AF361655

Air Quality Monitoring for Smart Cities and Industrial Applications

Air Quality Monitoring for Smart Cities and Industrial Applications

Editors

Daniele Sofia
Paolo Trucillo

Basel • Beijing • Wuhan • Barcelona • Belgrade • Novi Sad • Cluj • Manchester

Editors

Daniele Sofia
DIIN-Department of
Industrial Engineering
University of Salerno
Fisciano, Italy

Paolo Trucillo
DICMAPI-Department of
Chemical Engineering,
Materials and Industrial
Production Design
University of Naples
Federico II
Napoli, Italy

Editorial Office
MDPI
St. Alban-Anlage 66
4052 Basel, Switzerland

This is a reprint of articles from the Special Issue published online in the open access journal *Processes* (ISSN 2227-9717) (available at: https://www.mdpi.com/journal/processes/special_issues/Quality_Air_Monitoring).

For citation purposes, cite each article independently as indicated on the article page online and as indicated below:

Lastname, A.A.; Lastname, B.B. Article Title. *Journal Name* **Year**, *Volume Number*, Page Range.

ISBN 978-3-0365-8742-4 (Hbk)
ISBN 978-3-0365-8743-1 (PDF)
doi.org/10.3390/books978-3-0365-8743-1

Cover image courtesy of Chris Leboutillier

Contents

About the Editors

Daniele Sofia

Daniele Sofia (1984) is a chemical engineer; he received his PhD in industrial engineering at the University of Salerno. He is a researcher at the University of Salerno and delivers lectures on environmental data analysis at the Department of Chemistry and Biology. He has expertise in particle technology in environmental monitoring and in new low-environmental-impact additive manufacturing techniques.

Paolo Trucillo

Paolo Trucillo (1988) is a chemical engineer; he received his PhD in industrial engineering at the University of Salerno, Italy. He is assistant professor at the University of Naples Federico II and delivers lectures on innovative materials for design for the master's degree of design for the built environment at the Department of Architecture. He has expertise in drug carrier production, chemical plant design and sustainability indicators for materials and processes.

 *processes*

Editorial

Editorial Overview of the Special Issue "Air Quality Monitoring for Smart Cities and Industrial Applications"

Daniele Sofia [1,2,*] **and Paolo Trucillo** [3]

[1] Department of Industrial Engineering, University of Salerno, Via Giovanni Paolo II, 132, 84084 Fisciano, Italy
[2] Research Department, Sense Square Srl, Corso Garibaldi 33, 84123 Salerno, Italy
[3] Department of Chemical, Material and Industrial Production Engineering, University of Naples Federico II, Piazzale Tecchio, 80, 80125 Napoli, Italy; paolo.trucillo@unina.it
* Correspondence: dsofia@unisa.it

Citation: Sofia, D.; Trucillo, P. Editorial Overview of the Special Issue "Air Quality Monitoring for Smart Cities and Industrial Applications". *Processes* **2023**, *11*, 2458. https://doi.org/10.3390/pr11082458

Received: 8 August 2023
Accepted: 10 August 2023
Published: 15 August 2023

The Special Issue entitled "Air Quality Monitoring for Smart Cities and Industrial Applications" addresses the pressing concern of environmental pollution, particularly air pollution, and its impact on global well-being. The primary focus is on monitoring air quality, which is pivotal for combating air-related health issues, such as obstructive pulmonary diseases, cardiovascular illnesses, lung cancers, asthma, or lower respiratory infections [1,2].

This Special Issue examines gaseous compounds released from industrial plants, considering the challenges posed by strict government regulations, trying to contain pollution that nowadays is felt as a global burden [3]. The concept of circular economy is being correctly forced into our daily lives, especially in manufacturing activities [4]. This is why humans are trying to reduce the generation of solid, liquid, and gaseous waste.

Therefore, the significance of investing in air quality research is emphasized, covering indoor and outdoor monitoring systems, pollution-related health reviews, and pollution's connection to human health. Specific themes include gaseous waste treatment, air quality monitoring equipment development, analysis of outdoor air quality and industrial emissions, and indoor air quality assessment in inadequately ventilated spaces. The impact of air pollutants on health, especially during the COVID-19 pandemic, is also explored.

Papers published within this Special Issue include research that developed a network of low-cost sensors to monitor PM2.5 particulate matter in Temuco City, Chile, assessing spatial and temporal air quality variations [5]. Another paper explores the evolution of research trends in ozone formation sensitivity using bibliometric analysis [6]. Among the studies presented in this issue, there is one that reports a crowdsensing-based vehicle sensor network system for real-time monitoring of urban PM levels, discussing its cost-effectiveness and benefits [7]. The impact of poor indoor air quality on work efficiency is examined through a case study of PM2.5 levels in a large shopping mall in Macao [8].

Moreover, the research delves into aircraft emissions within Spain's domestic aviation market, emphasizing the importance of balancing economic profitability and environmental impact [9]. In the aims of sustainability, nowadays it is fundamental to satisfy simultaneously the conditions of profitability, environmental impact, and social impact. Only after balancing these three conditions can we address our research efforts to the path of global Earth sources regeneration.

The effects of the COVID-19 lockdown measures on criteria pollutants are analyzed, particularly in relation to Australia's prescribed burns during the first worldwide lockdown period due to the COVID-19 pandemic illness [10]. A study focuses on blood heavy metal absorption in areas with varying environmental impacts, assessing the connection between air quality and health [11]. An investigation predicts PM2.5 concentrations using multi-time scale fusion, contributing to environmental protection efforts [12]. The global burden of disease due to air pollution is assessed, revealing trends in disease attribution over the years [13], revealing the same problems all over the world [14]. Additionally, research simulates the spread of the novel Coronavirus in an aircraft cabin [15] and in other different transportation moods [16], highlighting infection risk.

In summary, this Special Issue addresses the multifaceted aspects of air quality monitoring, pollution's effects on health, and the potential implications for policy and public awareness. The published papers delve into various facets of air quality, pollution sources, monitoring technologies, and health impacts, aiming to contribute to a deeper understanding of these critical environmental and health issues.

Acknowledgments: The co-guest editors thank the authors for providing their excellent papers, sharing their knowledge and experience.

Conflicts of Interest: The authors declare no conflict of interest.

References

1. Iloeje, U.H.; Redlich, C.A. Indoor air pollution: An update for the clinician. *Clin. Pulm. Med.* **2000**, *7*, 128–133. [CrossRef]
2. Raziani, Y.; Raziani, S. The effect of air pollution on myocardial infarction. *J. Chem. Rev.* **2001**, *3*, 83–96.
3. Aigbe, G.O.; Cotton, M.; Stringer, L.C. Global gas flaring and energy justice: An empirical ethics analysis of stakeholder perspectives. *Energy Res. Soc. Sci.* **2023**, *99*, 103064. [CrossRef]
4. Beaurain, C.; Chembessi, C.; Rajaonson, J. Investigating the cultural dimension of circular economy: A pragmatist perspective. *J. Clean. Prod.* **2023**, *417*, 138012. [CrossRef]
5. Muñoz, C.; Huircan, J.; Jaramillo, F.; Boso, A. Calibration of Sensor Network for Outdoor Measurement of PM2.5 on High Wood-Heating Smoke in Temuco City. *Processes* **2023**, *11*, 2338. [CrossRef]
6. Javed, Z.; Mehmood, K.; Liu, C.; Zheng, X.; Xu, C.; Tanvir, A.; Khan, M.A.; Siddique, N.; Du, D. Examining Current Research Trends in Ozone Formation Sensitivity: A Bibliometric Analysis. *Processes* **2023**, *11*, 2240. [CrossRef]
7. Diviacco, P.; Iurcev, M.; Carbajales, R.J.; Viola, A.; Potleca, N. Design and Implementation of a Crowdsensing-Based Air Quality Monitoring Open and FAIR Data Infrastructure. *Processes* **2023**, *11*, 1881. [CrossRef]
8. Lei, T.M.; Chan, Y.W.; Mohd Nadzir, M.S. Monitoring PM2.5 at a Large Shopping Mall: A Case Study in Macao. *Processes* **2023**, *11*, 914. [CrossRef]
9. Martínez Raya, A.; Segura de la Cal, A.; González Díaz, R.E. An Empirical Analysis of the Aircraft Emissions by Operating from Scheduled Flights within the Domestic Market in Spain. *Processes* **2023**, *11*, 741. [CrossRef]
10. Verma, P.; Sisodiya, S.; Banait, S.K.; Chowdhury, S.; Dwivedi, G.; Zare, A. The Impact of Coronavirus Disease of 2019 (COVID-19) Lockdown Restrictions on the Criteria Pollutants. *Processes* **2023**, *11*, 296. [CrossRef]
11. Lotrecchiano, N.; Montano, L.; Bonapace, I.M.; Giancarlo, T.; Trucillo, P.; Sofia, D. Comparison Process of Blood Heavy Metals Absorption Linked to Measured Air Quality Data in Areas with High and Low Environmental Impact. *Processes* **2022**, *10*, 1409. [CrossRef]
12. Zhang, J.; Xia, W. Prediction of PM2.5 Concentration on the Basis of Multi-Time Scale Fusion. *Processes* **2022**, *10*, 171. [CrossRef]
13. Dhimal, M.; Chirico, F.; Bista, B.; Sharma, S.; Chalise, B.; Dhimal, M.L.; Ilesanmi, O.S.; Trucillo, P.; Sofia, D. Impact of air pollution on global burden of disease in 2019. *Processes* **2021**, *9*, 1719. [CrossRef]
14. Lotrecchiano, N.; Trucillo, P.; Barletta, D.; Poletto, M.; Sofia, D. Air pollution analysis during the lockdown on the city of Milan. *Processes* **2021**, *9*, 1692. [CrossRef]
15. Zhang, M.; Yu, N.; Zhang, Y.; Zhang, X.; Cui, Y. Numerical simulation of the novel coronavirus spread in commercial aircraft cabin. *Processes* **2021**, *9*, 1601. [CrossRef]
16. Zhang, Y.; Yu, N.; Zhang, M.; Ye, Q. Particulate matter exposures under five different transportation modes during spring festival travel rush in China. *Processes* **2021**, *9*, 1133. [CrossRef]

Article

Calibration of Sensor Network for Outdoor Measurement of PM2.5 on High Wood-Heating Smoke in Temuco City

Carlos Muñoz [1,*], Juan Huircan [1], Francisco Jaramillo [2] and Álex Boso [3]

[1] Department of Electrical Engineering, Faculty of Engineering and Sciences, Universidad de La Frontera, Av. Francisco Salazar, Temuco 01145, Chile; juan.huircan@ufrontera.cl

[2] Department of Electrical Engineering, Faculty of Physical and Mathematical Sciences, University of Chile, Av. Tupper, Santiago 2007, Chile; francisco.jaramillo@ing.uchile.cl

[3] Department of Environment, CIEMAT, Avenida Complutense 40, 28040 Madrid, Spain; alex.boso@ciemat.es

* Correspondence: carlos.munoz@ufrontera.cl

Abstract: In order to ascertain the spatial and temporal changes in the air quality in Temuco City, Chile, we created and installed a network of inexpensive sensors to detect PM2.5 particulate matter. The 21 measurement points deployed were based on a low-cost Sensiron SPS30 sensor, complemented with temperature and humidity sensors, an Esp32 microcontroller card with LoRa and WiFi wireless communication interface, and a solar charging unit. The units were calibrated using an airtight combustion chamber with a Grimm 11-E as a reference unit. The calibration procedure fits the parameters of a calibration model to map the raw low-cost particle-material measurements into reliable calibrated values. The measurements showed that the concentrations of fine particulate material recorded in Temuco present a high temporal and spatial variability. In critical contamination episodes, pollution reaches values as high as 354 $\mu g/m^3$, and at the same time, it reaches 50 $\mu g/m^3$ in other parts of the city. The contamination episodes show a similar trend around the city, and the peaks are in the time interval from 07:00 PM to 1:00 AM. In the winter, this time of day coincides with when families are usually home and there are low temperatures outside.

Keywords: air pollution; low-cost sensor; particle-matter calibration; spatial–temporal distribution

Citation: Muñoz, C.; Huircan, J.; Jaramillo, F.; Boso, Á. Calibration of Sensor Network for Outdoor Measurement of PM2.5 on High Wood-Heating Smoke in Temuco City. *Processes* **2023**, *11*, 2338. https://doi.org/10.3390/pr11082338

Academic Editor: Paolo Trucillo

Received: 12 May 2023
Revised: 26 June 2023
Accepted: 27 June 2023
Published: 3 August 2023

1. Introduction

In southern Chile, burning wood is widely valued as a primary energy source in the domestic sphere due to its low price compared to other heating alternatives [1]. However, its massive use and inefficient combustion generate high emissions of air pollutants. The city of Temuco, in recent years, has increased its levels of environmental pollution, with the highest levels of PM2.5 and PM10 [2–4] exceeding in both cases 300 $\mu g/m^3$. These reports show that every winter, the city of Temuco experiences numerous critical episodes due to poor air quality, with particulate matter that exceeds the national regulations and international standards, impacting public health and the quality of life of its residents. The leading cause of pollution by PM2.5 and PM10 in this city is the widespread use of wood-fired heaters due to their low cost and high availability in the area [1].

The city of Temuco is medium-sized in terms of population (approximately 304,000 inhabitants), with an urban area of 50 km^2 and a riverbed topography with surrounding hills. This city has two FEM (Federal Equivalent Methods) stations installed and operated by the Ministry of the Environment [5]. These two FEM stations are located in residential sectors separated by 4.5 km from each other. The air-quality measurements obtained through these FEM stations allow the adoption of measures to restrict the use of wood heating [3].

Becnel et al. [6] and Johnston et al. [7] state that FEMs can deliver reliable and accurate data to support decision-making; however, they are expensive to maintain. Due to the high costs, the number of stations is usually sparse, lacking the spatial resolution necessary to assess community exposure to PM2.5 particulate matter.

According to Johnston et al. [7], the impact of air pollution levels on health is dependent on pollutant concentrations and exposure levels. These factors vary at fine spatio-temporal scales in urban environments, driving the need for more data by increasing the number of sensor deployments and improving the sampling frequency. Alfano et al. [8] present an extensive review of the low-cost particulate matter sensors (LCPMSs) currently available on the market, their electronic characteristics, and their applications in the published literature and from specific tests. These LCPMSs are proposed as a complement to the measurements provided by the FEMs [9].

Among the LCPMSs described in the literature, there is a sensor node developed by Johnston et al. [10] that uses four low-cost devices (Alphasense OPC-N2, Plantower PMS5003 and PMS7003, Honeywell HPMA115S0). Johnston et al. [10] compared these LCPMSs to certified government stations, resulting in an RMS error of 6.065 and a Pearson coefficient of 0.878. Sayahi et al. [11] carried out a field evaluation of Plantower PMS1003 and PMS5003 sensors with reference equipment under different conditions, showing a good correlation in the winter season ($R^2 > 0.858$). Sayahi et al. [11] concluded that having various calibration factors for the same sensor model requires a systematic laboratory tuning or an on-site calibration strategy. Badura et al. [12] evaluated a network of 20 PMS5003 sensors, and the average values of the coefficient of determination (R^2) calculated between the sensor nodes and the government FEM stations were around 0.89 (with a range of 0.82 to 0.91, depending on the sensing device and station). Becnel et al. [6] proposed a platform with PMS3003 sensors made up of 50 nodes, which they deployed in an area of 100 km^2, obtaining a correlation of $R^2 = 0.88$ between the sensor nodes and the FEM. Tagle et al. [13] built a small network based on SDS011 particulate-matter sensors, which were compared with reference FEM stations, obtaining a correlation coefficient (R^2) between 0.47 and 0.86.

Ferrer-Cid et al. [14] state that PM2.5 particulate matter sensors, among others, are generally not calibrated by the manufacturer. If they have been, they lack calibration under the environmental conditions in which they will provide measurement data. To calibrate LCPMSs in the laboratory, Papapostolou et al. [15] and Sayahi et al. [16] developed calibration chambers that allow stable and reproducible gas and aerosol concentrations at low, medium, and high levels. Malings et al. [17] then uses an instrument laboratory standard, which has a combustion chamber to generate characteristic curves that allow adjustment of a specific calibration polynomial for each LCPMS. Magi et al. [18] proposed improving the correction with compensation of the measurements for humidity and temperature variations. Zaidan et al. [19] developed calibration systems to enhance the accuracy of LCPMSs, incorporating calibration models based on machine learning and virtual sensors. The methodology requires the construction of a database using continuous measurements of low-cost sensors and reference instruments. Then, it is analyzed to obtain information on atmospheric characteristics and air pollutants, which allows an understanding of the performance of the sensors in terms of consistency (relative to other sensors) and accuracy by comparison with the reference instruments. According to Ferrer-Cid et al. [14], incorporating low-cost sensors is a viable alternative to complement the measurements made at the FEM stations.

According to Datta et al. [20], using multiple LCPMSs allows obtaining information on intra-urban space–time variation, which allows extraction of sectorized information on PM2.5 pollution. In this sense, Schneider et al. [21] and Liu et al. [22] accounted for the non-uniformity of pollution through updated high-resolution maps of particulate pollution in the environment using 24 LCPMSs in the city of Oslo, Norway. A space-time analysis of PM2.5 particulate matter using multiple sensors in Dezhou City, China, Cao et al. [23] identified four stages of daily PM2.5 variation: accumulation, continuous pollution, dispersion, and cleaning, which is consistent with the pollution episodes generated by the use of heating.

Using mobile units makes achieving a flexible system to acquire information in a broad spatial spectrum possible. Still, comparing measurements taken in different parts of the city makes it difficult, since these are captured at deferred times. On the other hand, using measurements through LCPMSs is one of many mechanisms to achieve higher-resolution pollution maps. Blanco et al. [24] and Quinteros et al. [25] carried out spatial characterization of particulate matter during winter nights in Temuco City using vehicles with GPS tracking that carried PM2.5 recording units. The investigation indicated that pollutants are distributed unevenly in Temuco City, revealing that in some neighborhoods, PM2.5 concentrations are almost twice as high as those measured at the government station.

The objective of the present study is to deploy and calibrate a low-cost sensor network platform in the city of Temuco, Chile. This sensor network allows for an analysis of the density of contaminants in the phases of accumulation, continuous contamination, dispersion, and cleaning during contamination episodes. The novelty of this paper and its main contribution is that we deployed a calibrated LCPM sensor network over a city with high-contamination episodes, allowing us to analyze the behavior of the pollutants over the entire high-contamination episode. This network will allow us to measure the pollutants in the city, evidencing the differences in particle-matter concentration within high-contamination episodes simultaneously in different sectors.

The paper is structured as follows: First, Section 1 furnishes an overview of the current research status on obtaining information on intra-urban space–time variation, which allows extraction of sectorized information on PM2.5 pollution, and depicts the objective of deploying and calibrating a low-cost sensor network platform in the city of Temuco, Chile. Then, in Section 2, we show the methods, and Section 3 focuses on depicting a novel calibration method for low-cost particle-matter sensors based on a nonlinear calibration model. Subsequently, in Section 4, we address the case study, presenting the deployment of the LCPM sensor network in Temuco City. As the main result, we show the spatio-temporal analysis of PM2.5 on three consecutive days, in which we can observe the accumulation of pollution and the washout due to rain. Lastly, in Section 5, the results and conclusions show the experience deploying this sensor network and calibration method, including limitations and future research directions.

2. Methods

Figure 1 depicts the method used, which involves two stages: calibration of the sensors in the laboratory with a reference instrument and deployment of the sensor nodes in the city. The following sections will describe each of the stages.

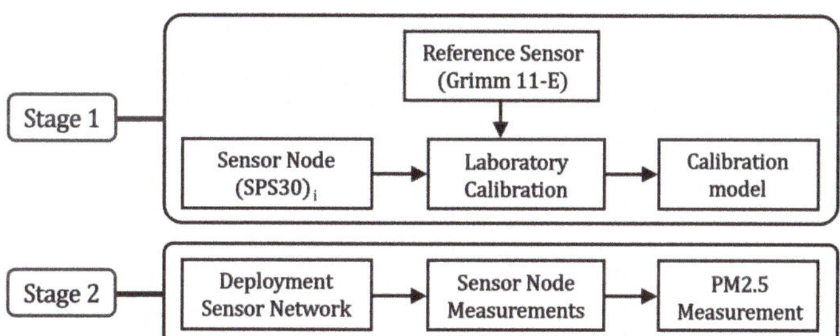

Figure 1. Methodology used in this work. In stage 1 the calibration setup was made and the calibration models were fitted. In stage 2, the sensor network was deployed and operated.

In the first stage, as the first step, the method uses a combustion chamber for plant material to compare the measurements of sensor nodes based on the Sensirion SPS30 device with a GRIMM 11-E reference instrument. Then, we apply calibration models to match the responses of the Sensiron SPS30 and those obtained with the GRIMM 11-E.

2.1. Particulate Matter IoT Devices

Each IoT device for measuring particulate matter has the structure indicated in Figure 2 and consists of the following components: LCPMS, temperature/humidity sensor, WiFi communication module, LoRa communication module, power for a 3.6 V Lithium battery,

7 W solar panel, and charge controller, all managed by an ESP32 microcontroller card. It has a storage unit type micro SD. All this is installed in a case with the IP67 standard.

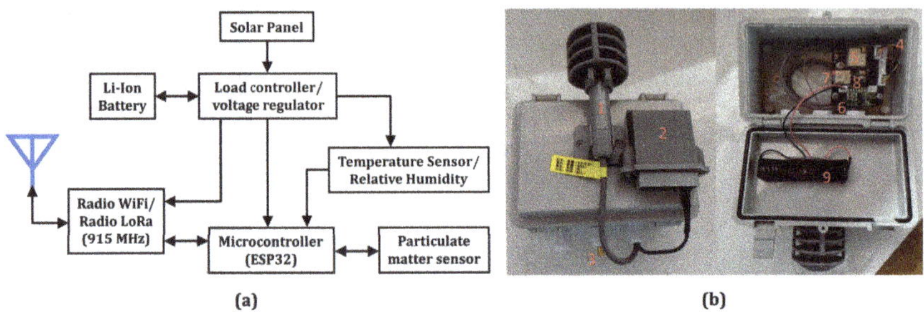

Figure 2. Sensor node of particulate matter. (**a**) Diagram. (**b**) Device, 1: Temperature Sensor / Relative Humidity, 2: Particulate Matter Sensor, 3: To the solar panel, 4: To the battery, 5: Microcontroller, Radio WiFi/Radio LoRa, 6: Load controller /Voltage regulator, 7: MicroSD Card , 8: Circuit board 9: Battery Holder for Li-Ion Battery

As the LCPMS, we used the Sensiron SPS30, a device for online measurements of particulate material PM1, PM2.5, PM4, and PM10 based on light scattering. This device presents an excellent performance Bulot et al. [26] and has a more compact package than the alternatives. The net measurement is converted into units of mass concentration (0–1000 µg/m^3) [27]. The device supports operating temperatures between -10 and 60 °C and humidity up to 95% RH, which is compatible with the range of temperatures measured in the City of Temuco (-5 to 38 °C) [28], and during periods of cold weather and without precipitation, the humidity reaches 95%. When precipitation occurs, the device turns off because there are no restrictions on using wood-fired heaters since the particulate material settles with the rain. According to the manufacturer, PM2.5 has a mass concentration accuracy of ± 10 µg/m^3 for a range of 0 to 100 µg/m^3. On the other hand, for ranges between 100 and 1000 µg/m^3, the accuracy is $\pm 10\%$ of the measured value. Furthermore, according to the manufacturer, the SPS030 has an estimated lifespan of 10 years and includes a self-cleaning mechanism.

Additionally, we programmed the sensor node with a data-acquisition application, including programmable sampling time, local storage in the micro SD memory, tools to update the software remotely, and an MQTT protocol to upload the collected data online to a mySQL database. We used the ThingsBoard open-source IoT platform for real-time data monitoring.

2.2. Reference Instrument

We used the GRIMM 11-E equipment to acquire the reference material particulate measurements. This instrument complies with the EN12341 [29] and EN14907 [30] standards for PM10 and PM2.5 measurement, respectively. It has a counting range of 1–2,000,000 particles/L and has a measurement range from 0.1 µg/m^3 to 6 mg/m^3 with reproducibility of $\pm 3\%$ over the entire measurement range.

3. Sensor-Node Calibration

With the purpose of calibrating the LCPMSs, we designed and implemented a controlled laboratory experiment. In this experiment, an airtight chamber with forced airflow was prepared to generate the combustion of a specific vegetable mass inside. Two LCPMSs and a reference instrument (GRIMM 11E) were installed to register PM2.5 measurements, as shown in Figure 3a.

The three sensors were configured to be synchronized with the start of the PM2.5 measurements. The sampling time for data acquisition was set at 6 s, where the recorded data were stored in the internal memory of the LCPMS and the reference instrument. In addition, the devices were always on during the experiment. Finally, the total length of the experiment was set at 8 h.

The data registered by the three sensors were preprocessed using a technique derived from Coleman and Meggers [31]. In our case, we considered a time window of 30 s (i.e., five measurements), where the median value for each time window was extracted, thus reducing the sampling time (to 30 s) and the quantity of measured data (930 data points). Figure 3b shows the variation in time of the PM2.5 concentration after preprocessing. The measured concentration exceeds the specified upper limit for each instrument to finally reach a minimum value at the end of the experiment. The yellow curve in Figure 3b represents the PM2.5 measurements acquired by the reference instrument used to calibrate both LCPMSs considered in this work (LCPMS 1, blue curve; LCPMS 2, red curve).

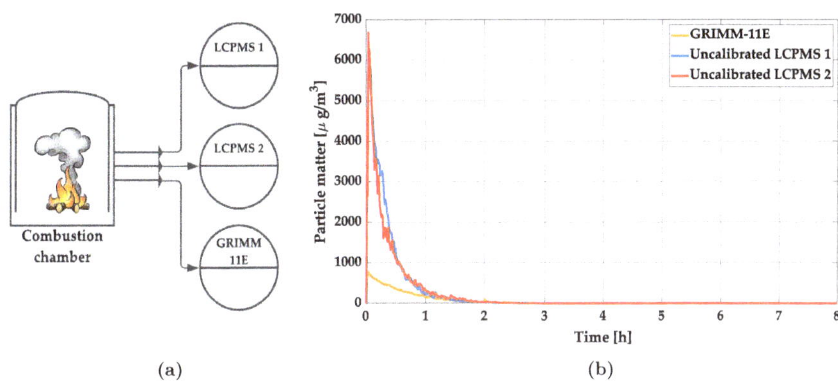

(a)   (b)

Figure 3. (**a**) Calibration experiment scheme. (**b**) Acquired signals from the reference instrument (yellow line) and the LCPMSs (blue and red lines).

The next step in the calibration procedure involves finding a calibration model $y_c(x, \theta)$ capable of mapping the raw LCPMS measurements onto reliable calibrated values. In this work, we defined a cost function that should be minimized using the data extracted from the laboratory experiment and a proposed calibration model to find the optimum set of parameters of $y_c(x, \theta)$. The cost function is detailed in Equation (1), where y_c is the calibration model and x_1, x_2, and y are the data related to LCPMS 1, LCPMS 2, and GRIMM 11E, respectively.

$$\theta = \arg\min_{\theta} \left(\sum_{i=1}^{N} [y_c(x_1(i), \theta) - y(i)]^2 + \sum_{i=1}^{N} [y_c(x_2(i), \theta) - y(i)]^2 \right) \quad (1)$$

3.1. Polynomial Calibration Model

In references [32,33], polynomial functions are one of the commonly proposed calibration models; due to this, we consider as $y_c(x, \theta)$ a polynomial model of order p, indicated in Equation (2).

$$y_c(x, \theta) = \sum_{i=1}^{p} \theta_i \cdot x^{i-1} \quad (2)$$

By using the MATLAB optimization toolbox, the cost function in Equation (1) was minimized considering different values for p to the proposed model in Equation (2). It is important to mention that, before the optimization, a second preprocessing step was applied to the experimental data for outlier removal. As a result, the optimum order of the polynomial was $p = 9$, and the optimum parameters are detailed in Table 1. Moreover, Figure 4 shows three graphs that complement the results achieved by the proposed calibration process. At first, Figure 4a displays the optimum calibration model for LCPMS

1/LCPMS 2. Then, Figure 4b presents the error histogram between LCPMS 1/LCPMS 2 and the calibrated model, where the standard deviation of the calibration error is equal to 7.5904. Finally, Figure 4c compares the experimental data registered from the reference instrument (GRIMM-11E) and the calibrated LCPMS 1/LCPMS 2.

Table 1. Optimum parameter set θ with 95% confidence bounds for the polynomial calibration model.

Coefficient (θ)	Value	95% Confidence Bounds
θ_1	111.9	(110.6, 113.1)
θ_2	411.6	(408.1, 415)
θ_3	-487	$(-504.2, -469.9)$
θ_4	418.8	(392.3, 445.3)
θ_5	-199.8	$(-217.3, -182.4)$
θ_6	54.99	(49.03, 60.95)
θ_7	-9	$(-10.15, -7.854)$
θ_8	0.8655	(0.7405, 0.9905)
θ_9	-0.04517	$(-0.0524, -0.03794)$
θ_{10}	0.000987	(0.0008149, 0.001159)

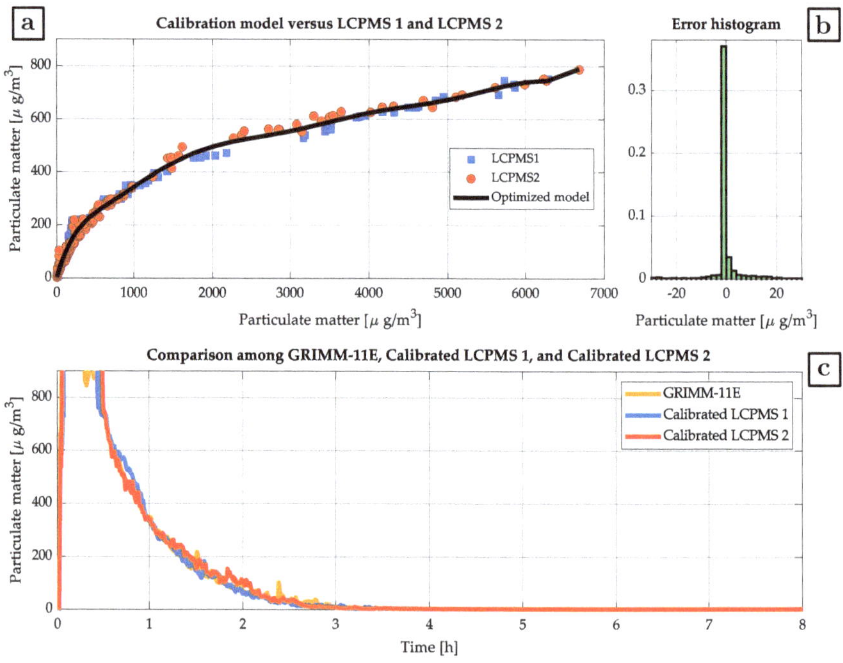

Figure 4. (**a**) Ninth-order polynomial calibration model for LCPMS 1 and LCPMS 2. (**b**) Calibration error histogram. (**c**) Comparison among GRIMM-11E and the calibrated LCPMS.

3.2. Nonlinear Function Calibration Model

To avoid overfitted and over-parameterized calibration models, we propose a nonlinear function composed of two exponential components and one square root term. This nonlinear model is detailed in Equation (3):

$$y_c(x, \theta) = a_0 \cdot e^{b_0 \cdot x} + a_1 \cdot e^{b_1 \cdot x} + c_0 \cdot \sqrt{x} \ , \tag{3}$$

where the parameter vector $\boldsymbol{\theta}$ is represented by $\theta = \begin{bmatrix} a_0 & b_0 & a_1 & b_1 & c_0 \end{bmatrix}$.

After applying the second preprocessing step for outlier removal and using the MATLAB optimization toolbox, the cost function in Equation (1) was minimized considering the nonlinear model (Equation (3)). The resultant optimum set of parameters $\boldsymbol{\theta}$ is detailed in Table 2.

Table 2. Optimum parameter set θ with 95% confidence bounds for the nonlinear calibration model.

Coefficient (θ)	Value	95% Confidence Bounds
a_0	193.8	(192.7, 194.8)
b_0	5.989×10^{-5}	$(5.901 \times 10^{-5}, 6.076 \times 10^{-5})$
a_1	-198.8	$(-199.8, -197.7)$
b_1	-0.001553	$(-0.001559, -0.001546)$
c_0	5.969	(5.939, 5.999)

In the same manner, as shown above, Figure 5 helps to understand better the scope of the results achieved by the proposed calibration process. Figure 5a displays the optimum calibration model for the LCPMS 1/LCPMS 2. Then, Figure 5b presents the error histogram between LCPMS 1/LCPMS 2 and the calibrated nonlinear model, where the standard deviation of the calibration error equals 6.4562. Finally, Figure 5c compares the experimental data registered from the reference instrument (GRIMM-11E) and the calibrated LCPMS 1/LCPMS 2.

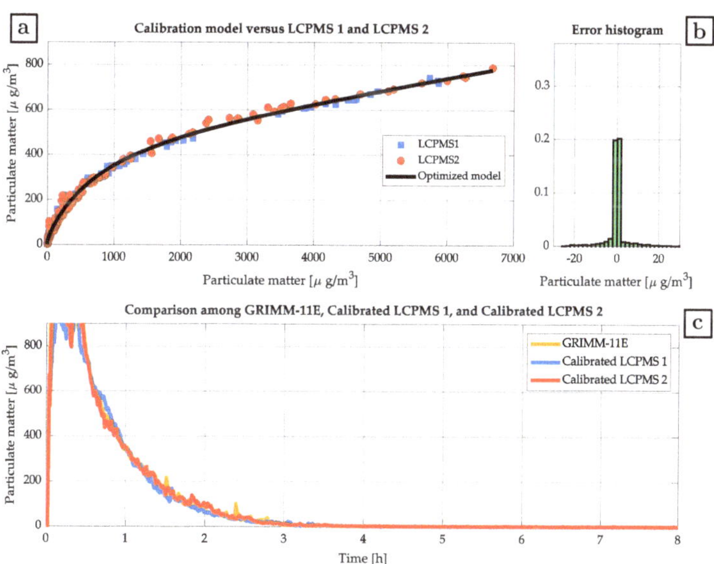

Figure 5. (**a**) Nonlinear calibration model for LCPMS 1 and LCPMS 2. (**b**) Calibration error histogram. (**c**) Comparison among GRIMM-11E and the calibrated LCPMSs.

Given the calibration results (Figures 4 and 5), both proposed models achieved an adequate mapping from the raw data to the calibrated data. However, the nonlinear calibration model was the one that obtained the best results as to the standard deviation of the calibration error and lower number of parameters. This last is crucial to obtain generalized models instead of overfitted models.

4. Case Study: IoT Network for Particle-Size Monitoring in Temuco City

4.1. Comparison of the LCPM with Other Solutions

For a comparative analysis of the LCPM device with respect to other sensor devices, we analyzed characteristics such as the type of use (indoor, outdoor), price, range, precision, and communication. The outdoor property is relevant because the sensor network is intended to acquire the PM10 particle-matter behavior outdoors. The communication is a requirement to real-time collect the data acquired in a central database. The cost per unit is relevant to deploy a significant number of units to give an insight into the pollution behavior in the city through time. Table 3 compares these sensors' characteristics, and as shown, the LCPM fulfills the requirements issued for the sensor network deployment. In the case of GRIMM 11-E, the precision was not available in the datasheet, but the vendor stated a reproducibility of ±3% over the total measuring range. We calibrated the LCPM devices in the laboratory, coded and installed the data-collection system with the respective dashboards, and finally proceeded with the next step: deploying the sensor network in the city.

Table 3. Comparison of particle-matter sensors.

Features	Sensor Type		
	Low-Cost Node Sensors Proposed	Low-Cost Node Sensors Commercially Available	GRIMM 11-E or Similar
Cost per unit USD\$	120	200–500 [34,35]	10,000–20,000 [35]
PM2.5	yes	yes	yes
PM10	yes	yes	yes
Range	0–1000 $\mu g/m^3$	0–1000 $\mu g/m^3$	0.1–6000 $\mu g/m^3$
Precision	<100 $\mu g/m^3$: ± 10 $\mu g/m^3$	<100 $\mu g/m^3$: ± 10 $\mu g/m^3$	N.A.
Particle Range Size	0.3 to 2.5 µm	N.A.	0.25–31µm
Wireless Connectivity	LoRa, WiFi	Global cellular 2G/3G/4G	Bluetooth
Use Type	Outdoor	Outdoor	Laboratory
Additional Measurement	Temperature, Humidity	Temperature, Humidity	N.A.

4.2. Node-Deployment Criteria

We deployed a network of 21 IoT devices with calibrated LCPMSs for measuring particulate matter in Temuco City, as shown in Figure 6.

For the location of the nodes, the following aspects were considered: that the place is far from a source of smoke emission, that there is access to the installation site, the availability of connectivity, the availability of energy, and the uniform distribution of the equipment. This last element is not necessarily actual since the x-distant point to an already installed node was not necessarily available to install the next node. In the case of residential nodes, we considered the participation of citizens, who agreed to provide space for the installation in the backyard of their homes and connectivity for the required nodes. In the case of the nodes installed in public distributions, the authority provided access to data and energy connectivity. We installed the nodes at outdoor sites more than 30 m from smoke emission sources (stove or boiler chimneys) and around 2 m in height. The selected 2 m is because the average size of people is between 1.5 m and 1.8 m. In addition, this height allows us to install the solar panel on a nearby roof.

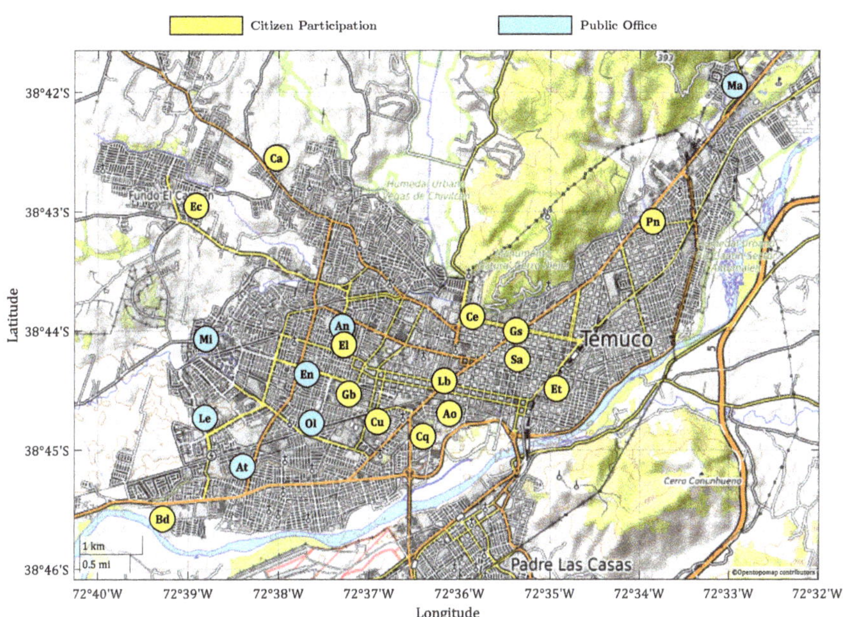

Figure 6. Temuco map, 14 July 2020 at 21:00 h (adapted from OpenStreetMap [36]).

Table 4. Node information.

Node Name	Abbreviation	Type
Altamira	At	Citizen Participation
Altos de Mirasur	Mi	Citizen Participation
Andes	An	Citizen Participation
Aseo y Ornato	Ao	Public Office
Bodega Droguería	Bd	Public Office
CECOSF Arquenco	Ca	Public Office
CECOSF Las Quilas	Cq	Public Office
Cementerio	Ce	Public Office
Ciencias Físicas UFRO	Cu	Public Office
El Carmen	Ec	Public Office
El Trencito	Et	Public Office
Entrelagos	En	Citizen Participation
Escuela Llaima	El	Public Office
Galo Sepúlveda	Gs	Public Office
German Becker	Gb	Public Office
La Lechería	Le	Citizen Participation
Las Mariposas	Ma	Citizen Participation
Liceo Bicentenario	Lb	Public Office
Olimpia	Ol	Citizen Participation
Pueblo Nuevo	Pn	Public Office
Smart Araucanía	Sa	Public Office

The calibrated LCPMSs were configured to capture information every 10 min. To compare the PM25 measurements in different sectors every hour, we computed the average hourly values in every station. To identify the areas in which the nodes have been installed, identifiers similar to those proposed in [37–39] were used. Since there is no single urban zoning identification, we associated the deployment of the nodes with two different types of locations, citizen participation and public office. Citizen participation refers to citizens who put the location to install sensor nodes at our disposition. In this case, the nodes were deployed in the backyards of residential homes powered by internal batteries and a small solar panel. These residential nodes are connected to the home's WiFi to deliver the sensed data. Public office refers to public schools, family health centers, and municipal warehouses. The nodes were plugged into the local electricity network and connected to the data centers via Ethernet wires. Figure 6 shows the deployment of the nodes by type in the city of Temuco, and Table 4 shows the characteristics of the nodes installed.

4.3. Results of Three Consecutive Days

The concentrations of fine particulate material recorded in Temuco show an evident temporal variability, with the maximum attention in the time slot from 7:00 p.m. to 1:00 a.m. This time of day coincides with when families are usually at home and temperatures outside are low. Regarding home emissions associated with firewood, various studies have shown a high dependence between the temperature, a high energy demand for heating in homes with poor thermal insulation, and the emission rate of pollutants [3]. This relationship is observed in the differences in levels of particulate matter found in our sensors. Some residential neighborhoods with the highest population density and poorest housing quality show a higher emission of air pollutants associated with firewood, especially during peak hours. On the other hand, the sensors indicate that the highest concentrations, with emergency levels for PM2.5, occur in areas around the geographic center of the city, as can be seen in Figures 6 and 7. From the center and moving towards the periphery, the average levels of particulate matter decrease.

This situation can be due to the geographical distribution of the city, which is in a valley encased by two hills and a river. When the wind blows, it tends to displace pollutants in the direction of the river.

Figures 7–9 show the results of monitoring on three consecutive days (14 June, 15 June, and 16 June). These figures show that days 14 and 15 were cold, with a minimum temperature below 5 °C and low humidity. However, on the 17th, it rained, the temperature rose, and pollution diminished throughout the city. We want to note that the 15th day was the most critical since the contamination had accumulated for two consecutive days, reaching critical thresholds in many stations at peak times (21:00–03:00). In some stations, these peaks were more extreme, reaching values as high as 350 μg/m^3, which greatly exceeded all permitted pollution standards for many consecutive hours. This situation could be explained due to many wood stoves heating homes when temperatures are freezing and people are home. However, given the geographical characteristics of the city, where a natural wind corridor along the river exists, there is an excellent cleaning capacity. The figures show that at some off-peak hours, the city shows medium to low pollution rates in all the stations. From Figures 7–9 we see that only some stations show extreme PM10 values, and the places that reach the highest pollution values are near the city's center, which is geographically located where the wind corridor enclosed by two hills narrows (see Figure 6). An analysis with more data is required to determine how the topography in different climatic conditions affects the distribution of particulate-matter pollution of PM2.5 and PM10 over the city.

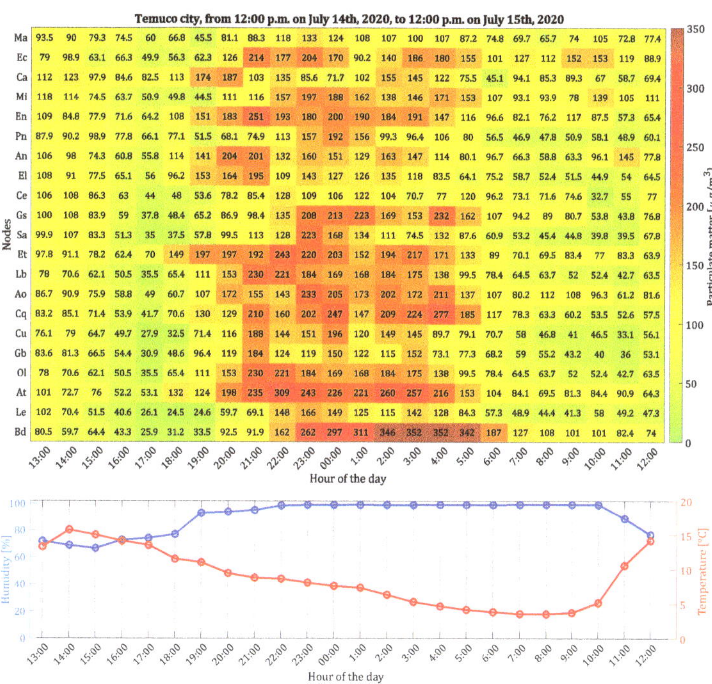

Figure 7. Nodes and particulate matter on 14 July 2020. Red line: Temperature, Blue line: Relative humidity.

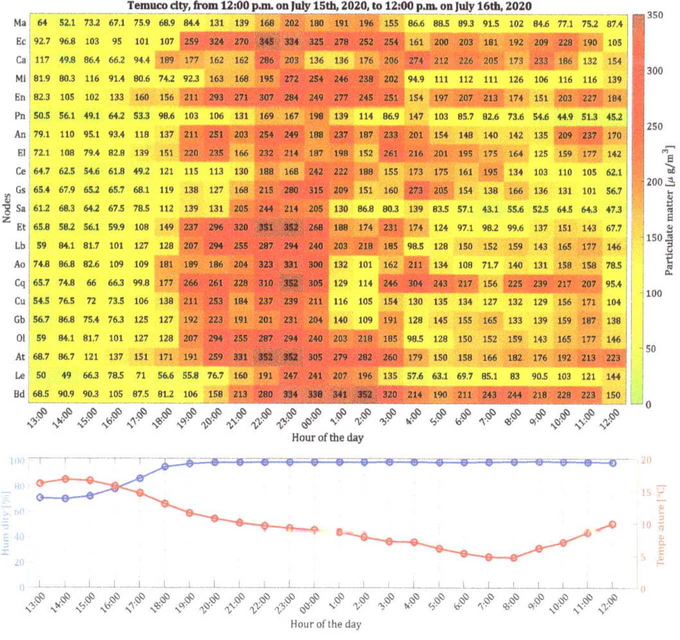

Figure 8. Nodes and particulate matter on 15 July 2020. Red line: Temperature, Blue line: Relative humidity.

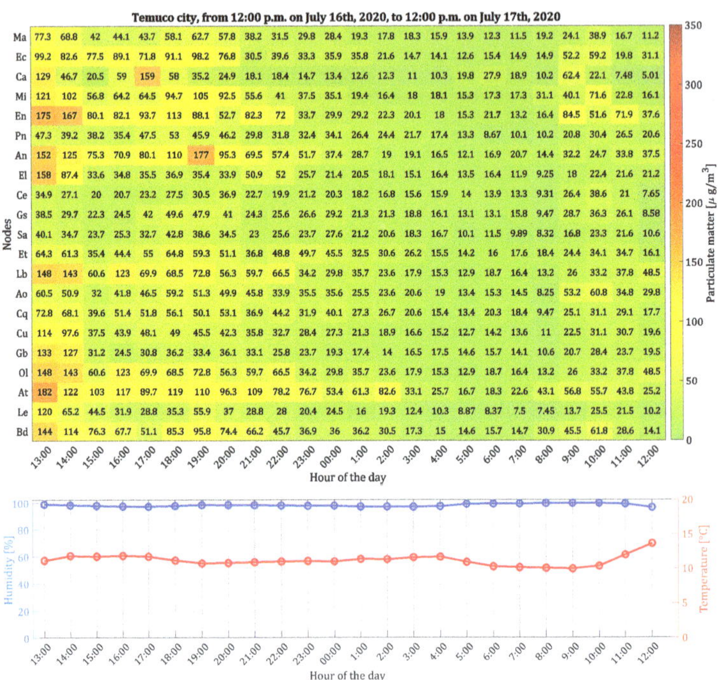

Figure 9. Nodes and particulate matter on 16 July 2020. Red line: Temperature, Blue line: Relative humidity.

5. Discussion and Conclusions

The objective of this study was to deploy and calibrate a low-cost particulate-matter sensor (LCPMS) network platform in Temuco City, Chile. This sensor network allows for analysis of the density of contaminants in the phases of accumulation, continuous contamination, dispersion, and cleaning during contamination episodes. The main contribution of this paper is to review the dynamic behavior of the pollutants in the whole city over the entire event of a high-contamination episode. This spatio-temporal analysis provides an insight into how the contaminants move in Temuco City over time during episodes of high contamination and cleansing due to rain. In Latin America and, in general, in low- and middle-income countries, low-cost sensors can revolutionize traditional monitoring systems since they provide air-quality information in real time, with excellent spatial precision and, when calibrated, with a remarkable level of validity and reliability. Their low-cost nature, especially when compared to official monitoring stations, gives this technology the potential to democratize air pollution information, making it accessible to a broad audience. The development of low-cost sensor-based monitoring systems constitutes an opportunity to give voice to, amplify, and represent local needs, especially those of socially vulnerable groups. To achieve this, it will be necessary that, in addition to technological progress such as those shown in this study, researchers and environmental authorities coordinate their work with volunteers, representatives of non-governmental organizations, and community groups.

Properly selecting the calibration methods for low-cost pollution sensors is relevant to acquire reliable data worth making a spatio-temporal analysis during contamination episodes.

Regarding the calibration methods shown in this work, we proposed a nonlinear calibration model, which was compared with a polynomial calibration model seen in the state of the art [40–43]. The proposed nonlinear model obtained a lower standard deviation of the calibration error (6.4562) than the polynomial model (7.5904). Furthermore, the nonlinear model includes five parameters instead of the nine parameters of the polynomial model. This smaller number of parameters may make the model easier to calibrate and less susceptible to overfitting, which is consistent with the goals stated by Warder and Piggott [44]. The LCPM sensors deployed have been working outdoors, enduring rough operating conditions such as rainy days and temperatures lower than minus 5 °C. The calibration can be affected by the intrusion of little spiders and dust to form tartar stuck inside the measuring chamber. We advise recalibrating the LCPM sensors at least once a year. The amount of LCPM sensors deployed depends on topography factors, population density, and the network coverage to deploy a sensor network that provides enough spatio-temporal information to understand the characteristics of the contamination episodes in the target location. This study showed that relevant data to understand the movement of spatio-temporal behavior of the contaminants could be acquired using calibrated LCPM sensor networks. The future challenge is to keep running this LCPM sensor network, increasing the number of nodes and replicating this setup in other cities with their features, such as population density, network coverage, and topography. It is relevant to study strategies to keep involving the city's inhabitants in the issue of deploying the sensor network with physical spaces for LCPM sensor installation and using their home's internet connectivity.

Author Contributions: Conceptualization, C.M., F.J., and J.H.; methodology, C.M., F.J., and J.H.; software, C.M., F.J., and J.H.; validation, C.M., F.J., Á.B, and J.H.; formal analysis, C.M., F.J., Á.B., and J.H.; investigation, C.M., F.J., and J.H.; resources, C.M., F.J., and J.H.; data curation, C.M., F.J., and J.H.; writing–original draft preparation, C.M., F.J., and J.H.; writing–review and editing, C.M., F.J., and J.H.; visualization, C.M., F.J. , Á.B., and J.H.; supervision, C.M., F.J., and J.H.; project administration, C.M., F.J., and J.H.; funding acquisition, C.M., F.J., and J.H. All authors have read and agreed to the published version of the manuscript.

Funding: This research was funded by grant number IDI18-0003, BID-FOMIN-LAB, SMARTCITY IN A BOX, the ANID/FONDECYT 1220178, and ANID—Basal funding for Scientific and Technological Center of Excellence, IMPACT (Center of Interventional Medicine for Precision and Advanced Cellular Therapy), FB210024.

Acknowledgments: The authors thank Rodrigo Fuentes Inzunza from Universidad de Concepción for allowing us to use the GRIMM 11-E equipment for acquiring the reference material particulate measurements used in the LCPM calibration procedures.

Conflicts of Interest: The authors declare no conflict of interest.

References

1. Jorquera, H.; Barraza, F.; Heyer, J.; Valdivia, G.; Schiappacasse, L.N.; Montoya, L.D. Indoor PM2.5 in an urban zone with heavy wood smoke pollution: The case of Temuco, Chile. *Environ. Pollut.* **2018**, *236*, 477–487. [CrossRef] [PubMed]
2. Quinteros, M.E.; Lu, S.; Blazquez, C.; Cárdenas-R, J.P.; Ossa, X.; Delgado-Saborit, J.M.; Harrison, R.M.; Ruiz-Rudolph, P. *Use of Data Imputation Tools to Reconstruct Incomplete Air Quality Datasets: A Case-Study in Temuco, Chile*; Elsevier Ltd.: Amsterdam, The Netherlands, 2019; Volume 200, pp. 40–49. [CrossRef]
3. Mardones, C.; Cornejo, N. Ex-post evaluation of a program to reduce critical episodes due to air pollution in southern Chile. *Environ. Impact Assess. Rev.* **2020**, *80*, 106334. [CrossRef]
4. Boso, A.; Álvarez, B.; Oltra, C.; Garrido, J.; Muñoz, C.; Hofflinger, A. Out of sight, out of mind: participatory sensing for monitoring indoor air quality. *Environ. Monit. Assess.* **2020**, *192*, 1–15. [CrossRef] [PubMed]
5. Ministerio_del_Medio_Ambiente_de_Chile. Sistema de Información Nacional de Calidad del Aire Home Page. 2020. Available online: https://sinca.mma.gob.cl/index.php/ (accessed on 16 April 2023)
6. Becnel, T.; Tingey, K.; Whitaker, J.; Sayahi, T.; Le, K.; Goffin, P.; Butterfield, A.; Kelly, K.; Gaillardon, P.E. A Distributed Low-Cost Pollution Monitoring Platform. *IEEE Internet Things J.* **2019**, *6*, 10738–10748. [CrossRef]
7. Johnston, S.J.; Basford, P.J.; Bulot, F.M.; Apetroaie-Cristea, M.; Easton, N.H.; Davenport, C.; Foster, G.L.; Loxham, M.; Morris, A.K.; Cox, S.J. City scale particulate matter monitoring using LoRaWAN based air quality IoT devices. *Sensors* **2019**, *19*, 209. [CrossRef]
8. Alfano, B.; Barretta, L.; Giudice, A.D.; De Vito, S.; Francia, G.D.; Esposito, E.; Formisano, F.; Massera, E.; Miglietta, M.L.; Polichetti, T. A review of low-cost particulate matter sensors from the developers' perspectives. *Sensors* **2020**, *20*, 6819. [CrossRef]
9. Levy Zamora, M.; Xiong, F.; Gentner, D.; Kerkez, B.; Kohrman-Glaser, J.; Koehler, K. Field and Laboratory Evaluations of the Low-Cost Plantower Particulate Matter Sensor. *Environ. Sci. Technol.* **2019**, *53*, 838–849. [CrossRef]
10. Johnston, S.J.; Basford, P.J.; Bulot, F.M.; Apetroaie-Cristea, M.; Fosterx, G.L.; Loxhamz, M.; Cox, S.J. IoT deployment for city scale air quality monitoring with Low-Power Wide Area Networks. In Proceedings of the 2018 Global Internet of Things Summit, GIoTS 2018, Bilbao, Spain, 4–7 June 2018. [CrossRef]

11. Sayahi, T.; Butterfield, A.; Kelly, K.E. Long-term field evaluation of the Plantower PMS low-cost particulate matter sensors. *Environ. Pollut.* **2019**, *245*, 932–940. [CrossRef]

12. Badura, M.; Sówka, I.; Szymański, P.; Batog, P. Assessing the usefulness of dense sensor network for PM2.5 monitoring on an academic campus area. *Sci. Total. Environ.* **2020**, *722*, 137867. [CrossRef]

13. Tagle, M.; Rojas, F.; Reyes, F.; Vasquez, Y.; Hallgren, F.; Linden, J.; Kolev, D.; Watne, A.K.; Oyola, P. Field performance of a low-cost sensor in the monitoring of particulate matter in Santiago, Chile. *Environ. Monit. Assess.* **2020**, *192*, 171. [CrossRef]

14. Ferrer-Cid, P.; Barcelo-Ordinas, J.M.; Garcia-Vidal, J.; Ripoll, A.; Viana, M. Multisensor Data Fusion Calibration in IoT Air Pollution Platforms. *IEEE Internet Things J.* **2020**, *7*, 3124–3132. [CrossRef]

15. Papapostolou, V.; Zhang, H.; Feenstra, B.J.; Polidori, A. Development of an environmental chamber for evaluating the performance of low-cost air quality sensors under controlled conditions. *Atmos. Environ.* **2017**, *171*, 82–90. [CrossRef]

16. Sayahi, T.; Kaufman, D.; Becnel, T.; Kaur, K.; Butterfield, A.E.; Collingwood, S.; Zhang, Y.; Gaillardon, P.E.; Kelly, K.E. Development of a calibration chamber to evaluate the performance of low-cost particulate matter sensors. *Environ. Pollut.* **2019**, *255*, 113131. [CrossRef] [PubMed]

17. Malings, C.; Tanzer, R.; Hauryliuk, A.; Saha, P.K.; Robinson, A.L.; Presto, A.A.; Subramanian, R. Fine particle mass monitoring with low-cost sensors: Corrections and long-term performance evaluation. *Aerosol Sci. Technol.* **2020**, *54*, 160–174. [CrossRef]

18. Magi, B.I.; Cupini, C.; Francis, J.; Green, M.; Hauser, C. Evaluation of PM2.5 measured in an urban setting using a low-cost optical particle counter and a Federal Equivalent Method Beta Attenuation Monitor. *Aerosol Sci. Technol.* **2020**, *54*, 147–159. [CrossRef]

19. Zaidan, M.A.; Motlagh, N.H.; Fung, P.L.; Lu, D.; Timonen, H.; Kuula, J.; Niemi, J.V.; Tarkoma, S.; Pet, T. Intelligent Calibration and Virtual Sensing for Integrated Low-Cost Air Quality Sensors. *IEEE Sens. J.* **2020**, *1*, 15. [CrossRef]

20. Datta, A.; Saha, A.; Zamora, M.L.; Buehler, C.; Hao, L.; Xiong, F.; Gentner, D.R.; Koehler, K. Statistical field calibration of a low-cost PM2.5 monitoring network in Baltimore. *Atmos. Environ.* **2020**, *242*, 117761. [CrossRef]

21. Schneider, P.; Castell, N.; Vogt, M.; Dauge, F.R.; Lahoz, W.A.; Bartonova, A. Mapping urban air quality in near real-time using observations from low-cost sensors and model information. *Environ. Int.* **2017**, *106*, 234–247. [CrossRef]

22. Liu, H.Y.; Schneider, P.; Haugen, R.; Vogt, M. Performance assessment of a low-cost PM 2.5 sensor for a near four-month period in Oslo, Norway. *Atmosphere* **2019**, *10*, 41. [CrossRef]

23. Cao, R.; Li, B.; Wang, Z.; Peng, Z.R.; Tao, S.; Lou, S. Using a distributed air sensor network to investigate the spatiotemporal patterns of PM2.5 concentrations. *Environ. Pollut.* **2020**, *264*, 114549. [CrossRef]

24. Blanco, E.; Rubilar, F.; Quinteros, M.E.; Cayupi, K.; Ayala, S.; Lu, S.; Jimenez, R.B.; Cárdenas, J.P.; Blazquez, C.A.; Delgado-Saborit, J.M.; et al. Spatial distribution of particulate matter on winter nights in Temuco, Chile: Studying the impact of residential wood-burning using mobile monitoring. *Atmos. Environ.* **2022**, *286*, 119255. [CrossRef]

25. Quinteros, M.E.; Blanco, E.; Sanabria, J.; Rosas-Diaz, F.; Blazquez, C.A.; Ayala, S.; Cárdenas-R, J.P.; Stone, E.A.; Sybesma, K.; Delgado-Saborit, J.M.; et al. Spatio-temporal distribution of particulate matter and wood-smoke tracers in Temuco, Chile: A city heavily impacted by residential wood-burning. *Atmos. Environ.* **2023**, *294*, 119529. [CrossRef]

26. Bulot, F.M.J.; Russell, H.S.; Rezaei, M.; Johnson, M.S.; Ossont, S.J.J.; Morris, A.K.R.; Basford, P.J.; Easton, N.H.C.; Foster, G.L.; Loxham, M.; et al. Laboratory comparison of low-cost particulate matter sensors to measure transient events of pollution. *Sensors* **2020**, *20*, 2219. [CrossRef]

27. Sensirion - Particulate Matter Sensor SPS30 Home Page. Available online: https://sensirion.com/products/catalog/SEK-SPS30 (accessed on 28 June 2023).

28. Reyes, F.; Ahumada, S.; Rojas, F.; Oyola, P.; Vásquez, Y.; Aguilera, C.; Henriquez, A.; Gramsch, E.; Kang, C.M.; Saarikoski, S.; et al. Impact of biomass burning on air quality in Temuco city, chile. *Aerosol Air Qual. Res.* **2021**, *21*, 210110. [CrossRef]

29. European Standard, E.. Ambient Air: Standard Gravimetric Measurement Method for the Determination of the PM10 or PM2.5 Mass Concentration of Suspended Particulate Matter. 2014. Available online: https://www.en-standard.eu/bs-en-12341-2014-ambient-air-standard-gravimetric-measurement-method-for-the-determination-of-the-pm-sub-10-sub-or-pm-sub-2-sub-d-sub-5-sub-mass-concentration-of-suspended-particulate-matter/ (accessed on 28 June 2023).

30. European Standard, B.E.. Ambient Air Quality. Standard Gravimetric Measurement Method for the Determination of the PM2.5 Mass Fraction of Suspended Particulate Matter. 2005. Available online: https://www.une.org/encuentra-tu-norma/busca-tu-norma/norma?c=N0036322 (accessed on 28 June 2023).

31. Coleman, J.R.; Meggers, F. Sensing of indoor air quality—characterization of spatial and temporal pollutant evolution through distributed sensing. *Front. Built Environ.* **2018**, *4*, 1–12. [CrossRef]

32. Giordano, M.R.; Malings, C.; Pandis, S.N.; Presto, A.A.; McNeill, V.F.; Westervelt, D.M.; Beekmann, M.; Subramanian, R. From low-cost sensors to high-quality data: A summary of challenges and best practices for effectively calibrating low-cost particulate matter mass sensors. *J. Aerosol Sci.* **2021**, *158*, 105833. [CrossRef]

33. Liang, L. Calibrating low-cost sensors for ambient air monitoring: Techniques, trends, and challenges. *Environ. Res.* **2021**, *197*, 111163. [CrossRef]

34. Clarity Node-S Home Page. Available online: https://www.clarity.io/products/clarity-node-s (accessed on 25 May 2023).

35. Ardon-Dryer, K.; Kelley, M.; Xueting, X.; Dryer, Y. The aerosol research observation station (AEROS). *Atmos. Meas. Tech.* **2022**, *15*, 2345–2360. [CrossRef]

36. OpenStreetMap Home Page. Available online: https://www.openstreetmap.org/ (accessed on 28 June 2023).

37. Song, J.; Lin, T.; Li, X.; Prishchepov, A.V. Mapping Urban Functional Zones by Integrating Very High Spatial Resolution Remote Sensing Imagery and Points of Interest: A Case Study of Xiamen, China. *Remote Sens.* **2018**, *10*, 1737. [CrossRef]

38. Sun, R.; Lü, Y.; Chen, L.; Yang, L.; Chen, A. Assessing the stability of annual temperatures for different urban functional zones. *Build. Environ.* **2013**, *65*, 90–98. [CrossRef]

39. Yu, Z.; Jing, Y.; Yang, G.; Sun, R. A New Urban Functional Zone-Based Climate Zoning System for Urban Temperature Study. *Remote Sens.* **2021**, *13*, 251. [CrossRef]

40. Kelly, K.E.; Whitaker, J.; Petty, A.; Widmer, C.; Dybwad, A.; Sleeth, D.; Martin, R.; Butterfield, A. Ambient and laboratory evaluation of a low-cost particulate matter sensor. *Environ. Pollut.* **2017**, *221*, 491–500. [CrossRef] [PubMed]

41. Li, J.; Li, H.; Ma, Y.; Wang, Y.; Abokifa, A.A.; Lu, C.; Biswas, P. Spatiotemporal distribution of indoor particulate matter concentration with a low-cost sensor network. *Build. Environ.* **2018**, *127*, 138–147. [CrossRef]

42. Rogulski, M.; Badyda, A. Investigation of Low-Cost and Optical Particulate Matter Sensors for Ambient Monitoring. *Atmosphere* **2020**, *11*, 1040. 10.3390/atmos11101040. [CrossRef]

43. Ghamari, M.; Soltanpur, C.; Rangel, P.; Groves, W.A.; Kecojevic, V. Laboratory and field evaluation of three low-cost particulate matter sensors. *IET Wirel. Sens. Syst.* **2022**, *12*, 21–32. [CrossRef]

44. Warder, S.C.; Piggott, M.D. Optimal experiment design for a bottom friction parameter estimation problem. *GEM Int. J. Geomath.* **2022**, *13*, 7. [CrossRef]

 *processes* 

Article

Examining Current Research Trends in Ozone Formation Sensitivity: A Bibliometric Analysis

Zeeshan Javed [1,†], Khalid Mehmood [1,†], Cheng Liu [2,3,4,5], Xiaojun Zheng [1,*], Chunsheng Xu [6], Aimon Tanvir [7], Muhammad Ajmal Khan [8], Nadeem Siddique [9] and Daolin Du [1]

[1] Institute of Environmental Health and Ecological Security, School of the Environment and Safety Engineering, Jiangsu University, Zhenjiang 212013, China; zeeshan@mail.ustc.edu.cn (Z.J.); ddl@ujs.edu.cn (D.D.)

[2] Department of Precision Machinery and Precision Instrumentation, University of Science and Technology of China, Hefei 230026, China

[3] Key Laboratory of Environmental Optics and Technology, Anhui Institute of Optics and Fine Mechanics, Hefei Institutes of Physical Science, Chinese Academy of Sciences, Hefei 230031, China

[4] Center for Excellence in Regional Atmospheric Environment, Institute of Urban Environment, Chinese Academy of Sciences, Xiamen 361021, China

[5] Key Laboratory of Precision Scientific Instrumentation of Anhui Higher Education Institutes, University of Science and Technology of China, Hefei 230026, China

[6] Jiangsu Yangjing Environmental Protection Service Co., Ltd., Lianyungang 222000, China; xcs13905136575@126.com

[7] Shanghai Key Laboratory of Atmospheric Particle Pollution and Prevention (LAP3), Department of Environmental Science and Engineering, Fudan University, Shanghai 200433, China

[8] Deanship of Library Affairs, Imam Abdulrahman Bin Faisal University, Dammam 1982, Saudi Arabia; makhan@iau.edu.sa

[9] Gad and Birgit Rausing Library, Lahore University of Management Sciences (LUMS), Lahore 54792, Pakistan; nadeem.siddique@lums.edu.pk

[*] Correspondence: xjzheng@ujs.edu.cn

[†] These authors contributed equally to this work.

Citation: Javed, Z.; Mehmood, K.; Liu, C.; Zheng, X.; Xu, C.; Tanvir, A.; Ajmal Khan, M.; Siddique, N.; Du, D. Examining Current Research Trends in Ozone Formation Sensitivity: A Bibliometric Analysis. *Processes* **2023**, *11*, 2240. https://doi.org/10.3390/pr11082240

Academic Editors: Daniele Sofia and Paolo Trucillo

Received: 3 June 2023
Revised: 5 July 2023
Accepted: 6 July 2023
Published: 26 July 2023

Abstract: The end of the 20th century brought about drastic changes in the tropospheric ozone (O_3) around the globe. It is, therefore, highly important to gain insight into O_3 formation mechanisms and their key precursors in order to assist policymaking to combat O_3 pollution. This article synthesizes a bibliometric analysis of O_3 formation sensitivity from 1965 to 2022, reported in English language journals available in the Web of Science Core Collection. This study shows that constant expansion in the number of publications has occurred since 2008, with the highest number occurring in 2021. Most publications are from the United States of America (USA), with 406 papers (42.7%), followed by China with 128 papers (13.5%), and the United Kingdom (UK) with 87 papers (9.1%). Citation burst analysis and significant and highly cited research work analysis are used to discover and assess evolving research tendencies. The thematic evolution of author-supplied keywords indicates that the terms "volatile organic compounds" and "ozone precursors" have recently emerged with a higher frequency. This suggests that there is a growing trend in research focused on these topics in the future. The objective of this study is to provide research primacies and future prospects for better analysis of O_3 sensitivity, thereby helping to manage O_3 pollution.

Keywords: O_3 formation sensitivity; O_3-NO_X-VOCs; O_3 precursors; bibliometrics; VOSviewer; Web of Science Core Collection

1. Introduction

Tropospheric ozone (O_3) is a harmful, invisible gas present in the atmosphere. This pollutant has detrimental effects on ecosystems and human health. It irritates the nose and eyes and causes respiratory problems [1–5]. Crop yield losses due to prolonged exposure to O_3 have been reported [6]. During the 1950s, it was discovered that O_3 is produced secondarily by the photochemical reactions of anthropogenic emissions of volatile organic compounds (VOCs) and nitrogen oxides (NO_x = NO + NO_2) [7].

O_3 formation from VOCs and NOx precursors occurs in a non-linear manner in the presence of sunlight [8–10]. NO_x concentration can increase O_3 formation, and, in other cases, VOCs can enhance O_3 formation [11–13]. Tropospheric O_3 formation is a complex process that depends on precursor species and several other factors. The COVID-19 pandemic and subsequent lockdown measures resulted in emission reductions of VOCs and NO_x, but O_3 concentrations still increased in many regions [14–16]. Recent studies suggest this is due to the various factors (e.g., aerosols, meteorology and pollution source distribution) that govern atmospheric O_3 formation [12,17].

O_3 formation in the atmosphere occurs in one of three ways: (a) VOC-limited; (b) NO_x-limited; and (c) transition regimes. In the VOC-limited regime, O_3 production relies primarily on VOC levels, whereas in the NO_x-limited regime, O_3 formation depends on NO_X concentrations [18,19]. Insufficient precursor emission reductions can lead to enhanced tropospheric O_3 production. Therefore, accurately identifying O_3 sensitivity regions is necessary to cultivate effective control strategies to combat O_3 pollution. The situation is more complex in China, particularly as the ozone concentration has not shown a reduction in its level since the implementation of China's National Air pollution prevention plan in 2013. This is due to the production of O_3 via complex heterogeneous processes [15,20,21].

Given the complex relationship between O_3 precursors and O_3 formation sensitivity, we provide a comprehensive overview of the current research trends of O_3 precursors and their formation sensitivity. A bibliometric on ozone formation sensitivity from 1965 to 2022 is presented to elucidate the patterns of research and offer perspective into the present advancement and forthcoming paradigm of research. Rapid developments in scientific research result in the increased importance of bibliometric analysis. This sort of investigation is exceptional, as it offers classified information on various aspects of a scientific discipline, including publications, keywords, citations, and collaborations. This study utilizes distinct methods compared to previous research on ozone formation sensitivity. These methodologies enable the construction of a comprehensive research framework and facilitate the identification, organization, and analysis of the principal elements within the topic. To the best of the authors' knowledge, this is the first bibliometric study on ozone formation sensitivity and its precursors. Thus, the objective is to study recent global research trends and present content and citation analyses in order to recognize the leading studies in this domain. Furthermore, this study examined a wide range of research topics and analyzed the current emerging trends to identify potential valuable directions for further research.

2. Materials and Methods

This section describes the data collection methods used in this bibliometric study of ozone formation sensitivity. Scientific literature can be systematically analysed using bibliometric techniques, retrieving data from online databases such as the Web of Science Core Collection (WoSCC) [22]. We have compiled a comprehensive list of keywords related to ozone formation sensitivity research to ensure maximum precision and recall. A query for WoSCC was formulated by connecting the keywords using Boolean operators. Specifically, the following query was used:

TS = ("ozone formation regimes" OR "ozone formation sensitivity" OR "O_3-NOX-VOCs Sensitivity" OR "ozone precursor*")

The search was performed through the 'Topic Field', which considers the 'Title', 'Abstract' and 'Keywords Plus®' of a record. "TS" stands for topic and restricts the search to the articles' title, abstract, and keywords and "OR" is a boolean operator that connects the various terms in the query. The initial query returned 972 documents. We examined the results from the publication of the first article to the date of the data retrieval (20 November 2022) to obtain a comprehensive, up-to-date understanding of the subject. We narrowed the search to 952 documents by restricting the document types to articles, proceeding papers, and review articles. We downloaded the bibliographic data and imported it into EndNote to ensure there were no duplicate titles, authors, or years. One duplicate document was found and removed. An outline of the bibliometric study road map is shown in Figure 1. Finally, all the screened documents were examined carefully for their contribution to ozone formation sensitivity. The study used MS Excel, MS Access, Power BI, an online visualization platform (https://flourish.studio/ accessed on 20 March 2022), BiblioAnalytics, Citespace, VOSviewer software (version 1.6.15), and Biblioshiny software for the analysis [23].

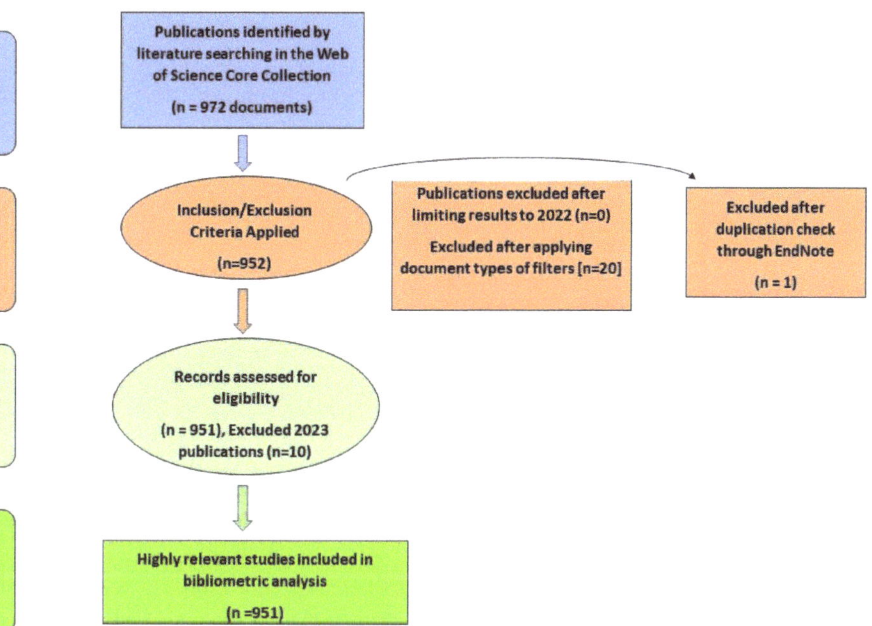

Figure 1. Flow chart of the steps of the study.

The analysis of bibliographic data is critical to identifying scientific research trends. The bibliometric analysis provides a quantitative means of evaluating scientific research and its impact by identifying the most influential authors and publications and finding research hotspots by analysing publication patterns and citation networks [24]. We systematically examine the existing literature on ozone formation sensitivity using bibliometric techniques. By identifying key research themes, we provide insights for future research directions.

3. Results

3.1. Temporal Analysis of Articles and Their Types

Scientific literature is usually evaluated on the basis of the number of publications and citations. The publication and citation trend can demonstrate the link between lifespan and citation per publication [22,25,26]. Figure 2 illustrates the growth in total publications (TP) and total citations (TC). The data show that the first research publication relating to ozone formation sensitivity was in 1965. The analysis shows little to no increase in publications in the early years, with only single-digit numbers and many years without any publications. Double-digit numbers appear between 1995 and 2008. After 2008, consistent publication growth was observed, with the highest number of publications occurring in 2021 (followed by 2020, 2012, and 2013). The results indicate that research into O_3 pollution gained attention relatively lately. This is due to the development of observation methods, including satellite and ground-based systems, over the past two decades, resulting in a vast range of continuous data with wide area coverage and high precision. The number of citations dropped after 2017 and the reasons for citations dropping after 2017 could be explored further.

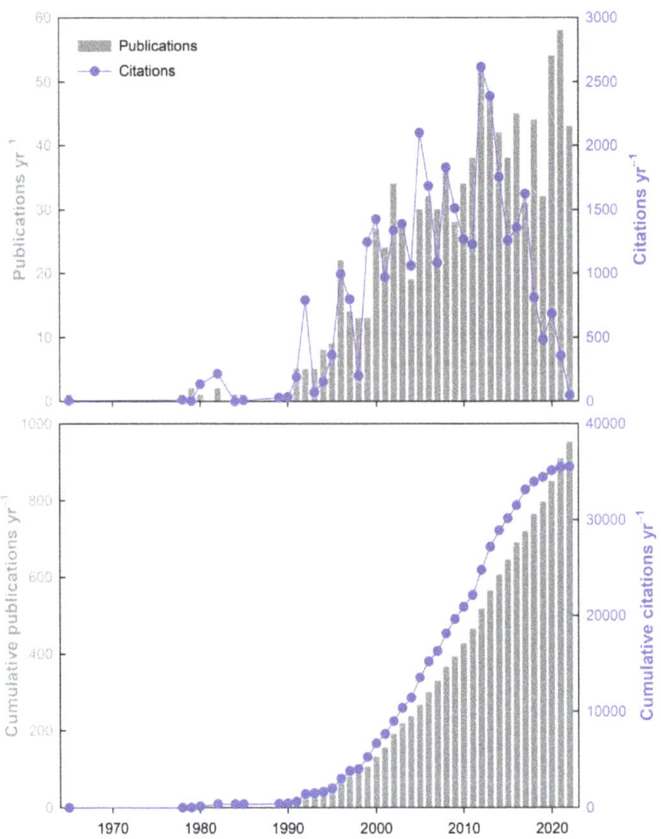

Figure 2. Yearly growth in ozone formation sensitivity research in terms of publications and citations.

The preferred document type for researchers working on ozone formation sensitivity is "article" with 876 TP, distantly followed by "proceedings paper" with 43 TP, and "review" with 32 TP. "Articles" also have the highest number of citations (29,274 TC), followed by "reviews" (4801 TC) and "proceedings papers" (1468 TC). The document type "review" has more impact (150 TC/TP) than "articles" (33.42 TC/TP) and "proceedings papers" (34.14 TC/TP).

3.2. Authorship Patterns and Most Prolific Scholars in Ozone Formation Sensitivity

Figure 3 shows the authorship patterns in ozone formation sensitivity research ranging from 1 to 41. The data analysis shows a collaborative research trend in this field of study. This analysis reveals the domination of two-author patterns with 139 publications, followed by five and four-author patterns with 136 and 132 publications, respectively. The citation analysis ranks four-author publications top, with 5042 citations, followed by six-author reports, with 4717 citations. Two-author papers have 4426 citations.

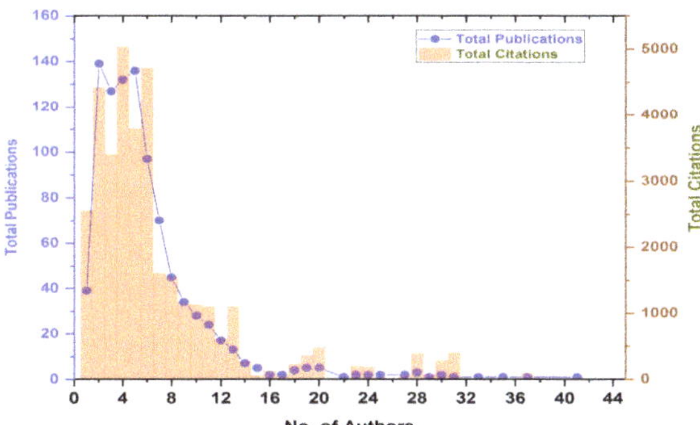

Figure 3. Authorship patterns in ozone formation sensitivity research in terms of publications and citations.

Table 1 gives the top 20 most prolific researchers in ozone formation sensitivity research. The data shows a trend for collaborative research among the most productive authors, as all authors among the top 20, except one (Shindell DT) published research which was the result of a joint effort. The most productive author, Wang JL, published five pieces of research on the topic as the first author. The second, Collins WJ, published five documents as the first author and 12 in other author categories. The third, Horowitz LW, maintained that position by contributing all publications in other author categories. The citation analysis ranks Shindell DT in the top position with the most citations, followed by Faluvegi G and Collins WJ.

Table 1. Top 20 most prolific authors.

Author	Publications by Authorship			TP	TC
	Single	First	Other		
WANG JL		5	18	23	776
COLLINS WJ		5	12	17	1692
HOROWITZ LW			16	16	1550
NAIK V		2	13	15	1400
DERWENT RG		6	8	14	572
PARRISH DD		3	10	13	842
THOURET V			13	13	743
CHANG CC		3	9	12	434
FIORE AM		1	11	12	715
NEDELEC P			12	12	750
STEVENSON DS		1	11	12	1174
SHINDELL DT	1	3	7	11	2081
SUDO K		1	10	11	1318
WEST JJ		2	9	11	767
ZENG G		2	9	11	1265
BEIG G		3	7	10	199
FALUVEGI G			10	10	1831
LAMARQUE JF		2	8	10	1488
PICKERING KE		2	8	10	494
THOMPSON AM		2	8	10	617

3.3. Most Productive Journals and Publishers

Identifying the distribution of journals is a crucial component of bibliometric analysis. This information is valuable for guiding future researchers in determining where to concentrate their efforts for finding literature and publishing their work. The top 20 most productive journals publishing on O_3 formation sensitivity are listed in Table 2. According to the analysis, Atmospheric Environment is ranked top for total publications (TP) with 169 and cited papers (CP) with 164. Atmospheric Chemistry and Physics, and the Journal of Geophysical Research-Atmospheres, take joint second place with a TP count of 100. The highest number of total citations are also from Atmospheric Environment, followed by the Journal of Geophysical Research-Atmospheres, and Atmospheric Chemistry and Physics. Geoscientific Model Development appears in 12th place; however, its impact (TC/TP) is the highest, followed by Environmental Science & Technology and Science of the Total Environment".

Table 2. Most productive journals.

Journal	CP	TP	TC	Impact (TC/TP)
ATMOSPHERIC ENVIRONMENT	164	169	7192	42.56
ATMOSPHERIC CHEMISTRY AND PHYSICS	96	100	4311	43.11
JOURNAL OF GEOPHYSICAL RESEARCH-ATMOSPHERES	98	100	5343	53.43
SCIENCE OF THE TOTAL ENVIRONMENT	31	31	1723	55.58
ATMOSPHERE	24	29	197	6.79
JOURNAL OF THE AIR & WASTE MANAGEMENT ASSOCIATION	27	27	704	26.07
GEOPHYSICAL RESEARCH LETTERS	24	24	915	38.13
ENVIRONMENTAL SCIENCE & TECHNOLOGY	19	21	1220	58.1
ENVIRONMENTAL POLLUTION	16	16	359	22.44
ATMOSPHERIC RESEARCH	14	14	555	39.64
ENVIRONMENTAL RESEARCH LETTERS	13	14	235	16.79
ENVIRONMENTAL MONITORING AND ASSESSMENT	11	12	219	18.25
JOURNAL OF ATMOSPHERIC CHEMISTRY	12	12	326	27.17
ENVIRONMENTAL SCIENCE AND POLLUTION RESEARCH	12	12	391	32.58
AIR QUALITY ATMOSPHERE AND HEALTH	9	10	133	13.3
INTERNATIONAL JOURNAL OF REMOTE SENSING	8	8	85	10.63
GEOSCIENTIFIC MODEL DEVELOPMENT	8	8	534	66.75
WATER AIR AND SOIL POLLUTION	8	8	84	10.5
JOURNAL OF APPLIED METEOROLOGY	7	7	132	18.86

The top publishers on the subject of O_3 formation are listed in Table 3, with Elsevier being on top, followed by the Amer geophysical union and Copernicus gesellschaft mbh. As well as the highest number of publications, Elsevier also has the highest number of citations, followed by Amer geophysical union and Copernicus gesellschaft mbh. The Nature Publishing Group has the highest impact (TC/TP), despite comparatively few publications, followed by Wiley-Blackwell and the Natl Acad Sciences.

Table 3. Most productive publishers.

Publisher	TP	TC	Impact (TC/TP)
ELSEVIER	312	11,866	38.03
AMER GEOPHYSICAL UNION	130	6279	48.3
COPERNICUS GESELLSCHAFT MBH	114	4931	43.25
SPRINGER	76	1849	24.33
MDPI	39	285	7.31
TAYLOR & FRANCIS LTD	35	706	20.17
AMER CHEMICAL SOC	24	1252	52.17
WILEY-BLACKWELL	17	1045	61.47
IOP PUBLISHING LTD	14	235	16.79
AIR & WASTE MANAGEMENT ASSOC	12	424	35.33
KLUWER ACADEMIC PUBL	12	394	32.83
NATURE PUBLISHING GROUP	11	1208	109.82
SCIENCE PRESS	11	224	20.36
AMER METEOROLOGICAL SOC	10	390	39
ROYAL SOC CHEMISTRY	9	459	51
NATL ACAD SCIENCES	7	400	57.14
TAIWAN ASSOC AEROSOL RES-TAAR	7	131	18.71
TURKISH NATL COMMITTEE AIR POLLUTION RES & CONTROL-TUNCAP	6	81	13.5
NATURE PORTFOLIO	6	334	55.67
INDERSCIENCE ENTERPRISES LTD	5	49	9.8

3.4. Countries and Organizational Productivity

Based on the author's affiliations, the regional research output on the subject of O_3 formation regime is presented. It is important to mention here that the number of publications in a particular area of interest is representative of the significance of that specific region for the subject. Figure 4 illustrates the contribution to research on the topic on a continental basis. Europe has the most publications (contributions from 29 countries), followed by North America (two countries) and Asia (23 countries). Europe also has the highest number of citations, followed by North America and Asia.

Figure 5 shows the most productive countries. The USA appears on top with 406 publications, followed by China (128) and the UK (87). The USA also has the highest number of citations. The UK, however, published the third-highest number of documents but is in second place for citations. Articles are regarded as publications from the United Kingdom (UK) if they are from Ireland, England, Scotland, Northern Ireland, and Wales. These countries are also among the most impactful in terms of the emission of gases and have greater resources to study the corresponding air pollution problems. There is greater mitigation concern for O_3, compared to other air pollutants, due to its complex non-linear photochemistry, especially in China [9]. Therefore, researchers in these countries have recently started working on this topic, resulting in a high number of publications.

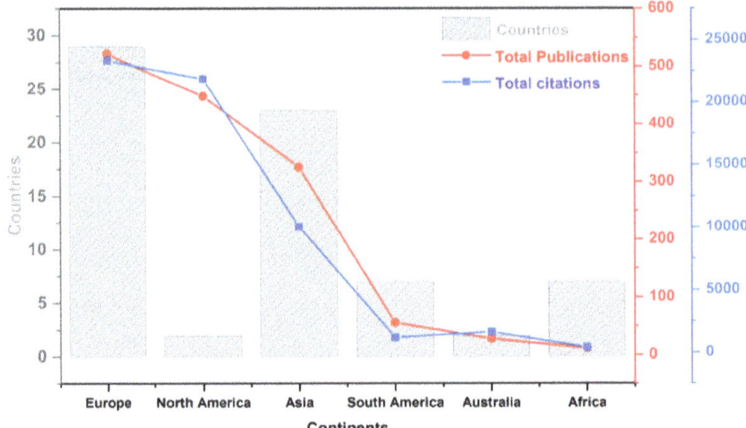

Figure 4. Most productive continents with number of countries and productivity (Publications and citations).

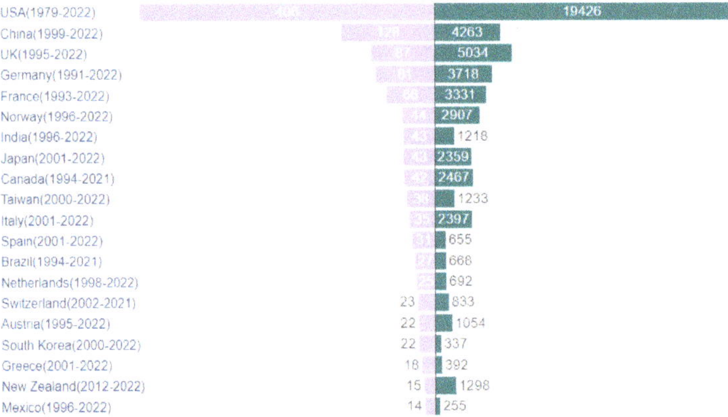

Figure 5. Most productive countries in terms of total publications and citations (Purple bar shows number of publications and green shows number of citations).

Bibliometric analysis offers insights into the leading institutions in various countries regarding the subject. This analysis is beneficial for researchers, as it helps them identify potential hubs for specific research, training opportunities, workshops, and post-doctoral study tours. By examining the most productive institutions in various countries, scholars can establish connections and collaborations to further their research endeavours [27].

Figure 6 shows the top organizations and their impact on publishing research on this topic. Ten of the top 20 organizations are located in North America, including the top three. The other ten come from Asia and Europe, with five from each. North American organizations also have higher impacts than those from other continents. The top four organizations, in terms of publications, are NASA (National Aeronautics and Space Administration), USA, NOAA (National Oceanic and Atmospheric Administration), USA, University of Colorado, USA and China Academy of Sciences, China. It can be inferred that these organizations are actively involved in studying the sensitivity of ozone formation.

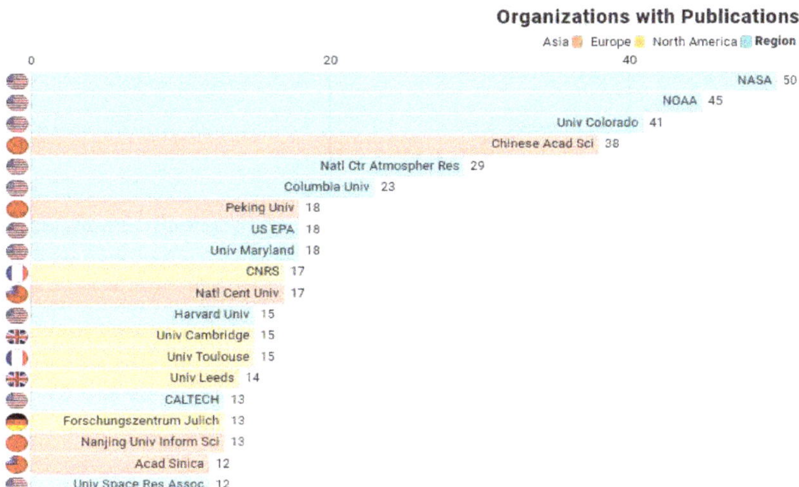

Figure 6. Most productive organizations in ozone formation sensitivity research publications.

Figure S1 shows the top organizations in terms of citations in research regarding O_3 formation sensitivity. Figure S2 shows collaborations among countries. The USA has the most collaborations and published research on the topic, collaborating with the highest number of countries. The UK is the second most collaborative country, followed by Germany, France, and China. The USA collaborates most with the UK, Germany, France, China, and Norway. The UK collaborates most with Germany, France, Norway, and Japan.

3.5. Frequently Used Author Keywords

Author keyword analysis generates information about the words that appear in the titles or abstracts of publications. This kind of analysis is crucial for exploring research topics, as authors use keywords to convey specific ideas [28]. Based on author keywords, we investigate the areas focused on and give a topical analysis with an emphasis on O_3 formation regimes over the years. We investigate the shifts in the focus of authors dealing with the topic. Keywords with higher recurrence are chosen carefully to ascertain the key research scopes and related evolving dimensions [29,30].

Figure 7 shows the author keyword map, and Figure S3 shows the author keywords as a word cloud. The size of the circles in Figure 7 indicates the keyword search frequency use, with bigger circles implying more significant usage. Likewise, the word size in Figure S3 indicates the keyword use frequency. In both Figures 7 and S3, the author keywords "ozone", "ozone precursors", "air quality", "surface ozone", "tropospheric ozone", "nitrogen oxides", "climate change", and "air pollution" appear most frequently. Figure S4 shows the evolution of author-supplied keywords in ozone research and the author-supplied keywords with the highest frequency in the corresponding year. The keyword flow between nodes shows the evolutionary direction, while the node size indicates keyword usage frequency. Themes that have no connection with the evolved themes are not shown.

The keyword "ozone" remained the same from 1965 to 2000 but then evolved into new keywords, as shown in Figure S4. The keyword "biogenic emissions" evolved into "volatile organic compounds" and "ozone precursors".

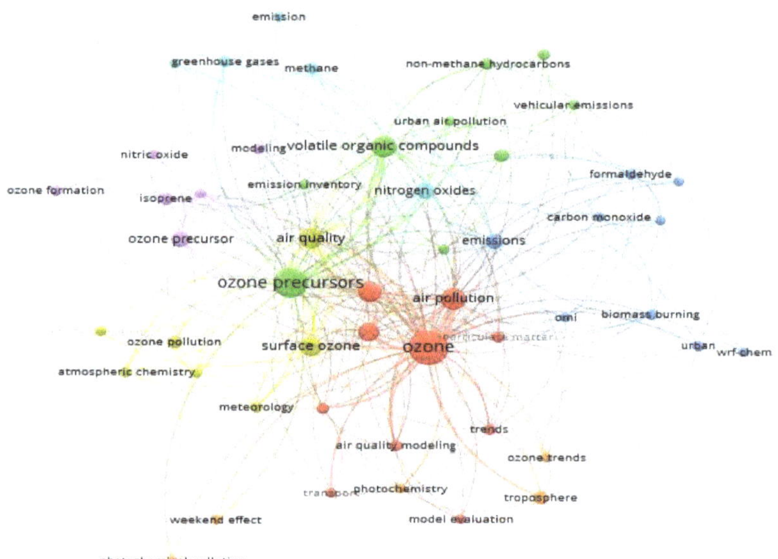

Figure 7. Frequently used author keywords network visualization in ozone formation sensitivity research.

3.6. Highly Cited Documents

Citation analysis is an essential bibliometric feature that demonstrates the significance and prestige of a research paper within the scientific community. A highly impactful article (HIA) is an indicator of a high-quality publication. Wang and Ho [28] propose that researchers should prioritize recent HIAs, taking into account their citation count. Sillman S [11] is the most cited document in ozone formation regime research, published in the journal Atmospheric Environment. This article provides an overview of the research into the relationship between O_3 and its primary precursors, volatile organic compounds and nitrogen oxides. The second most cited article is Wang et al. [31], published in the journal Science of the Total Environment. This study provides a review of studies on tropospheric O_3 in China and insight into the impact of meteorological processes and precursors on ozone formation. The third most cited document is Chameides et al. [32], published in the Journal of Geophysical Research. This study analyses concentrations of O_3, NOx and non-methane hydrocarbons to show the relationship between ozone and its precursors.

4. Implications and Limitations of This Study

The results of this study have vital implications for the scientific productivity surrounding the subject. The analyses pave the way toward a better understanding of how specific countries, institutions, authors and research hotspots contribute to ozone formation research. This study highlights the extensive work published over the years based on various metrics and illustrates the standing of various research institutions in terms of research output.

It is noteworthy that, except for China, Asian countries lag behind European countries in terms of research productivity. A detailed insight into current and prospective positions on the subject is provided, with an in-depth literature review of pioneering research articles and key research hotpots. A comprehensive analysis with detailed information on each parameter is conducted. A country's scientific productivity is, for instance, not only an indicator of the quality of research but of the potential for future collaboration networks. The high impact of published research is suggestive of the vitality of the subject for its specific audience. An important implication of the current analysis is recognizing faltering deviations in the research dynamics, which is crucial for policymaking in this domain.

Our study reveals some significant and interesting findings; however, there are certain limitations. Firstly, it is difficult to extract full information from keyword analysis maps generated by the bibliometric technique. A more comprehensive investigation could expand the information content to include the titles, abstracts and main text, not relying merely on keywords. Furthermore, we show that the subject lacks quantitative studies, which should be more of a focus in the future. However, such studies require objective data, which is either lacking or difficult to obtain. There is still uncertainty in some data sources and about data collection tools' ability to provide the desired datasets for analysis and research. Finally, this study only includes articles from the WOSCC search engine, which covers most articles, but some articles from other databases and search engines may be missing. Therefore, an analysis based on cross-comparison and multi-source searching would be more accurate and conclusive.

5. Conclusions

The current study uses a bibliometric technique to analyse the growth in the subject domain by finding recent trends and hotspots in O_3 formation sensitivity research. In order to meet the objectives, we perform an analysis of 951 publications and visualize the prospects of these publications in terms of prominent countries, research categories, institutions, authors, and journals. This study is a crucial indicator of the importance and contribution of research institutions that employ different research metrics across the globe. In terms of research output, the USA and China emerge on top. The USA has the greatest impact, followed by the UK, Germany, France, and China. On a territory basis, Europe is the most productive continent, followed by North America and Asia. Based on the analysis of the thematic evolution using author-supplied keywords, it is observed that the terms "volatile organic compounds" and "ozone precursors" have recently gained prominence with an increased frequency. This indicates a growing trend in research focused on these topics in the future. This analysis is crucial, as it has a substantial tendency to monitor indeterminate variations in research inclinations reflected in the literature, which act as crucial keys to assist policymaking. Using recent citation trends, the articles with the highest impact are identified.

Supplementary Materials: The following supporting information can be downloaded at: https://www.mdpi.com/article/10.3390/pr11082240/s1, Figure S1: Most productive organizations in terms of citation regarding ozone formation sensitivity research; Figure S2. The collaborations network among the countries regarding ozone formation sensitivity research; Figure S3. The frequently used author keywords represented as word clouds; Figure S4. Thematic evolution of Author keywords in Ozone formation sensitivity research.

Author Contributions: Conceptualization, Z.J. and K.M.; methodology, Z.J. and K.M.; software, K.M.; validation, C.X., A.T. and Z.J.; formal analysis, K.M. and Z.J.; investigation, X.Z.; resources, Z.J., X.Z. and D.D.; data curation, A.T., M.A.K. and N.S.; writing—original draft preparation, Z.J.; writing—review and editing, C.L., X.Z. and D.D.; visualization, M.A.K. and N.S.; supervision, X.Z. and K.M.; project administration, Z.J. and X.Z.; funding acquisition, Z.J., X.Z. and D.D. All authors have read and agreed to the published version of the manuscript.

Funding: This study was supported by the Jiangsu Province funding program for Excellent Postdoctoral Talent (Grant No. 2022ZB651), the National Natural Science Foundation of China (Grant No. 32271587), the Carbon Peak and Carbon Neutrality Technology Innovation Foundation of Jiangsu Province (Grant No. BK20220030), the Scientific Research Foundation for Senior Talent of Jiangsu University, China (Grant No. 20JDG067), and the Jiangsu Province "Double Innovation PhD" Grant.

Institutional Review Board Statement: Not applicable.

Informed Consent Statement: Not applicable.

Data Availability Statement: The datasets used during the study are available from the corresponding author on request.

Conflicts of Interest: The authors declare no conflict of interest.

References

1. Khaniabadi, Y.O.; Hopke, P.K.; Goudarzi, G.; Daryanoosh, S.M.; Jourvand, M.; Basiri, H. Cardiopulmonary mortality and COPD attributed to ambient ozone. *Environ. Res.* **2017**, *152*, 336–341. [CrossRef] [PubMed]
2. Huang, J.; Li, G.; Xu, G.; Qian, X.; Zhao, Y.; Pan, X.; Huang, J.; Cen, Z.; Liu, Q.; He, T.; et al. The burden of ozone pollution on years of life lost from chronic obstructive pulmonary disease in a city of Yangtze River Delta, China. *Environ. Pollut.* **2018**, *242*, 1266–1273. [CrossRef] [PubMed]
3. Yu, S.; Yin, S.; Zhang, R.; Wang, L.; Su, F.; Zhang, Y.; Yang, J. Spatiotemporal characterization and regional contributions of O_3 and NO_2: An investigation of two years of monitoring data in Henan, China. *J. Environ. Sci.* **2020**, *90*, 29–40. [CrossRef] [PubMed]
4. Belis, C.A.; Van Dingenen, R.; Klimont, Z.; Dentener, F. Scenario analysis of PM2.5 and ozone impacts on health, crops and climate with TM5-FASST: A case study in the Western Balkans. *J. Environ. Manag.* **2022**, *319*, 115738. [CrossRef]
5. Juráň, S.; Grace, J.; Urban, O. Temporal Changes in Ozone Concentrations and Their Impact on Vegetation. *Atmosphere* **2021**, *12*, 82. [CrossRef]
6. Chaudhary, I.J.; Rathore, D. Relative effectiveness of ethylenediurea, phenyl urea, ascorbic acid and urea in preventing groundnut (*Arachis hypogaea* L.) crop from ground level ozone. *Environ. Technol. Innov.* **2020**, *19*, 100963. [CrossRef]
7. Haagen-Smit, A.J. Chemistry and physiology of Los Angeles smog. *Ind. Eng. Chem.* **1952**, *44*, 1342–1346. [CrossRef]
8. Ma, M.; Yao, G.; Guo, J.; Bai, K. Distinct spatiotemporal variation patterns of surface ozone in China due to diverse influential factors. *J. Environ. Manag.* **2021**, *288*, 112368. [CrossRef]
9. Sun, J.; Shen, Z.; Wang, R.; Li, G.; Zhang, Y.; Zhang, B.; He, K.; Tang, Z.; Xu, H.; Qu, L.; et al. A comprehensive study on ozone pollution in a megacity in North China Plain during summertime: Observations, source attributions and ozone sensitivity. *Environ. Int.* **2021**, *146*, 106279. [CrossRef]
10. Du, Y.; Zhao, K.; Yuan, Z.; Luo, H.; Ma, W.; Liu, X.; Wang, L.; Liao, C.; Zhang, Y. Identification of close relationship between large-scale circulation patterns and ozone-precursor sensitivity in the Pearl River Delta, China. *J. Environ. Manag.* **2022**, *312*, 114915. [CrossRef]
11. Sillman, S. The relation between ozone, NOx and hydrocarbons in urban and polluted rural environments. *Atmos. Environ.* **1999**, *33*, 1821–1845. [CrossRef]
12. Lu, H.; Lyu, X.; Cheng, H.; Ling, Z.; Guo, H. Overview on the spatial–temporal characteristics of the ozone formation regime in China. *Environ. Sci. Process. Impacts* **2019**, *21*, 916–929. [CrossRef] [PubMed]
13. Zhang, K.; Xu, J.; Huang, Q.; Zhou, L.; Fu, Q.; Duan, Y.; Xiu, G. Precursors and potential sources of ground-level ozone in suburban Shanghai. *Front. Environ. Sci. Eng.* **2020**, *14*, 92. [CrossRef]
14. Javed, Z.; Wang, Y.; Xie, M.; Tanvir, A.; Rehman, A.; Ji, X.; Xing, C.; Shakoor, A.; Liu, C. Investigating the impacts of the COVID-19 lockdown on trace gases using ground-based MAX-DOAS observations in Nanjing, China. *Remote Sens.* **2020**, *12*, 3939. [CrossRef]
15. Tanvir, A.; Javed, Z.; Jian, Z.; Zhang, S.; Bilal, M.; Xue, R.; Wang, S.; Bin, Z. Ground-based MAX-DOAS observations of tropospheric NO_2 and HCHO during COVID-19 lockdown and spring festival over Shanghai, China. *Remote Sens.* **2021**, *13*, 488. [CrossRef]
16. Javed, Z.; Bilal, M.; Qiu, Z.; Li, G.; Sandhu, O.; Mehmood, K.; Wang, Y.; Ali, M.A.; Liu, C.; Wang, Y.; et al. Spatiotemporal characterization of aerosols and trace gases over the Yangtze River Delta region, China: Impact of trans-boundary pollution and meteorology. *Environ. Sci. Eur.* **2022**, *34*, 86. [CrossRef] [PubMed]
17. Kroll, J.H.; Heald, C.L.; Cappa, C.D.; Farmer, D.K.; Fry, J.L.; Murphy, J.G.; Steiner, A.L. The complex chemical effects of COVID-19 shutdowns on air quality. *Nat. Chem.* **2020**, *12*, 777–779. [CrossRef]
18. Duncan, B.N.; Yoshida, Y.; Olson, J.R.; Sillman, S.; Martin, R.V.; Lamsal, L.; Hu, Y.; Pickering, K.E.; Retscher, C.; Allen, D.J.; et al. Application of OMI observations to a space-based indicator of NOx and VOC controls on surface ozone formation. *Atmos. Environ.* **2010**, *44*, 2213–2223. [CrossRef]
19. Hong, Q.; Zhu, L.; Xing, C.; Hu, Q.; Lin, H.; Zhang, C.; Zhao, C.; Liu, T.; Su, W.; Liu, C. Inferring vertical variability and diurnal evolution of O3 formation sensitivity based on the vertical distribution of summertime HCHO and NO_2 in Guangzhou, China. *Sci. Total Environ.* **2022**, *827*, 154045. [CrossRef]
20. Dai, H.; Zhu, J.; Liao, H.; Li, J.; Liang, M.; Yang, Y.; Yue, X. Co-occurrence of ozone and PM2.5 pollution in the Yangtze River Delta over 2013–2019: Spatiotemporal distribution and meteorological conditions. *Atmos. Res.* **2021**, *249*, 105363. [CrossRef]
21. Xue, J.; Zhao, T.; Luo, Y.; Miao, C.; Su, P.; Liu, F.; Zhang, G.; Qin, S.; Song, Y.; Bu, N.; et al. Identification of ozone sensitivity for NO_2 and secondary HCHO based on MAX-DOAS measurements in northeast China. *Environ. Int.* **2022**, *160*, 107048. [CrossRef] [PubMed]
22. Mehmood, K.; Bao, Y.; Saifullah Cheng, W.; Khan, M.A.; Siddique, N.; Abrar, M.M.; Soban, A.; Fahad, S.; Naidu, R. Predicting the quality of air with machine learning approaches: Current research priorities and future perspectives. *J. Clean. Prod.* **2022**, *379*, 134656. [CrossRef]
23. Ye, N.; Kueh, T.B.; Hou, L.; Liu, Y.; Yu, H. A bibliometric analysis of corporate social responsibility in sustainable development. *J. Clean. Prod.* **2020**, *272*, 122679. [CrossRef]
24. Wong, S.; Nyakuma, B.; Wong, Y.; Lee, T.; Lee, H.; Lee, H. Microplastics and nanoplastics in global food webs: A bibliometric analysis (2009–2019). *Mar. Pollut. Bull.* **2020**, *158*, 111432. [CrossRef] [PubMed]
25. Chiu, W.-T.; Ho, Y.-S. Bibliometric analysis of tsunami research. *Scientometrics* **2007**, *73*, 3–17. [CrossRef]

26. Usman, M.; Ho, Y.-S. A bibliometric study of the Fenton oxidation for soil and water remediation. *J. Environ. Manag.* **2020**, *270*, 110886. [CrossRef]
27. Vanzetto, G.V.; Thomé, A. Bibliometric study of the toxicology of nanoescale zero valent iron used in soil remediation. *Environ. Pollut.* **2019**, *252*, 74–83. [CrossRef]
28. Wang, C.-C.; Ho, Y.-S. Research trend of metal–organic frameworks: A bibliometric analysis. *Scientometrics* **2016**, *109*, 481–513. [CrossRef]
29. Hamidi, A.; Ramavandi, B. Evaluation and scientometric analysis of researches on air pollution in developing countries from 1952 to 2018. *Air Qual. Atmos. Health* **2020**, *13*, 797–806. [CrossRef]
30. Mao, G.; Hu, H.; Liu, X.; Crittenden, J.; Huang, N. A bibliometric analysis of industrial wastewater treatments from 1998 to 2019. *Environ. Pollut.* **2021**, *275*, 115785. [CrossRef]
31. Wang, T.; Xue, L.; Brimblecombe, P.; Lam, Y.F.; Li, L.; Zhang, L. Ozone pollution in China: A review of concentrations, meteorological influences, chemical precursors, and effects. *Sci. Total Environ.* **2017**, *575*, 1582–1596. [CrossRef] [PubMed]
32. Chameides, W.L.; Fehsenfeld, F.; Rodgers, M.O.; Cardelino, C.; Martinez, J.; Parrish, D.; Lonneman, W.; Lawson, D.R.; Rasmussen, R.A.; Zimmerman, P.; et al. Ozone precursor relationships in the ambient atmosphere. *J. Geophys. Res. Atmos.* **1992**, *97*, 6037–6055. [CrossRef]

Article

Design and Implementation of a Crowdsensing-Based Air Quality Monitoring Open and FAIR Data Infrastructure

Paolo Diviacco *, Massimiliano Iurcev, Rodrigo José Carbajales, Alberto Viola and Nikolas Potleca

National Institute of Oceanography and Applied Geophysics, Borgo Grotta Gigante 42/C, 34010 Sgonico, Italy
* Correspondence: pdiviacco@ogs.it; Tel.: +39-040-2140380

Abstract: This work reports on the development of a real-time vehicle sensor network (VSN) system and infrastructure devised to monitor particulate matter (PM) in urban areas within a participatory paradigm. The approach is based on the use of multiple vehicles where sensors, acquisition and transmission devices are installed. PM values are measured and transmitted using standard mobile phone networks. Given the large number of acquisition platforms needed in crowdsensing, sensors need to be low-cost (LCS). This sets limitations in the precision and accuracy of measurements that can be mitigated using statistical methods on redundant data. Once data are received, they are automatically quality controlled, processed and mapped geographically to produce easy-to-understand visualizations that are made available in almost real time through a dedicated web portal. There, end users can access current and historic data and data products. The system has been operational since 2021 and has collected over 50 billion measurements, highlighting several hotspots and trends of air pollution in the city of Trieste (north-east Italy). The study concludes that (i) this perspective allows for drastically reduced costs and considerably improves the coverage of measurements; (ii) for an urban area of approximately 100,000 square meters and 200,000 inhabitants, a large quantity of measurements can be obtained with a relatively low number (5) of public buses; (iii) a small number of private cars, although less easy to organize, can be very important to provide infills in areas where buses are not available; (iv) appropriate corrections for LCS limitations in accuracy can be calculated and applied using reference measurements taken with high-quality standardized devices and methods; and that (v) analyzing the dispersion of measurements in the designated area, it is possible to highlight trends of air pollution and possibly associate them with traffic directions. Crowdsensing and open access to air quality data can provide very useful data to the scientific community but also have great potential in fostering environmental awareness and the adoption of correct practices by the general public.

Keywords: particulate matter; crowdsensing; citizen science; low-cost sensors; infrastructure

Citation: Diviacco, P.; Iurcev, M.; Carbajales, R.J.; Viola, A.; Potleca, N. Design and Implementation of a Crowdsensing-Based Air Quality Monitoring Open and FAIR Data Infrastructure. *Processes* **2023**, *11*, 1881. https://doi.org/10.3390/pr11071881

Academic Editors: Hsin Chu, Daniele Sofia and Paolo Trucillo

Received: 6 May 2023
Revised: 11 June 2023
Accepted: 20 June 2023
Published: 23 June 2023

1. Introduction

The World Health Organization defines air pollution as the "contamination of the indoor or outdoor environment by any chemical, physical or biological agent that modifies the natural characteristics of the atmosphere" [1]. Epidemiological evidence suggests that polluted air is one of the leading factors associated with the development of respiratory illness, cardiovascular disease and lung cancer [2]. At the same time, air pollution directly and indirectly affects the climate and damages buildings and cultural heritage [3,4]. Many countries have introduced specific legislation setting strict objectives for air quality. In the United States, this was implemented in the 1970s with the Air Quality Act [5] and later in Europe with the 2008/50/EC directive [6]. Notwithstanding the fact that, in general, air quality has improved a lot since then, there are still several hotspots of air pollution in most of the Western countries [7,8]. Recently, several analysts highlighted the risks that, due to the current geopolitical situation and the shortage of natural gas, the resulting increase in the use of solid fuels poses to worsen the situation [9]. Indeed, while the combustion of natural gas contributes to the formation of smog and acid rain, particulate matter (PM) emissions are generally low [10]. In contrast, liquid and solid fuel combustion produces large quantities of PM and high concentrations of sulfur and heavy metals [11].

Air quality monitoring is generally performed by government agencies using standardized methods and devices at specific fixed locations in order to have reliable long time series that could be considered as reference measurements. The position of such monitoring stations is linked to the specific problem to be considered, be it traffic congestion in an urban area or an industrial site or any other issue that could be relevant to public health.

Reference methods are intrinsically expensive and need well-trained personnel. As a consequence, there are limitations in the possible number of stations to install. In order to reconstruct the geographic distribution of the quality of air in an area, measurements at the sparse stations can be inter/extrapolated using statistical [12] or modeling techniques [13]. Although generally very accurate, these methods can be problematic where high gradients are present. In such cases, the possibility to increase the geographic and temporal coverage and resolution of phenomena could be very helpful.

In this perspective, a new paradigm can be introduced that has already been successfully applied in several scientific fields. Since the seminal work of Irwin [14], a large number of initiatives have, in fact, flourished that aim at enrolling resources from outside the scientific community and employing them within several research activities [15–19]. This new approach is generally referred to as citizen science, although slightly different definitions may be more suited to each specific application. In this work, we prefer to use the term 'crowdsensing', which refers to a technique where a large number of volunteers offer their help in acquiring and sharing measurements taken with devices or software provided by the project designer. Within this perspective, a large literature exists on the use of air quality sensors [20] for several air quality parameters [21–23], both indoors [24–26] and outdoors [27,28]. Within this work, we will focus on mobile crowdsensing [29], where PM acquisition devices are installed on vehicles such as cars, vans or public buses. As a result, the availability of a large number of vectors has the potential to radically increase the amount of data available and improve geographic and temporal coverage and resolution.

Moreover, participative research activities such as crowdsensing and citizen science have the possibility to deliver benefits well beyond scientific outcomes. Starting from the importance of understanding how the general public is informed about environmental topics, such as climate change or air quality, it is easy to understand that if the discussions remain confined within the scientific community, it is very unlikely that the general public will be able to take informed actions in order to mitigate those phenomena. Effective and correct communication via the mass media and the Internet is therefore necessary to spread correct messages, which, unfortunately, is often not the case because mass media and, in particular, social media may at best be partial if not altogether manipulative.

In contrast, active participation in research activities by volunteers, together with the possibility to freely and easily access reliable data and information on environmental issues, can have a wide range of positive effects. These can span from an increase in trust in the scientific community, to the improvement of awareness and engagement of citizens in environmental issues [30] up to their empowerment in steering political and economic decisions [31,32]. While improvement in subject-matter knowledge and stronger scientific literacy is generally easy to be traced as a participative research outcome, the actual impact on policy making of such initiatives is not always easy to fully understand and is sometimes a matter of debate [33,34].

To support citizen science and crowdsourcing activities in the field of environmental monitoring, several, often intermingled, aspects have to be considered. Each of them can constitute on their own a topic for a considerable analysis. In previous works [35–39], we have considered some of these aspects and the scientific results that we were able to obtain by exploiting them. Here, we describe the technological aspects of our work in the hope that our experience could prove beneficial to others intending to replicate or eventually improve what we have been able to build so far.

2. Materials and Methods

Within this work, we will describe a system and infrastructure we developed that, leveraging the crowdsensing paradigm, allows the monitoring of air quality and represents its geographic distribution on a web portal in real time. The initiative is named "COCAL" after the dialectal term used for seagulls in the city of Trieste (Italy), where it has been developed and first deployed. The reason for using a seabird name comes from the fact that Trieste is a coastal city where applications of the crowdsensing paradigm can be envisaged in multiple environments. As a matter of fact, the first trials of COCAL were focused on monitoring marine parameters such as temperature, pH, salinity and dissolved oxygen. Further information on such trials can be found in [35,36].

Trieste is located at the north-east tip of the Adriatic Sea and occupies an NW-SE trending elongated area of approximately 90 square kilometers between the sea and the Karst plateau, which acts as a barrier to air masses (Figure 1). The city center lies in a restricted area that is characterized by economic activities linked to the tertiary sector and tourism and separated from an industrial area located in an SE sector. The wind regime is characterized in winter by a strong NE wind called Bora, which can effectively move polluted masses from the city to the sea and other nearby regions [13,40]. In summer, the sea breezes are the prevalent factor conditioning the behavior and position of the polluted air masses [41].

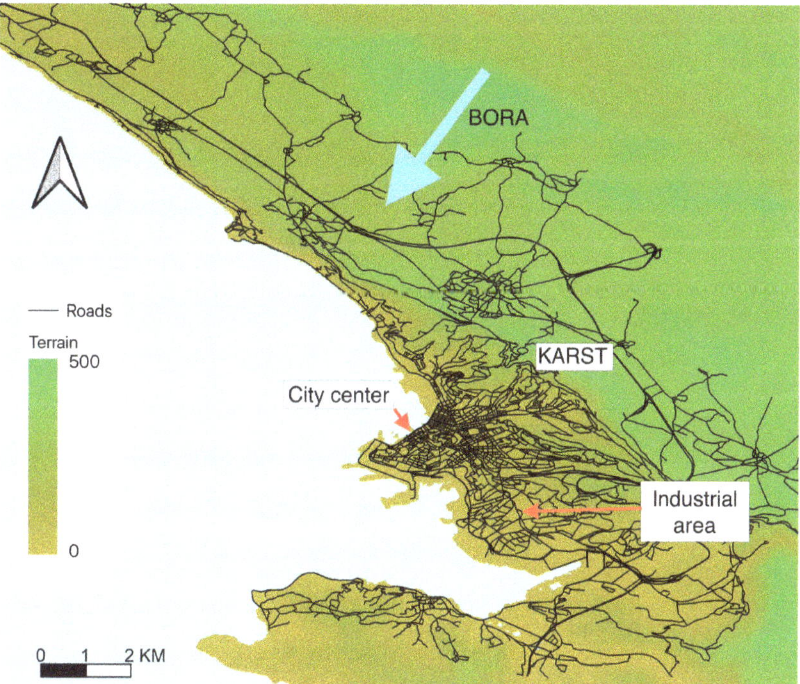

Figure 1. The area where the COCAL system is installed.

Within this work, we focus on monitoring PM only, but we are currently working on extending the method to other pollutants. It is worth highlighting that all the data acquired within this initiative and in other crowdsensing initiatives are gathered in an integrated database and managed within a fully FAIR-compliant perspective following international standards as mandated by ISO and OGC.

2.1. PM Sensors

Given the importance in crowdsensing of using a large number of acquisition devices, it is evident that increasing their number will inevitably imply increasing the overall cost of the initiative. Under this approach, in fact, it is not possible to use conventional PM monitoring techniques such as filters and gravimetric mass detection, which are very expensive and based on standardized procedures that can only be performed by trained personnel. Therefore, low-cost sensors (LCS) are needed. New technologies have emerged that use laser scattering, which relates the waveform of the scattered light to the diameter and number of particles, enabling real-time and continuous measurements of particulate matter.

A detailed description of the technologies behind PM sensors is beyond the scope of this work and can be found in other works such as, for example, [42,43]. Suffice it to say that these PM sensors consist of a fan, generally connected to a small tube, that pushes air into the sensing box. Light from a laser diode is scattered by the particles. This scattered light is received by a photodiode, which can estimate the concentration of each type of particle by classifying and counting the number of pulses detected.

Within COCAL, we use the SDS011 PM sensor from Nova Fitness Co., which enables simultaneously measuring both $PM_{2.5}$ and PM_{10} levels at a very low cost.

2.1.1. LCS Performances

The major advantages of LCS in terms of price and portability come at the expense of limitations in precision and accuracy [43–45].

Many LCS manufacturers and models are available on the market, and detailed comparisons between them can be found, for example, in [12,24,42]. These works highlight that, in addition to the limitations in precision and accuracy, it is very important to consider the environmental conditions in which LCS sensors operate. For example, because these sensors do not have sample conditioning equipment, they are susceptible to drifts due to relative humidity (RH), which can affect the hygroscopic growth of particles and distort measurements [27,46]. The results of these studies demonstrate that among LCS, the issue of quality and precision of the specific brand and model of sensors can be as relevant as the intrinsic limitations of the technology employed, the issues related to the deployment in the designated environment and the environmental conditions.

In this perspective, to monitor the environmental conditions in which the acquisition takes place, together with the PM LCS, we also use a Dallas Semiconductor DS18B20 one-wire communication sensor with waterproof protection outside the acquisition box, while internally, we use a Bosch BME280 temperature, pressure and RH sensor connected to the board via the I2C bus.

In [42], useful references can be found to understand the performance of the SDS011 sensor and other similar sensors under controlled laboratory conditions. The results of that work confirm that the SDS011 sensor is suitable for use within the COCAL project since it performs reasonably well in comparison with similar or even more expensive sensors. At the same time, downsides have been identified such as a general trend to underestimate PM values and the presence of a delay in the timing of measurements.

2.1.2. Statistical Analysis

Following [44], it is particularly important to understand the behavior of the LCS used in this work under real-world conditions. In order to devise possible mitigation strategies for the issues introduced by the use of LCS, following the protocols suggested by US EPA and [43], we designed two experiments, namely: (i) a study of the behavior of LCS in a highly variable PM level environment to evaluate precision and (ii) a co-location-based evaluation of LCS with a reference measurement station managed by the regional environmental agency ARPA-FVG, in order to evaluate accuracy. It should be noted that during these tests, the data completeness of the COCAL system, meaning the ability to avoid gaps in measurements and data transmission, has always been very high, with almost negligible glitches and well above the 75% threshold recommended by US EPA.

LCS Precision

To assess the precision of LCS, we studied the recordings of three co-located LCS in a highly variable PM concentration environment. Following US EPA standards and procedures mentioned by [43], precision can be estimated using the standard deviation (SD) and the coefficient of variation (CV). The SD shows a value of 2.15 for PM_{10} and a value of 1.17 for $PM_{2.5}$, while the CV shows a value of 24.70% for PM_{10} and 22.77% for $PM_{2.5}$.

US EPA recommends a target SD less or equal to 5 $\mu g/m^3$ and a CV less or equal to 30%. The tests we conducted therefore assess that LCSs used in COCAL match the recommendations for precision.

LCS Accuracy

To test the accuracy of the selected sensors, we placed a COCAL box at a short distance from a certified reference air quality station (ARPA-FVG station 'Rosmini'). The reference PM values of this station are made available through an API on the official ARPA-FVG website as daily average values only. The measurements and comparison took place from mid-March 2022 to the end of April 2023.

In Figure 2, one can compare measurements from the reference station and those taken during the same period with the COCAL box located close to the reference station.

Figure 2 is divided into two boxes. The upper box shows data and statistics for PM_{10}, while the lower one focuses on $PM_{2.5}$. In each box, the first graph (n.1) shows in red the time series of the reference daily average values as made available from the ARPA-FVG website, while the daily average values of the COCAL box located close to the reference station are plotted in blue.

High PM values were measured at the end of July 2022. Unfortunately, these were not outliers but the effects of a large forest fire that occurred for several days at a distance of about 20 km from the test site.

It is possible to note that COCAL measurements are generally lower than the reference measurements; however, during the first months of 2023, in three specific events (identified by boxes A, B and C), this behavior reverses. To better understand the performance of LCSs, Figure 2 also shows the difference between the reference and the COCAL time series (graph n.3), the RH time series (graph n.3) and the time series of the standard deviation of all COCAL measurements acquired near the reference station, calculated on a daily basis.

It is interesting to note that during the A, B and C events, the standard deviation of COCAL measurements increases. In two of these cases (A and C), this can be understood as owing to the sensitivity of the LCS to RH. In fact, the time series of the RH in those periods exceeded 60%, while elsewhere, when LCS underestimated the reference measurements, the RH remains below this threshold. In the case of event B, however, where RH is low, a different explanation is needed. This could be found in the different technologies and rate of sampling of the LCS and the reference acquisition system. COCAL boxes acquire data every ten seconds, which means that rapid variations in the actual PM concentrations can effectively be captured, increasing at the same time the standard deviation of the set of daily measurements. Reference systems sample much more slowly so that, even if very reliable, they can overlook rapid phenomena so that, in the comparison, LCSs data appear to be dispersed.

Following US EPA standards and procedures mentioned by [32], accuracy can be estimated using the coefficient of determination (R^2), slope (m), intercept (b), root mean square error (RMSE) and the normalized root mean square error (NRMSE). Results for the mentioned tests for PM_{10} and $PM_{2.5}$ are shown in Table 1, while Figure 3 provides a snapshot of the comparison of the LCS and reference measurements.

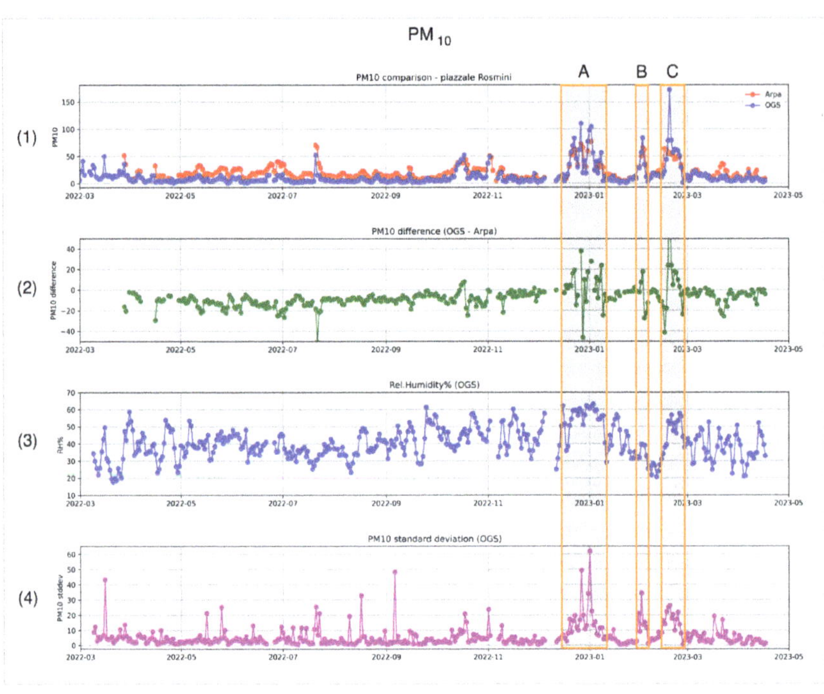

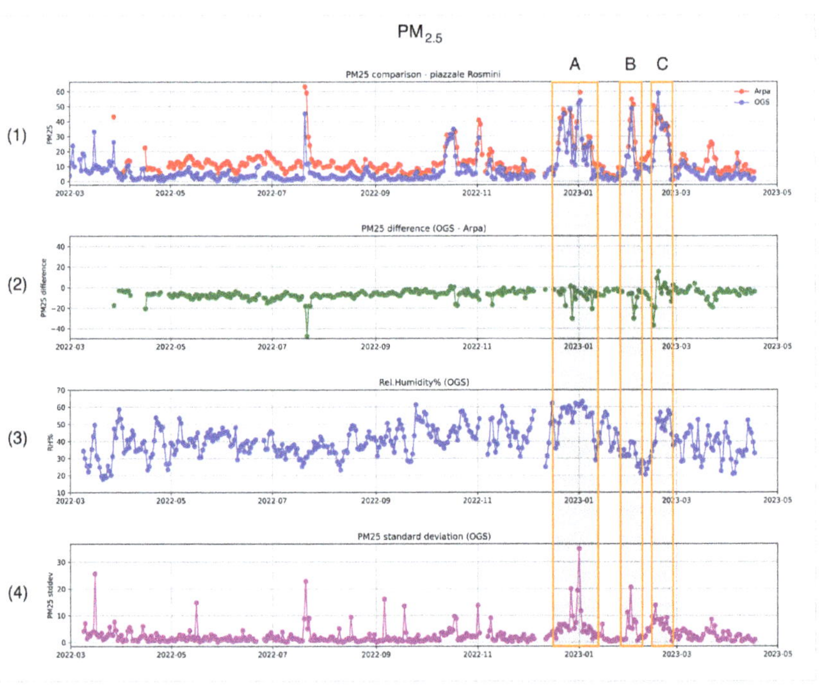

Figure 2. Comparison of 15 months of PM measurements from a certified reference station with measurements taken during the same period with a COCAL system located near the reference station. The upper set of graphs reports PM_{10} measurements, while the lower one reports $PM_{2.5}$ measurements.

In both cases, graph no. (**1**) shows the daily average COCAL measurements in blue, while the reference daily value is plotted in red. Graph no. (**2**) shows the difference between the two time series. Graph no. (**3**) shows the RH time series. Graph no. (**4**) shows the standard deviation of COCAL measurements calculated day by day. As can be seen, COCAL measurements are generally lower than the reference measurements. During the first months of 2023 in three specific events (identified in the graph by the boxes A, B and C), this behavior reverses. It is interesting to note that during these events, the standard deviation of measurements also becomes high. In two of these cases (A and C), this can be understood as owing to the sensitivity of the LCS to RH, but in the case of event B, the RH is low.

Table 1. LCS accuracy metrics.

Parameter	PM$_{10}$	PM$_{2.5}$
Coefficient of Determination (R^2)	0.45	0.25
Slope	1.06	0.80
Intercept (b)	−8.73	−4.09
Root Mean Square Error (RMSE)	13.76	8.73
Normalized Root Mean Square Error (NRMSE)	66.04	60.37

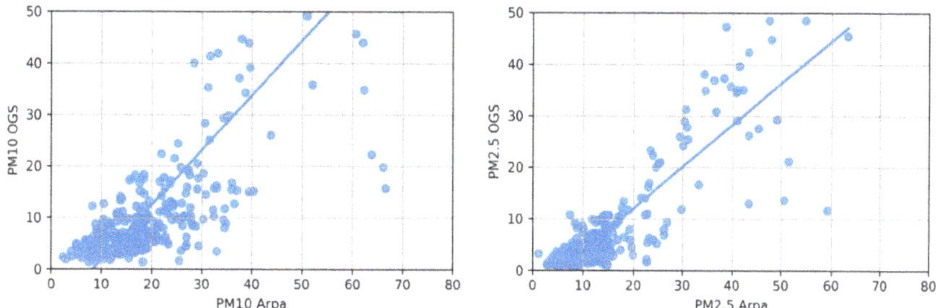

Figure 3. Scatter plot comparing Reference PM measurements and COCAL measurements.

Considering US EPA recommendations, these results can be problematic. In fact, considering R^2, this parameter is recommended to exceed 0.70, while the analysis reveals lower values. The target slope should be approximately 1 ± 0.35, a condition that is instead respected by the LCSs. Similarly, the intercept parameter performs relatively well for PM$_{2.5}$ sensors, while PM$_{10}$ sensors do not fall within the recommended range since EPA recommends a value between −5 and +5. Following EPA standards, RMSE should be lower than 7 μg/m^3, and again, also here, PM$_{2.5}$ scores rather well, whereas we measured a value twice the threshold for PM$_{10}$. In addition, unfortunately, the NRMSE results are also too high, being around 60%, while EPA recommends a value less than 30%.

It can therefore be said that accuracy-wise, the LCSs perform rather poorly, both for PM$_{10}$ and PM$_{2.5}$, and that a correction mechanism is needed to obtain results that could be comparable with the official ones. At the same time, it is possible to say that, since the precision is reasonably good, a geographic distribution of LCS measurements built on the integration of multiple COCAL boxes should reasonably be capable of highlighting general trends and pinpointing local anomalies.

2.1.3. LCS Performances Improvement

The results of the tests are consistent with a wide literature on LCS performance. Several authors [47–49] highlighted the impact of the environmental conditions and, in particular, of the RH in the deviation between reference measurements and LCS. To address such problems, reference stations are equipped with a device that, on heating air samples, induces the water vapor condensed onto the particle to evaporate. This, of course, is not possible with LCSs. To compensate for this effect, several RH correction approaches exist such as, for example, κ-Köhler theory-derived factors and various types of regressions or machine learning methods. After an extensive survey of the existing literature, Ref. [44] maintains that such corrections are applied very seldomly, with a simple linear regression, in that case, being the most used method. The same authors underline difficulties in accurately defining local parameters and accumulating knowledge from different cases and areas. It is also worth noting that RH itself can be a problematic parameter to measure and that, since COCAL is a VSN system, RH measurements taken with it can have further limitations.

Taking these considerations into account, and since the tests performed in Sections "LCS Precision" and "LCS Accuracy" with the LCSs we use in this work revealed that they perform reasonably well in terms of precision but unfortunately not well enough in terms of accuracy, we devised a specific and pragmatic two-step mitigation strategy to improve their performance.

The first step consists in filtering all measurements made in problematic conditions, for example, when RH is more than 60%. These data are automatically flagged and are not sent to the following processing flow.

The second step consists in calculating an accuracy correction to a reference station using a COCAL box located in its proximity. Since sensors proved to behave consistently among them, following [39], we apply the same accuracy correction to all the other sensors. As abovementioned, given that, in the designated area, only one value per day is currently available from the reference station, we calculate the difference between that reference value and the daily average value of all the measurements taken by an LCS co-located in the proximity of the reference station. Corrections are inter/extrapolated in all designated areas by means of the technique described in [39].

An example of the method's results can be seen in Figure 4, where on the left, the geographic distribution of LCS measurements before corrections is shown, while on the right, corrected values that are more consistent with reference measurements are shown. It is to be noted that this method can be problematic since applying the correction to areas far from where the reference station is located can unpredictably bias the final values. In the case proposed in Figure 4, measurements taken in the village of Opicina (upper part of the map) generally depict rather different conditions from the city center (lower part of the map). Opicina is, in fact, located uphill, is characterized by a different climatic setting and is less subject to vehicular traffic. No reference station is available in that area such that the only reference measurements available are those taken in the city center. In the example of Figure 4 (left), while the un-corrected data report a polluted city center and a much better situation uphill, after the corrections (Figure 4 right), the revised air quality also degrades notably in the hills. This could be an artifact that needs careful consideration when interpreting the data.

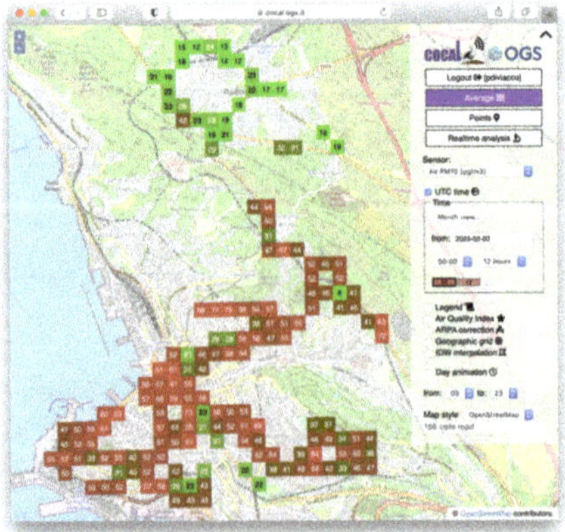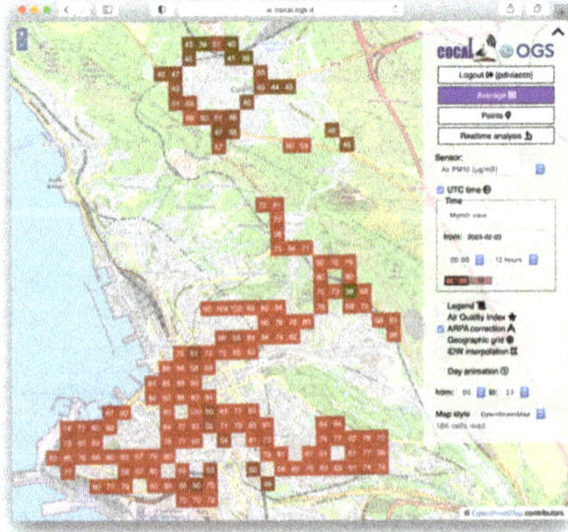

Figure 4. Distribution of raw PM_{10} values (**left**); distribution of data after correction to improve LCS accuracy (**right**).

2.2. Deployment on Mobile Platforms

As mentioned above, sensors were installed on two different platform types, namely (i) buses and (ii) cars. In both cases, we developed a tailor-made waterproof box that can easily be installed on the platform and where all the acquisition and transmission electronics can be safely protected while air inlets and outlets could effectively bring air samples to the LCS.

Bus deployment has been developed with key help from the local transportation authority, TPL Trieste Trasporti, which kindly offered to host several COCAL systems. The boxes were installed on the roof of the buses (Figure 5) in a closed compartment with a specific air inlet passing through a syphon in order to prevent rain from entering the box. The power supply is obtained from the bus using a temporized relay to minimize the impact of the COCAL box on the normal functioning of the vehicles.

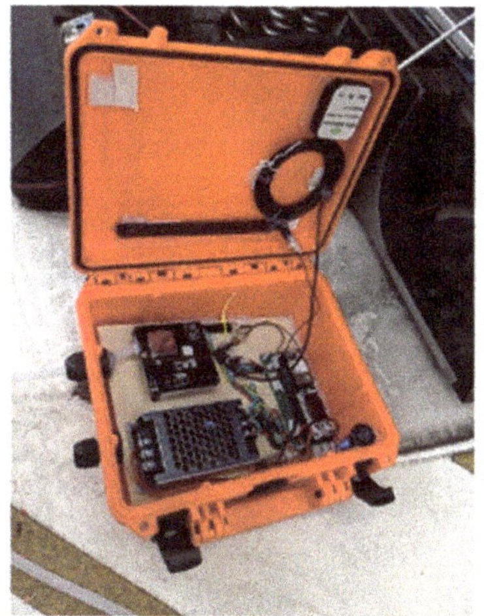

Figure 5. Installing a COCAL box on buses.

Buses are a very convenient acquisition platform because each unit can be redirected to several routes throughout the day, thereby covering a large portion of the urban area. On the other hand, bus routes tend to follow the main directions of the traffic in a city, which may somehow bias the coverage of the designated area.

Cars have several advantages over buses, one being that they generally do not follow predefined routes. This makes cars a good means to provide infills in areas where buses are not available. At the same time, cars introduce other constraints that depend mainly on volunteer drivers. Issues may arise, in fact, in order to motivate them to cover areas that are not within their daily routines. In our experience, this often meant also using vehicles belonging to our institute.

COCAL boxes for cars have been designed entirely by us and 3D-printed autonomously with the help of the ICTP FabLab laboratory (Figure 6). The boxes were conceived to be fully autonomous and least invasive as possible. This forced us to make some design choices; for example, since connecting the boxes to the car's power supply can be problematic, they are powered on batteries only. Autonomy is approximately one full day, although it can be longer depending on the rate of data transmission. Battery recharge can be completed in a few hours. Another choice was to avoid taking up space inside the vehicle or in the trunk, so we decided to position the box on the roof. To affix the box to the roof surface, we added magnetic plates on the bottom and, for further security, we decided to install it on cars with roof bars only, to which COCAL boxes are secured using Velcro strips. The air inlet passes through the curved white roof so that it remains dry in case of rain (Figure 3 lower left).

The air outlet is located on the back of the box. The roof can easily be removed to access the electronics inside.

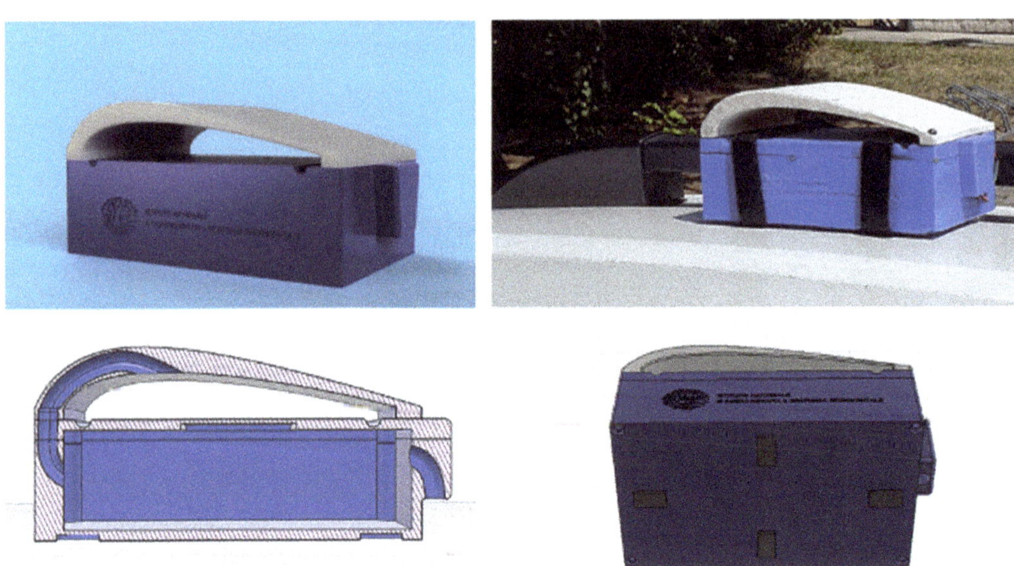

Figure 6. Design and installation of COCAL boxes for cars: Rendering of the box (**upper left**). Lateral section of the box (**lower left**): the air inlet passes through the curved white roof to remain dry in case of rain. The air outlet is located on the back of the box. The roof can easily be removed to access the electronics inside. Actual deployment on a car (**upper right**): boxes are located on car roofs secured with Velcro stripes to roof bars. COCAL boxes have magnetic plates to better adhere to the roofs (**lower right**).

Limitations of Mobile Platforms

Besides the already mentioned limitations in accuracy, we were also concerned about the possible effects of the deployment of LCSs on moving platforms. While it is known that platform speed influences the measurement, to the best of our knowledge, there is no specific study on this topic since most of the existing literature is based on fixed-position deployments. We therefore set up a test, where, passing multiple times in the same area at different speeds during a restricted period with stable meteorologic conditions, we collected a large dataset of measurements. The results of the experiment can be seen in Figure 7. These show an inverse relationship between PM and platform speed. Considering how the COCAL box is built, this is probably due to a depression induced by the platform movement on the inlet of the box. This increases with speed, reducing the quantity of air that reaches the detection device and therefore reducing the estimates of PM values. The drift is relatively small and below the precision of sensors for velocities lower than 50 km/h, while higher speed values tend to be more problematic. Since the system has been installed mostly in an area where the speed limit is below 50 km/h, we can safely say that data collection is not particularly affected by this issue. As a measure of further security, during data processing, measurements associated with a speed higher than 50 km/h are automatically filtered out of the calculations.

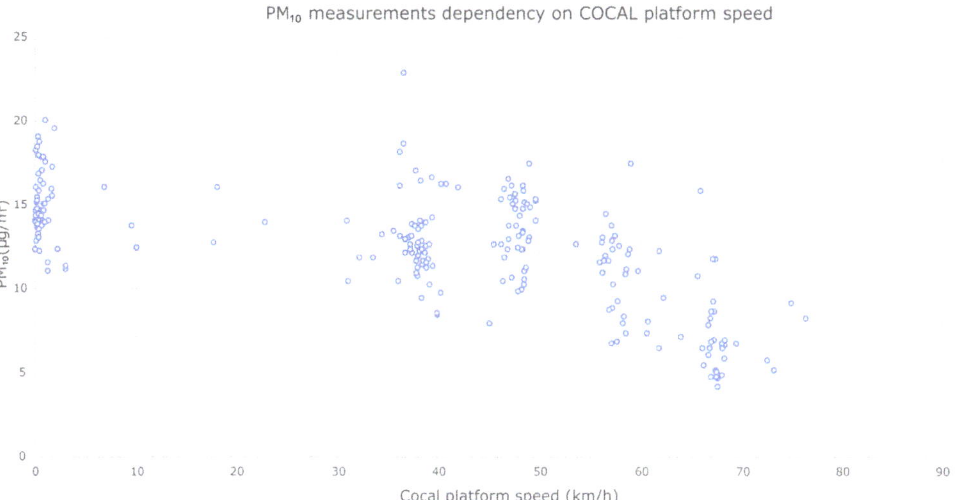

Figure 7. Measurements of PM_{10} concentration as a function of platform speed.

2.3. Data Acquisition and Transmission

The acquisition system (Figure 8) is based on a low-cost ESP32 microcontroller with WiFi and Bluetooth connectivity. We selected a Heltec LoRa 32 v2 board, which has an embedded OLED display and battery charger together with a LoRa chip and Wi-Fi and Bluetooth connectivity. These are used for testing and short-distance connectivity, while LoRa is used for long-distance connectivity [28]. To this, we added GSM and GPS functionalities using an A9G development board, designed by Ai-Thinker, which, with an active SIM card, allows data transmission using the GSM telephonic network where coverage is available. Data transmission using Wi-Fi and GSM stores data directly in an InfluxDB database, while, using Lora, we rely on The Things Network (TTN) LoRaWAN infrastructure in order to retrieve data transmitted using LoRa and store data into the database. Telegraf, a server-based agent, oversees retrieving data from TTN using the MQTT protocol and storing data into the database.

Figure 8. Heltec and A9G board on a PCB designed for COCAL.

2.4. Data Management

Figure 9 describes the general architecture of the COCAL system. The flux of incoming data transmitted through LoRaWAN flows into an InfluxDB table filled by the TTN service. A server script manages to reroute the data into the main InfluxDB time-series tables after proper conversion. The final storage and processing server is based on a Postgres database, with a PostGIS extension for dealing with georeferenced objects, geographic projections and geometric objects such as polylines. This storage/processing server (SPS) is built on an open-source architecture: Linux Ubuntu, Apache, PHP, Python and Postgres. It currently manages the database, as well as several scripts responsible for the processing and the web front-end. A PHP script periodically synchronizes the InfluxDB with the Postgres database, inserting in the latter one every valid measurement from a sensor with a time marker (UTC), WGS84 coordinates, all other GPS info (such as altitude or speed), the type of transmission (e.g., GSM), a device ID, a sensor ID (e.g., atmospheric pressure) and the measured value (e.g., 1007 mBar).

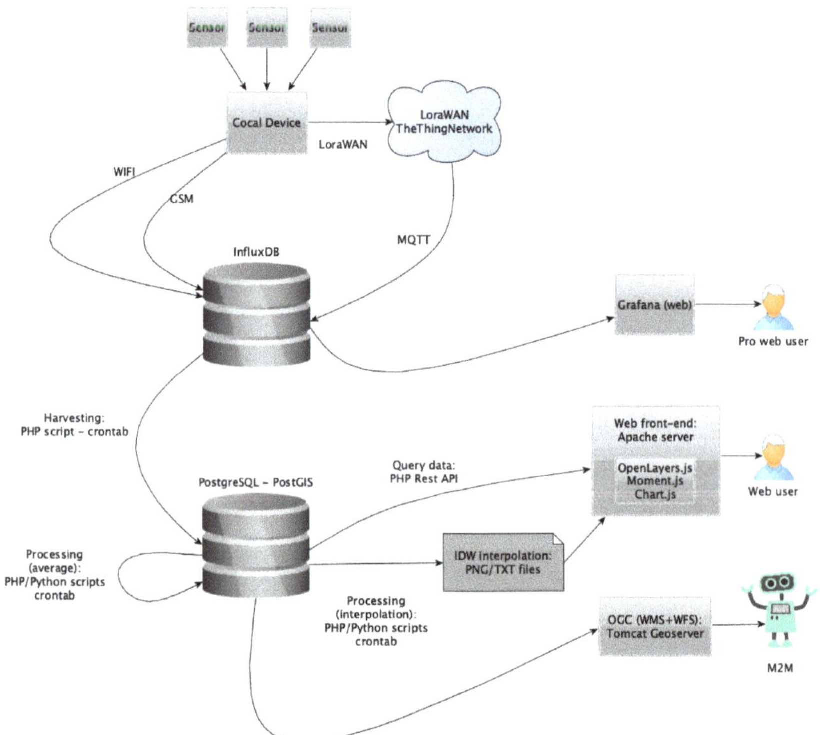

Figure 9. General architecture of the COCAL system.

2.5. Data Processing

The SPS performs different activities by means of PHP scripts, which are scheduled with crontab. The most demanding analyses are encoded in Python with its standard libraries such as NumPy, SciPy, Matplotlib or PIL.

All processed products are made available in near real-time and stored permanently for better performance.

2.5.1. Window Averaging

Window averaging is necessary to assimilate the large amount of data acquired by many devices spread across a wide area. As in [35], we define a geographical grid of

200 m per 200 m wide cells, based on a local projection. In addition, we subdivide the timeline into 1 h intervals. Every set of data spanning a spatial cell and a time interval is a datacube, including measurements from different devices but mounting the same kind of sensor (e.g., PM_{10}). The choice of a local projection provides good accuracy when the area of interest is limited, and, in the case of this work, we used WGS84/UTM zone 33N (EPSG:32633). In order to obtain values that are smoother and more representative of the physical phenomenon, reducing the outliers and providing a uniform subdivision of space and time, considering the good results obtained in [26], we adopted a similar approach by averaging data (calculating the median) over space and time datacubes. Larger time intervals (for example 2, 3, 4 or 8 h) can be analyzed by selecting the specific datacube. These are processed once a day and made available the next day. A discussion on the advantages of window averaging and the shape of the cells can also be found in [39].

Every averaged datacube is stored into the database, marked with a start time, end time and a polyline describing the square cell.

2.5.2. Correction of LCS Data

As mentioned in Section "LCS Accuracy", the accuracy of LCS can be problematic. The technologies used within these sensors and the rate of sampling together with the effects of environmental parameters such as the RH often induce drifts in the LCS measurements. In Section 2.1.3, we introduced a pragmatic method that can mitigate such effects. This is based on applying a correction value to the LCS measurements that are calculated daily as the difference between the value provided by a reference station and a fixed LCS located in the vicinity of the reference station. The correction is applied server-side one day after the LCS data are actually collected since the reference value is available only with such a delay. The resulting grid of data is then made available to the web portal.

2.5.3. Interpolation and Contouring

In order to provide a more intuitive insight into the measured phenomenon, interpolation is a useful tool. Following [39], there are many aspects to take into account when spatial interpolation is applied:

(i)   The accuracy of the method and how far the interpolated values from the samples are still meaningful, i.e., a consideration on "extrapolation". This issue can be partially solved by the definition of an area of interpolation, such as the bounding box or (better) the convex hull of the samples as the first approximation.

(ii)   The computational complexity and the relative speed of the interpolation method, which must comply with the near real-time requirement. In our implementation, we chose a very quick and sufficiently accurate method, inverse distance weighting (IDW). IDW interpolation is defined as follows. Assuming that $\{x_1 \ldots x_N\}$ are the interpolating points (samples) and x a generic point, the interpolant is:

$$f_{IDW}(x) = \frac{\sum_{k=1}^{N} w_k(x) f(x_k)}{\sum_{k=1}^{N} w_k(x)} \qquad (1)$$

where the weights are:

$$w_k(x) = \| x - x_k \|^{-p} \qquad (2)$$

(iii) In addition, there is an epistemological aspect to be further considered: all processing is automatic and cannot be assisted by human intervention. This fact excludes algorithms such as Kriging, which involve many discretional models and parameters. An excellent alternative solution is natural neighbor interpolation (Sibson's method), which is based only on the geometrical properties of the dataset and is approximately ten times slower than IDW [50]. Lastly, it is necessary to define the levels for the contouring in some adaptive way to improve the readability and also the color map for the interpolation, which must be coherent with all other data visualizations. The interpolation/contouring is implemented on the SPS with a Python script that reads the averaged values, applies the IDW method (with exponent $p = 3$), generates the contour and produces a transparent PNG image with a small text file for the georeferentiation. The image is clipped around the convex hull of the dataset, excluding the external area.

2.5.4. Near-Real-Time Web-Based Visualization

The visualization of environmental data is a topic that raises several questions: Are our data time series or spatial distributions? How do we represent time-varying phenomena? Which colors and graphic patterns are more effective and representative? To what extent does the computation have an impact on near-real-time web interaction? There are many different answers, of course, and much research involving mathematical, computational or psychological aspects (see, for example, [39]).

We implemented a set of visualizations in the web front-end, which allows the end user to browse through spatial and temporal coordinates and to select and analyze both single acquisitions and averaged maps (Figure 10).

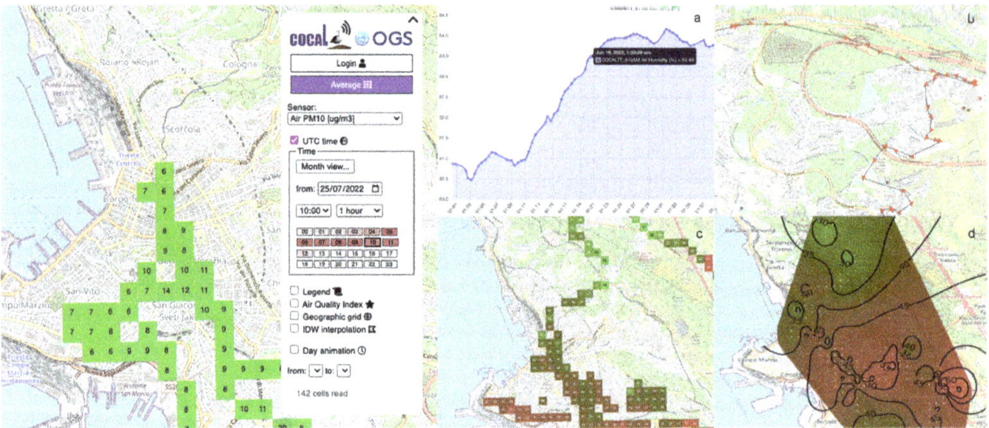

Figure 10. COCAL web-based GUI (**left**) and near-real-time visualizations: (**a**) time series; (**b**) single acquisitions; (**c**) near-real-time window averaging; (**d**) interpolation and contour.

All services are available at the web portal https://cocal.ogs.it (accessed on 1 June 2023).

The web interface allows easy navigation through the datasets by means of a simple window (Figure 10 left) where the user can select a single device, the acquisition sensor, the time interval and many different options. The time selection can be made in local time or in UTC, and a simplified view of the day shows the density of available data as shades of red, providing one-click access to time selection.

A calendar (month view button) shows the data density day by day by using the same principle. The acquisitions of a single device are represented as an interactive chart of the time series (a) or as a collection of connected points on a map (b). In the latter case, an arrow shape can show the GPS direction, and the point colors are mapped to the measure scale and corresponding legend.

The averaged data (on a rectangular grid) are represented as colored and labeled polygons (c). The interpolated data use the same color coding but are represented as continuous within the data convex hull, with superimposed contour lines (d). All graphic elements are responsive, showing all data details.

Additionally, we implemented a functionality that allows the user to follow cumulative data as an animation, cycling from a starting to an ending hour, in order to dynamically represent the temporal evolution of each parameter.

2.5.5. Advanced Analysis

The adoption of UTM33N as the map projection is a disadvantage when the acquisitions are beyond the limits of the range from 12° E to 18° E. Moreover, the implemented processing mechanism, which computes the averages periodically (in the background), is very efficient for a quick response and a fluid user experience but, on the downside, is rather fixed and rigid. A more flexible interface for data analysis has been tested, based on the global map projection "Web Mercator" (EPSG:3857, Pseudo-Mercator/Spherical Mercator). This spherical projection is used by most GIS systems such as Google Maps, Bing, ESRI, etc., has an increasing distortion at high latitudes and is not conformal but is the de facto standard for web applications and allows global coverage for processing. The web page shown in Figure 11 provides a wider range of query parameters and builds an analysis "on the fly" (windowed average or IDW interpolation). The computation is restricted to the visible bounding box and requires some computational time, favoring the extended query flexibility.

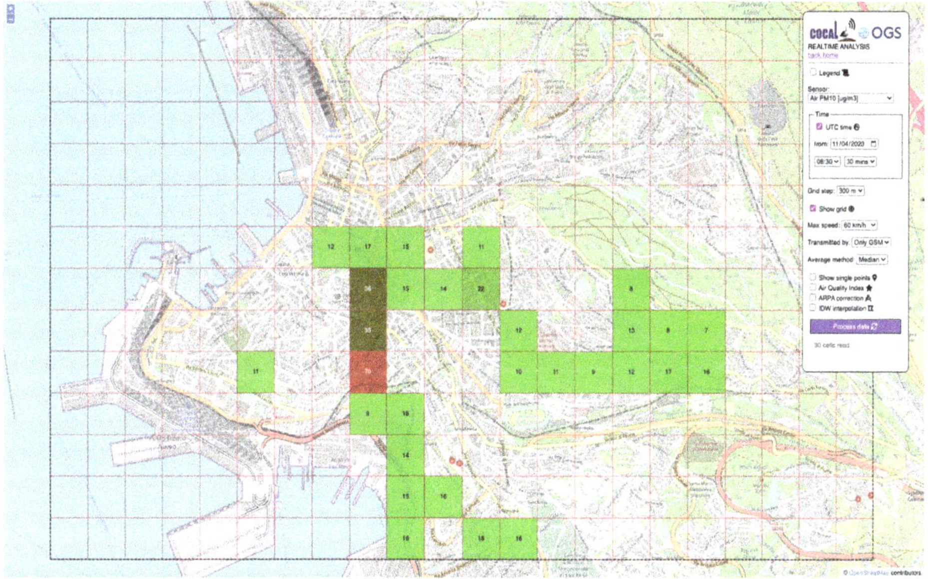

Figure 11. COCAL web "real time analysis" with advanced queries.

2.5.6. Open and FAIR Data Access

According to the FAIR principles, data must be findable, accessible, interoperable and reusable. COCAL deploys different protocols and implementations aiming to provide Open and FAIR data in accordance with well-established and official standards. In order to achieve discoverability, the initiative handles the standard ISO 19115-3 metadata profile through the Geonetwork catalog application. To ensure interoperability, such as machine-to-machine data flows, data harvesting or archive federations, the COCAL geospatial database is compliant with OGC (Open Geospatial Consortium) standards, deployed as Web HTTP services: (i) WMS (Web Map Service), which provides georeferenced map images of the requested features or (ii) WFS (Web Feature Service), which provides detailed and fine-grained information about features or general capabilities of the dataset in a structured text format (XML, JSON, etc.). All OGC services are implemented on a Geoserver platform, an open-source Apache Tomcat extension linked to the main database. In order to achieve accessibility, data products are fully open and accessible, while a download of raw data in CSV format is available on the COCAL web portal, after authentication with a trusted account.

3. Results

The technology behind COCAL has been under constant development since 2020. During the initial phase, trials took place using multiple simultaneous acquisition and transmission platforms mounted on vehicles operated by our institution. This allowed us to extensively test the system, its scalability, precision and accuracy during specific targeted surveys. Once the system was finalized, we were able to deploy a fully operative system on vehicles of the local transportation authority (Trieste Trasporti) and on some voluntary cars. COCAL entered into service in February 2021 and has been fully operative ever since.

Up to April 2023, the system acquired and processed a remarkable amount of data, both "points" (single measurements) and "cells" (averaged results). In Table 2, it is possible to see the approximated number of records per year during the period from January 2020 to April 2023.

Table 2. Approximated number of records per year.

Record Type	Total Count	Year	Count
Points	53M	2020	3.8M
		2021	13M
		2022	27M
		2023 (partial)	9.2M
Cells	18M	2020	0.9M
		2021	5.8M
		2022	9.4M
		2023 (partial)	1.8M

Data are fully public and can be accessed using standard procedures from the COCAL web portal (https://cocal.ogs.it, accessed on 1 June 2023). The main results of the work are, on one hand, the COCAL system itself, which has proved to be efficient, robust and easy to install and maintain, allowing a very high throughput of environmental data that strongly support the paradigm of low-cost participative systems in monitoring the environment. On the other hand, a very important result is the availability of a very large quantity of environmental measurements, allowing us to significantly increase the spatial and time coverage of the distribution of air pollutants in the designated area. Initial analysis of the dataset acquired enabled identifying several interesting features of the air quality in the area of Trieste. Some of these observations have already been published in the papers mentioned above.

Figure 12 shows the geographic distribution of the standard deviation of all measurements made by the COCAL systems in the designated area. For every possible grid cell, we consider all PM measurements during the whole considered period (. . . 2021 . . . 2022). Acquisitions within each cell are divided into time windows of 1 h, and some statistical parameters are calculated for every window, such as the number of samples, average, median and standard deviation. Eventually, the maximum standard deviation over the period is calculated for every cell.

This map should be interpreted with some caution. On one hand, the standard deviation could provide an idea of the amount of variation in measurements during a specific period of time. Where the standard deviation is high, this could mean that higher levels of PM have been recorded in that area compared to areas where the standard deviation is lower. On the other hand, there is a risk that if a biased coverage of data has been used, then the standard deviation could also be biased. Indeed, in Figure 12, it is easy to identify an NW-SE alignment (highlighted by the light blue line) where high values of the standard deviation seem to gather. This correlated almost perfectly with the main traffic direction of the city. It could be tempting to associate this trend with the traffic, concluding that those areas are the most polluted in the city. We think that some caution should be taken because this direction also corresponds to the more frequently followed routes of the buses used in the COCAL initiative. Most of the data, in fact, have been acquired in those areas so that, in comparison to other areas, measurements made in the former could have had the opportunity to detect all pollution events, while measurements made in other areas may just have overlooked them.

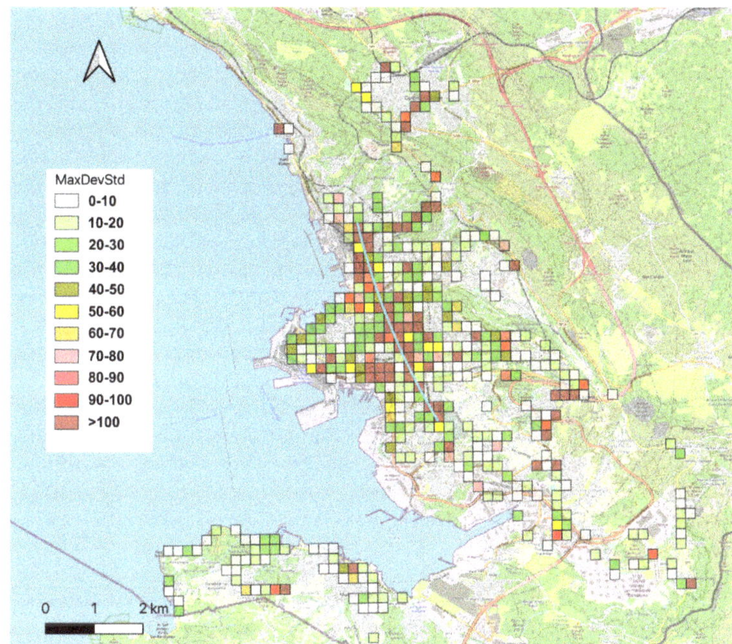

Figure 12. Map of maximum standard deviation of COCAL PM$_{10}$ measurements in the designated area. In light blue, the alignment of the higher values correlates with the main traffic direction of the city.

An interesting and surprising result we obtained is related to the possible deterioration of the sensors after long usage. After approximately 2 years of activity, we decided to substitute the hardware of the deployed systems and discovered that they practically had not accumulated any dirt inside and that their precision remained (considering the intrinsic limitations of the LCS) almost unaltered (Figure 13).

Figure 13. An LCS case opened after almost two years of activity. The interiors show only very small quantities of dirt.

4. Conclusions and Future Work

This work describes a crowdsensing-based air monitoring system following all of the technological segments of a path that starts from the actual measurement using LCS, to data transmission, to processing and a FAIR-compliant web-based representation and access to the reconstructed data products. The main conclusion of this work is, therefore, that the implementation of all segments of the system can be achieved using low-cost and open-source technology only. At the same time, the acquisition of data does not need trained personnel but can be performed with the help of volunteers and especially of the local transportation authorities. The results of the experience we propose here suggest that such systems can be trustworthy from the point of view of the precision of measurements, while it is necessary to rely on a reference value to correct the deviation of measurements due to the intrinsic limitations of the LCSs. Currently, within the proposed system, this correction is calculated daily, since the reference values are made available by the local environmental agency only as daily averages and one day after the actual measurements took place. We demonstrated that, in some cases, this can be problematic and that, when available, corrections should be calculated with a higher temporal resolution. Other limitations of the system were taken into consideration such as the speed of the acquisition vehicle. We showed that this speed should not exceed 60 km/h; otherwise, the variation in air pressure could bias the measurements. We also demonstrated that after almost two years of continuous field operations, the amount of dirt accumulated within the acquisition box in the case of the designated area was minimal. This, of course, can depend on the levels of pollution in the specific cases where the system is applied.

The amount of data acquired raised important questions from the viewpoint of the ITC system to be used. We understood that, for example, a modular approach in separating each activity on different virtual machines helps considerably in monitoring the performance of the system and in understanding where it may be necessary to increase dedicated resources.

After two years of testing the system with five COCAL boxes installed on local transportation authority buses, we understood that since this means of transportation can often be under maintenance or rerouted, five units of vectors is the minimum set of installation for a city of approximately 200,000 inhabitants and an area of approximately 100,000 square meters. In this perspective, much depends on the actual urban configuration. In fact, while installations on cars can cover the urban area almost randomly, buses follow the bus line distribution, which is generally concentrated in specific areas while neglecting others. The resulting data can be biased and limit the reconstruction of the distribution of air quality.

In the coming months, the number of COCAL installations on the public bus network of Trieste will be doubled in order to achieve a broader and more homogeneous coverage of the city.

The designated area being a coastal area, the system developed so far allows reconstructing the on-land area only. This does not allow studying in depth the phenomena related to the movements of polluted air masses due to sea breezes. In this perspective, an extension of the system is planned at sea using volunteers' recreational sailing ships, while the system will also be installed on a certain number of sea buoys managed by OGS.

Author Contributions: Conceptualization, P.D.; methodology, P.D., M.I. and R.J.C.; software, M.I., R.J.C.; validation, P.D., M.I., R.J.C., A.V. and N.P.; resources, R.J.C., N.P. and A.V.; data management, M.I. and N.P.; writing—original draft preparation, P.D., M.I. and R.J.C.; writing—review and editing, P.D.; visualization, M.I.; supervision, P.D.; project administration, P.D. All authors have read and agreed to the published version of the manuscript.

Funding: This research received no external funding.

Institutional Review Board Statement: Not applicable.

Informed Consent Statement: Not applicable.

Data Availability Statement: All the data used in this work are freely and openly available at the website https://COCAL.ogs.it, accessed on 3 May 2023.

Acknowledgments: The authors would like to thank Trieste Trasporti TPL, the local transportation authority of Trieste, for its support, without which this work would not have been possible. In particular, we would also like to thank, for their great help, Giuseppe Zottis, Fabrizio Godinich and Fabio Vidotto. We are also very grateful to the ICTP FabLAb for their effective help in designing and 3D printing the COCAL boxes.

Conflicts of Interest: The authors declare no conflict of interest.

References

1. Air Pollution. Available online: https://www.who.int/health-topics/air-pollution (accessed on 20 February 2023).
2. Almetwally, A.A.; Bin-Jumah, M.; Allam, A.A. Ambient air pollution and its influence on human health and welfare: An overview. *Environ. Sci. Pollut. Res.* **2020**, *27*, 24815–24830. [CrossRef]
3. Di Turo, F.; Proietti, C.; Screpanti, A.; Fornasier, M.F.; Cionni, I.; Favero, G.; De Marco, A. Impacts of air pollution on cultural heritage corrosion at European level: What has been achieved and what are the future scenarios. *Environ. Pollut.* **2016**, *218*, 586–594. [CrossRef]
4. Cachier, H.; Sarda-Estève, R.; Oikonomou, K.; Sciare, J.; Bonazza, A.; Sabbioni, C.; Greco, M.; Saiz-Jimenez, C.; Hermosin, A.; Reyes, J. Aerosol Characterization and Sources in Different European Urban Atmospheres: Paris, Seville, Florence and Milan. In *Air Pollution and Cultural Heritage*; Taylor and Francis Group: Abingdon, UK, 2004. Available online: https://scholar.google.com/scholar_lookup?title=Aerosol%20characterization%20and%20sources%20in%20different%20European%20urban%20atmospheres%3A%20Paris%2C%20Seville%2C%20Florence%20and%20Milan&publication_year=2004&author=H.%20Cachier&author=R.%20Sarda-Est%C3%A8ve&author=K.%20Oikonomou&author=J.%20Sciare&author=A.%20Bonazza&author=C.%20Sabbioni&author=M.%20Greco&author=J.%20Reyes&author=B.%20Hermosin&author=C.%20Saiz-Jimenez (accessed on 20 February 2023).
5. United States Environmental Protection Agency. Clean Air Act Text. Available online: https://www.epa.gov/clean-air-act-overview/clean-air-act-text (accessed on 21 February 2023).
6. Directive 2008/50/EC of the European Parliament and of the Council of 21 May 2008 on Ambient Air Quality and Cleaner Air for Europe. *Off. J. Eur. Union* **2008**, *152*, 169–212. Available online: http://data.europa.eu/eli/dir/2008/50/oj/eng (accessed on 21 February 2023).
7. European Environment Agency. European City Air Quality Viewer. Available online: https://www.eea.europa.eu/themes/air/urban-air-quality/european-city-air-quality-viewer (accessed on 22 February 2023).
8. United States Environmental Protection Agency. Interactive Map of Air Quality Monitors. Available online: https://www.epa.gov/outdoor-air-quality-data/interactive-map-air-quality-monitors (accessed on 22 February 2023).
9. Strzelecki, M.; Stezycki, K. "Burn Everything": Poland Chokes on the Smog of War. *Reuters*, 8 December 2022. Available online: https://www.reuters.com/world/europe/burn-everything-poland-chokes-smog-war-2022-12-08/ (accessed on 22 February 2023).
10. United States Environmental Protection Agency. EPA AP-42: Compilation of Air Emissions Factors. Available online: https://www3.epa.gov/ttnchie1/ap42/ch01/ (accessed on 12 August 2022).
11. Pudasainee, D.; Kurian, V.; Gupta, R. 2—Coal: Past, Present, and Future Sustainable Use. In *Future Energy*, 3rd ed.; Letcher, T.M., Ed.; Elsevier: Amsterdam, The Netherlands, 2020; pp. 21–48. [CrossRef]
12. Singh, D.; Dahiya, M.; Kumar, R.; Nanda, C. Sensors and systems for air quality assessment monitoring and management: A review. *J. Environ. Manag.* **2021**, *289*, 112510. [CrossRef]
13. Bonafè, G.; Montanari, F.; Stel, F. Air quality in Trieste, Italy—A Hybrid Eulerian-Lagrangian-Statistical Approach to Evaluate Air Quality in a Mixed Residential-Industrial Environment. 37p. Available online: https://pdfs.semanticscholar.org/72da/4005028680576c99a26eefc6d68a4df25c78.pdf (accessed on 5 May 2023).
14. Irwin, A. *Citizen Science: A Study of People, Expertise and Sustainable Development*; Routledge: London, UK, 1995.
15. Froeling, F.; Gignac, F.; Hoek, G.; Vermeulen, R.; Nieuwenhuijsen, M.; Ficorilli, A.; De Marchi, B.; Biggeri, A.; Kocman, D.; Robinson, J.A.; et al. Narrative review of citizen science in environmental epidemiology: Setting the stage for co-created research projects in environmental epidemiology. *Environ. Int.* **2021**, *152*, 106470. [CrossRef]
16. Fraisl, D.; Campbell, J.; See, L.; Wehn, U.; Wardlaw, J.; Gold, M.; Moorthy, I.; Arias, R.; Piera, J.; Oliver, J.L.; et al. Mapping citizen science contributions to the UN sustainable development goals. *Sustain. Sci.* **2020**, *15*, 1735–1751. [CrossRef]
17. Stewart, C.; Labrèche, G.; González, D.L. A Pilot Study on Remote Sensing and Citizen Science for Archaeological Prospection. *Remote Sens.* **2020**, *12*, 2795. [CrossRef]
18. Kasperowski, D.; Hillman, T. The epistemic culture in an online citizen science project: Programs, antiprograms and epistemic subjects. *Soc. Stud. Sci.* **2018**, *48*, 564–588. [CrossRef]
19. Kullenberg, C.; Kasperowski, D. What Is Citizen Science?—A Scientometric Meta-Analysis. *PLoS ONE* **2016**, *11*, e0147152. [CrossRef]

20. Ward, F.; Lowther-Payne, H.J.; Halliday, E.C.; Dooley, K.; Joseph, N.; Livesey, R.; Moran, P.; Kirby, S.; Cloke, J. Engaging communities in addressing air quality: A scoping review. *Environ. Health* **2022**, *21*, 89. [CrossRef]
21. Höhne, A.; Schulte, R.A.A.; Kulicke, M.; Huynh, T.-T.; Telgmann, M.; Frenzel, W.; Held, A. Assessing the Spatial Distribution of NO2 and Influencing Factors in Urban Areas—Passive Sampling in a Citizen Science Project in Berlin, Germany. *Atmosphere* **2023**, *14*, 360. [CrossRef]
22. De Craemer, S.; Vercauteren, J.; Fierens, F.; Lefebvre, W.; Meysman, F.J.R. Using Large-Scale NO2 Data from Citizen Science for Air-Quality Compliance and Policy Support. *Environ. Sci. Technol.* **2020**, *54*, 11070–11078. [CrossRef]
23. Ellenburg, J.A.; Williford, C.J.; Rodriguez, S.L.; Andersen, P.C.; Turnipseed, A.A.; Ennis, C.A.; Basman, K.A.; Hatz, J.M.; Prince, J.C.; Meyers, D.H.; et al. Global Ozone (GO3) Project and AQTreks: Use of evolving technologies by students and citizen scientists to monitor air pollutants. *Atmos. Environ. X* **2019**, *4*, 100048. [CrossRef]
24. Baldelli, A. Evaluation of a low-cost multi-channel monitor for indoor air quality through a novel, low-cost, and reproducible platform. *Meas. Sens.* **2021**, *17*, 100059. [CrossRef]
25. Bousiotis, D.; Alconcel, L.-N.S.; Beddows, D.C.S.; Harrison, R.M.; Pope, F.D. Monitoring and apportioning sources of indoor air quality using low-cost particulate matter sensors. *Environ. Int.* **2023**, *174*, 107907. [CrossRef]
26. Yang, G.; Zhou, Y.; Yan, B. Contribution of influential factors on PM2.5 concentrations in classrooms of a primary school in North China: A machine discovery approach. *Energy Build.* **2023**, *283*, 112787. [CrossRef]
27. Zusman, M.; Schumacher, C.S.; Gassett, A.J.; Spalt, E.W.; Austin, E.; Larson, T.V.; Carvlin, G.; Seto, E.; Kaufman, J.D.; Sheppard, L. Calibration of low-cost particulate matter sensors: Model development for a multi-city epidemiological study. *Environ. Int.* **2020**, *134*, 105329. [CrossRef]
28. Johnston, H.J.; Mueller, W.; Steinle, S.; Vardoulakis, S.; Tantrakarnapa, K.; Loh, M.; Cherrie, J.W. How Harmful Is Particulate Matter Emitted from Biomass Burning? A Thailand Perspective. *Curr. Pollut. Rep.* **2019**, *5*, 353–377. [CrossRef]
29. Ganti, R.K.; Ye, F.; Lei, H. Mobile crowdsensing: Current state and future challenges. *IEEE Commun. Mag.* **2011**, *49*, 32–39. [CrossRef]
30. Fraisl, D.; Hager, G.; Bedessem, B.; Gold, M.; Hsing, P.-Y.; Danielsen, F.; Hitchcock, C.B.; Hulbert, J.M.; Piera, J.; Spiers, H.; et al. Citizen science in environmental and ecological sciences. *Nat. Rev. Methods Primer* **2022**, *2*, 64. [CrossRef]
31. Wehn, U.; Gharesifard, M.; Ceccaroni, L.; Joyce, H.; Ajates, R.; Woods, S.; Bilbao, A.; Parkinson, S.; Gold, M.; Wheatland, J. Impact assessment of citizen science: State of the art and guiding principles for a consolidated approach. *Sustain. Sci.* **2021**, *16*, 1683–1699. [CrossRef]
32. Aristeidou, M.; Herodotou, C. Online citizen science: A systematic review of effects on learning and scientific literacy. *Citiz. Sci. Theory Pract.* **2020**, *5*, 11. [CrossRef]
33. Luger, T.M.; Hamilton, A.B.; True, G. Measuring Community-Engaged Research Contexts, Processes, and Outcomes: A Mapping Review. *Milbank Q.* **2020**, *98*, 493–553. [CrossRef]
34. Friedman, A.J. (Ed.) *Framework for Evaluating Impacts of Informal Science Education Projects*; Report from a National Science Foundation Workshop; The National Science Foundation: Alexandria, VA, USA, 2008.
35. Diviacco, P.; Nadali, A.; Iurcev, M.; Carbajales, R.; Busato, A.; Pavan, A.; Burca, M.; Grio, L.; Nolich, M.; Molinaro, A.; et al. MaDCrow, a Citizen Science Infrastructure to Monitor Water Quality in the Gulf of Trieste (North Adriatic Sea). *Front. Mar. Sci.* **2021**, *8*, 619898. [CrossRef]
36. Diviacco, P.; Nadali, A.; Nolich, M.; Molinaro, A.; Iurcev, M.; Carbajales, R.; Busato, A.; Pavan, A.; Grio, L.; Malfatti, F. Citizen science and crowdsourcing in the field of marine scientific research—The MaDCrow project. *J. Sci. Commun.* **2021**, *20*, A09. [CrossRef]
37. Diviacco, P.; Iurcev, M.; Carbajales, R.J.; Potleca, N. First Results of the Application of a Citizen Science-Based Mobile Monitoring System to the Study of Household Heating Emissions. *Atmosphere* **2022**, *13*, 1689. [CrossRef]
38. Diviacco, P.; Iurcev, M.; Carbajales, R.J.; Potleca, N.; Viola, A.; Burca, M.; Busato, A. Monitoring Air Quality in Urban Areas Using a Vehicle Sensor Network (VSN) Crowdsensing Paradigm. *Remote Sens.* **2022**, *14*, 5576. [CrossRef]
39. Iurcev, M.; Pettenati, F.; Diviacco, P. Improved automated methods for near real-time mapping—Application in the environmental domain. *Bull. Geophys. Oceanogr.* **2021**, *62*, 427–454.
40. Horvath, K.; Ivatek-Šahdan, S.; Ivančan-Picek, B.; Grubišić, V. Evolution and Structure of Two Severe Cyclonic Bora Events: Contrast between the Northern and Southern Adriatic. *Weather Forecast.* **2009**, *24*, 946–964. [CrossRef]
41. Orlic, M.; Penzar, B.; Penzar, I. Adriatic Sea and Land Breezes: Clockwise Versus Anticlockwise Rotation. *J. Appl. Meteorol.* **1988**, *27*, 675–679. [CrossRef]
42. Bulot, F.M.J.; Russell, H.S.; Rezaei, M.; Johnson, M.S.; Ossont, S.J.J.; Morris, A.K.R.; Basford, P.J.; Easton, N.H.C.; Foster, G.L.; Loxham, M.; et al. Laboratory Comparison of Low-Cost Particulate Matter Sensors to Measure Transient Events of Pollution. *Sensors* **2020**, *20*, 2219. [CrossRef]
43. Zimmerman, N. Tutorial: Guidelines for implementing low-cost sensor networks for aerosol monitoring. *J. Aerosol Sci.* **2022**, *159*, 105872. [CrossRef]
44. Liang, L. Calibrating low-cost sensors for ambient air monitoring: Techniques, trends, and challenges. *Environ. Res.* **2021**, *197*, 111163. [CrossRef]
45. Yi, W.Y.; Lo, K.M.; Mak, T.; Leung, K.S.; Leung, Y.; Meng, M.L. A Survey of Wireless Sensor Network Based Air Pollution Monitoring Systems. *Sensors* **2015**, *15*, 31392–31427. [CrossRef]

46. Mead, M.I.; Popoola, O.A.M.; Stewart, G.B.; Landshoff, P.; Calleja, M.; Hayes, M.; Baldovi, J.J.; McLeod, M.W.; Hodgson, T.F.; Dicks, J.; et al. The use of electrochemical sensors for monitoring urban air quality in low-cost, high-density networks. *Atmos. Environ.* **2013**, *70*, 186–203. [CrossRef]
47. Crilley, L.R.; Shaw, M.; Pound, R.; Kramer, L.J.; Price, R.; Young, S.; Lewis, A.C.; Pope, F.D. Evaluation of a low-cost optical particle counter (Alphasense OPC-N2) for ambient air monitoring. *Atmos. Meas. Technol.* **2018**, *11*, 709–720. [CrossRef]
48. Crilley, L.R.; Singh, A.; Kramer, L.J.; Shaw, M.D.; Alam, M.S.; Apte, J.S.; Bloss, W.J.; Hildebrandt Ruiz, L.; Fu, P.; Fu, W.; et al. Effect of aerosol composition on the performance of low-cost optical particle counter correction factors. *Atmos. Meas. Technol.* **2020**, *13*, 1181–1193. [CrossRef]
49. Malm, W.C.; Day, D.E.; Kreidenweis, S.M.; Collett, J.L.; Lee, T. Humidity-dependent optical properties of fine particles during the Big Bend Regional Aerosol and Visibility Observational Study. *J. Geophys. Res. Atmos.* **2003**, *108*, 4279. [CrossRef]
50. Sathiyanarayanan, M.; Varadarajan, V.; Pradeep, K.V. Visual analytics on spatial time series for environmental data. *Int. J. Recent Technol. Eng.* **2019**, *8*, 1173–1181.

Article

Monitoring PM$_{2.5}$ at a Large Shopping Mall: A Case Study in Macao

Thomas M. T. Lei [1,*], Yan W. I. Chan [1] and Mohd Shahrul Mohd Nadzir [2]

[1] Institute of Science and Environment, University of Saint Joseph, Macau 999078, China
[2] Department of Earth Sciences and Environment, Faculty of Science and Technology, Universiti Kebangsaan Malaysia, Bangi 43600, Selangor, Malaysia
* Correspondence: thomas.lei@usj.edu.mo

Abstract: Employees work long hours in an environment where the ambient air quality is poor, directly affecting their work efficiency. The concentration of particulate matters (PM) produced by the interior renovation of shopping malls has not received particular attention in Macao. Therefore, this study will investigate the indoor air quality (IAQ), in particular of PM$_{2.5}$, in large-scale shopping mall renovation projects. This study collected on-site PM data with low-cost portable monitoring equipment placed temporarily at specific locations to examine whether the current control measures are appropriate and propose some improvements. Prior to this study, there were no measures being implemented, and on-site monitoring to assess the levels of PM$_{2.5}$ concentrations was non-existent. The results show the highest level of PM$_{2.5}$ recorded in this study was 559.00 µg/m^3. Moreover, this study may provide a reference for decision-makers, management, construction teams, design consultant teams, and renovation teams of large-scale projects. In addition, the monitoring of IAQ can ensure a comfortable environment for employees and customers. This study concluded that the levels of PM$_{2.5}$ concentration have no correlation with the number of on-site workers, but rather were largely influenced by the processes being performed on-site.

Keywords: particulate matter; construction dust; health exposure; indoor air quality; air pollution

Citation: Lei, T.M.T.; Chan, Y.W.I.; Mohd Nadzir, M.S. Monitoring PM$_{2.5}$ at a Large Shopping Mall: A Case Study in Macao. *Processes* **2023**, *11*, 914. https://doi.org/10.3390/pr11030914

Academic Editors: Daniele Sofia and Paolo Trucillo

Received: 13 February 2023
Revised: 4 March 2023
Accepted: 10 March 2023
Published: 17 March 2023

1. Introduction

According to the United Nation's (UN) News, 55% of the world's population lived in urban areas in 2018, estimated to rise to 68% by 2050 [1]. The high population density in urban areas is often associated with air pollution and carbon emission problems [2]. As urban populations and the number of motor vehicles simultaneously increase, and countries become more industrialized, air pollution becomes more and more of a serious problem. The urban population spends their time in the indoor environment every day. Indoor air pollution directly affects people's health. Studies showed that more than 4.3 million deaths each year are caused by poor IAQ [3]. Poor indoor air can lead to human discomfort, health problems, and even disease. Poor IAQ also has a negative impact on work efficiency. Poor indoor air will make staff emotionally irritable and physically tired and can even cause disease epidemics. Macao is a highly urbanized region with a very high population density of 27,259 inhabitants/km^2 [4], and the total population of Macao reached 683,200 inhabitants in 2021 [5]. Therefore, IAQ is a serious issue for a high-density environment such as Macao [6–8]. There have been many related technical studies as the impact of IAQ on health has attracted more and more attention. The current understanding of the synergistic and antagonistic effects of multiple pollutants at low concentrations for a long time is not comprehensive.

This work focuses on the IAQ of shopping malls in Macao with frequent renovation projects. These projects of different sizes are carried out in operating buildings. However, the emission of PM$_{10}$ and PM$_{2.5}$ in the indoor environment caused by interior decoration is not thoroughly evaluated. More importantly, people spend about 90% of their time in indoor buildings, such as in schools, offices, and shopping centers [9–12]. Therefore, having a comfortable living environment and good IAQ are important factors that affect human health and work efficiency. The outbreak of COVID-19 in recent years has led to adopting social distancing measures such as school closures and working from home [13]. This has led to an increase in indoor activities and the growing problem of indoor air pollution, which has different levels of impact on the human body, so providing a good IAQ environment can minimize health impacts and economic losses.

In recent years, many buildings in Macao have been undergoing renovation work which causes an increase in particulate matters (PM) in indoor building. For example, Sands China invested \$470 million in Hong Kong Dollars (HKD) in renovation projects of their premises in 2021. The concessionaire has invested more than \$100 billion HKD over the years [14]. Nevertheless, there was minimal attention to indoor $PM_{2.5}$ emissions in numerous interior renovation projects. On-site operators are exposed to high concentrations of $PM_{2.5}$ for extended periods of time. The health of operators may be affected by the polluted air in the buildings. The emission of indoor $PM_{2.5}$ is a serious threat to human health.

The concept of IAQ was first proposed in the 1990s in Nordic Countries and North America, focusing on indoor air in non-production places, such as residences, offices, schools, hospitals, and hotels. Due to people who stayed indoors for a long time often feeling uncomfortable, new problems and concepts, such as SBS and Legionnaires' disease, gradually emerged [15]. People also realized that indoor air pollution was closely related to different diseases, such as asthma and lung cancer. SBS is a condition in which people experience acute health or comfort-related effects within a poor indoor environment, and the complaint may be for a specific room, area, or entire building, and research found that symptoms subsided or disappeared when the environmental elements causing SBS were removed [16]. The symptoms of SBS include headaches, dizziness, nausea, eye, nose, or throat irritation, dry cough, dry or itchy skin, difficulty concentrating, fatigue, reduced sensitivity to smells, hoarseness, allergies, colds, flu-like symptoms, increased incidence of asthma attacks, and personality changes [17]. The cause of these symptoms is unclear, but most victims said their symptoms could be relieved by leaving the property [18].

Some possible main contributors to SBS include chemical pollutants, biological pollutants, insufficient ventilation, poor air circulation, psychological factors, insufficient and inappropriate lighting, and poor acoustics [19–26]. Symptoms of BRI include coughing, chest pains, shortness of breath with mild exertion, edemas, palpitations, nosebleeds, cancer, pregnancy problems, and miscarriages. In addition, extrinsic allergic alveolitis, Legionnaires' disease, humidifier fever, pneumonia, and occupational asthma are also known to occur due to BRI [27]. A large amount of formaldehyde was released inside buildings during the renovation process, which led to many residents experiencing acute irritation and poisoning symptoms, and some even suffered toxic hepatitis or allergic purpura [28,29]. These studies discovered that volatile organic compounds (VOCs) have a high contribution to indoor pollution sources for human exposure.

Systematic studies around IAQ initially focused on the relationship between indoor and outdoor air quality and the impact of indoor air pollutants on human health. The world's first systematic and large-scale study of the relationship between indoor and outdoor air quality was conducted in 1965 [30]. The study measured the relationship between indoor and outdoor SO_2, smoke, and dust, determining the relatively safe levels of IAQ during air pollution episodes. The relationship between smoking and indoor aerosol generation, indoor SO_2 attenuation, and building age was also explored. The results showed significant differences between indoor and outdoor air quality and that IAQ might have more of an impact on human health than outdoor air quality [31].

This work hopes to establish a comprehensive management system for pre-assessment and to discuss IAQ management from the aspects of building material management, architectural design management, and facility management and to manage and control the level of IAQ in public places and civic buildings. In addition, this work aims to explore the correlation between the number of workers and the levels of $PM_{2.5}$ concentrations in the shopping mall. Prior to this study, there were no measures being implemented, and on-site monitoring to assess the levels of $PM_{2.5}$ concentrations was non-existent on this specific premises, which is a very concerning situation for both the workers and visitors. This study is the first assessment of the relationships between the level of $PM_{2.5}$ concentrations, number of workers, and the construction processes in a large shopping mall in Macau.

2. Materials and Methods

In recent years, the government has made great efforts to promote Macao as a "World Center of Tourism and Leisure." Shopping malls of different sizes and themes, each with its own unique architectural style, were built, large complexes combining food, drink, and entertainment, providing one-stop diversified facilities and attracting many tourists and locals, especially during weekends and holidays. In this study, fixed-point measurement was carried out on-site using low-cost portable monitoring sensors. As of now, there are no specific indoor air quality (IAQ) guidelines established or being enforced in the indoor buildings of Macao. Therefore, the IAQ standard would be the same as the ambient air quality standard established by the Macao Meteorological and Geophysical Bureau (SMG), which adopted the WHO Interim Target 2 (IT-2) standard.

The collected data were obtained using TSI Sidepak AM510 Personal Aerosol Monitor calibrated to factory settings, temporarily set up in two monitored areas, consisting of 2 h of data collection during the peak periods of the day when most shopping mall employees worked in their respective areas. The duration of monitoring campaign lasted for 14 days in each area. The collected data showed that the levels of $PM_{2.5}$ concentration were different between these three collection points in two monitored areas, including Construction Site A and B and Renovation Area. During the construction phase, the level of $PM_{2.5}$ concentration reached its peak because many processes were being performed, and different operation processes produced different levels of $PM_{2.5}$ concentration. The construction site had a suitable hoarding, forming a barrier between the construction and public areas. In comparison, the renovation site was adjacent to the public area, in which the area was small and also had a suitable hoarding to prevent customers from entering by mistake. The measurement points in each area were chosen for the following reasons:

The construction area was large, with collection points A and B being core areas with complicated processes and dense flows of people. For this study, the monitored construction site was located on the upper floor of a shopping mall, with an area of about $11,000\ m^2$. Hence, it was important to monitor these locations for the IAQ of the mall. The demolition process caused the worst air pollution at the construction site, which generally emits a high concentration of $PM_{2.5}$, followed by the grinding and the drilling installation tasks. Figure 1 shows the layout of the construction site.

The renovation area measurement point was chosen because the scope of the store renovation was relatively small, and the collection point was chosen in the center of the district. For this study, the monitored renovation area was about $200\ m^2$ and located in the center of the shopping mall, near many shops and restaurants. Figure 2 shows the collection point of the public area next to the store renovation area, and the measurement point was where the public gathered. The collected data reflect the impact of IAQ on the public and study the impact of $PM_{2.5}$ on the health of the operators. The peak levels of $PM_{2.5}$ concentration during the refurbishment were recorded during the cement-sand task when the cement and sand were mixed. The second highest levels recorded were during the grinding process, with a large amount of $PM_{2.5}$ emitted during the grinding of ceiling surface defects.

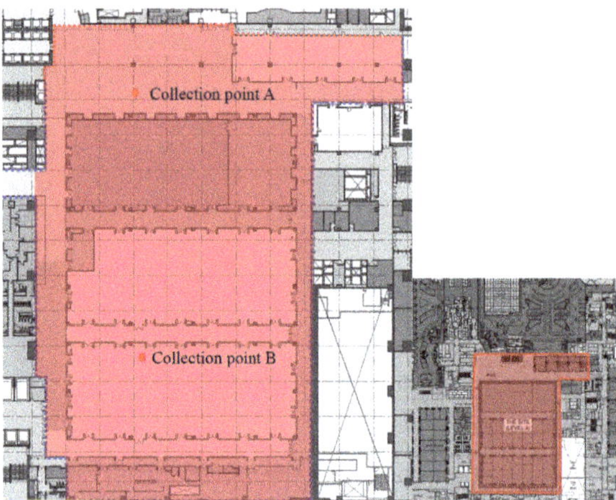

Figure 1. Layout of the construction site.

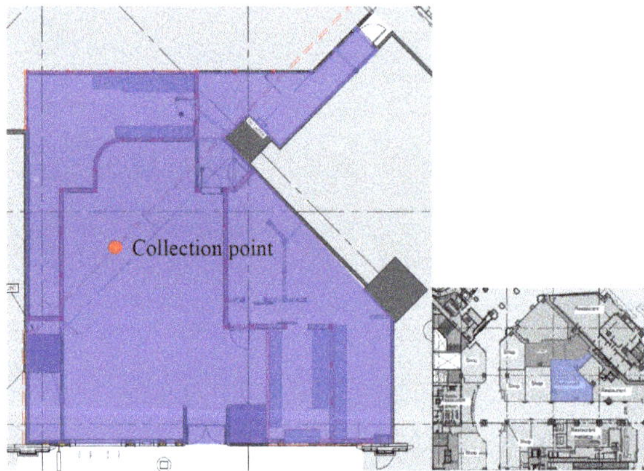

Figure 2. Layout of the renovation area.

The public area of the shopping mall is where customers gather for a short period of time. Customers may go to the stores or restaurants through the high-traffic corridor, which is very congested during public holidays and festivals. As shown in Figure 3, the collection point was located in a densely populated area. This study monitored the public area near the renovation area and collected the PM levels.

As part of this study, a survey on dust prevention measures was conducted. All 29 surveys received were completed by working professionals who applied at least one dust control measure in their working environment. The survey was voluntary and designed to allow the participants multiple answers to each question.

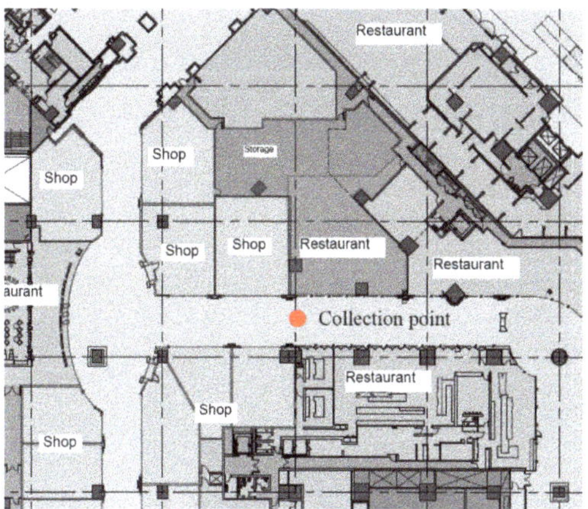

Figure 3. Layout of public area.

3. Results and Discussion

This section will discuss the levels of PM$_{2.5}$ concentrations during the 14-day monitoring campaign in the construction and renovation areas, with the public area being the control of this study. This section also explores the correlation between the number of workers and the observed values for PM$_{2.5}$ concentration in different locations, and the correlation between the levels of PM$_{2.5}$ concentrations and the meteorological parameters such as temperature and humidity. A previous study shows that each construction activity demonstrates different pattern and level of PM$_{2.5}$ concentration, and exposure to high concentration of PM$_{2.5}$ has caused impairment of lung function performance and led to respiratory health problems [32]. In addition, the construction site needs to control and monitor its emission of PM$_{2.5}$ with a proper construction time management schedule on site [33].

3.1. Construction Area

Figure 4 shows that the level of PM$_{2.5}$ concentration at the construction site of collection point A was significantly higher at 307.00 μg/m^3 on day 8 due to the spray-painting work. The level of PM$_{2.5}$ concentration at collection point B was significantly higher at 332.00 μg/m^3 on day 11 due to the drilling work. Painting and drilling work produce high levels of PM$_{2.5}$ concentration, so it is necessary to check whether the control measures on-site are sufficient. It is necessary to improve the personal protective equipment (PPE) of the on-site personnel, such as by wearing N95 filter masks.

As shown in Table 1, descriptive statistics were obtained for the mean, median mode, minimum, and maximum concentration of PM$_{2.5}$. Point B has a higher mean value for PM$_{2.5}$ compared to point A at the construction site due to the demolition and drilling work. Since there are many elderly people and children in the shopping malls, the use of a blower fan is recommended to manage PM$_{2.5}$ inside the shopping malls.

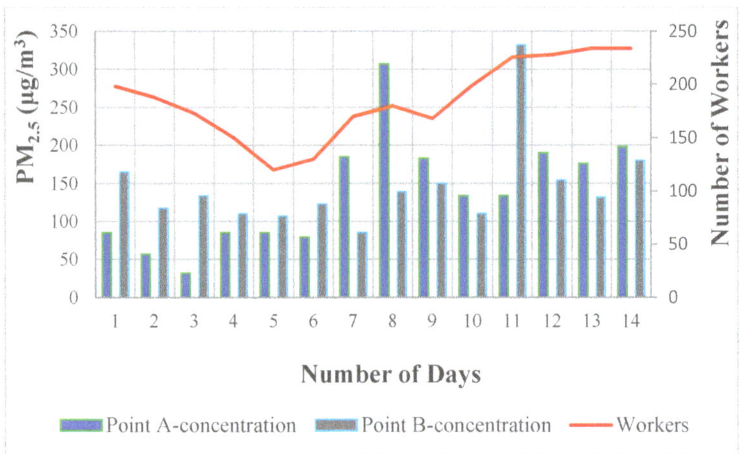

Figure 4. The levels of PM$_{2.5}$ concentration and the number of workers at a construction area throughout a 14-day monitoring campaign in a large shopping mall.

Table 1. Descriptive statistics for PM$_{2.5}$ concentration at the construction site (in µg/m^3).

PM$_{2.5}$—Construction Point A		PM$_{2.5}$—Construction Point B	
Mean	137.86	Mean	145.57
Median	134.00	Median	133.00
Mode	85.00	Mode	110.00
Minimum	32.00	Min	85.00
Maximum	307.00	Maximum	332.00

3.2. Renovation Area

Figure 5 shows the extremely high level of PM$_{2.5}$ concentration from days 7 to 9 because of the cement-sand mixing task, with day 7 reaching a peak of 559 µg/m^3. After the mixing task was completed, the level of PM$_{2.5}$ concentration slowly dropped to a more reasonable level.

The data collected on site showed no significant relationship between the number of workers and the level of PM$_{2.5}$ concentration. The high level of PM$_{2.5}$ concentration was recorded during special tasks, such as cement-sand mixing, drilling, and demolition work. The collected data showed the high levels of PM$_{2.5}$ concentration produced during these special tasks. Thus, corresponding control measures should be developed to reduce the high levels of PM$_{2.5}$ concentration.

Table 2 shows a descriptive analysis of the renovation area. The level of PM$_{2.5}$ concentration at the renovation site has a high peak, exceeding the levels in the construction area. The high PM$_{2.5}$ levels were caused by the cement-sand mixing task. Thus, the use of cement slurry mixers should be adopted by the shopping mall, and protective measures should be taken in advance. Hence, it is clear that the shopping mall staff should consider each renovation carefully to ensure a safe level of IAQ is guaranteed.

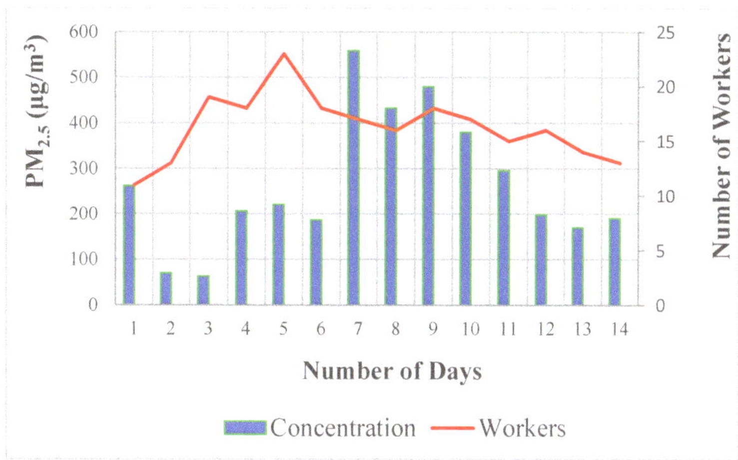

Figure 5. The levels of PM$_{2.5}$ concentration and the number of workers in the renovation area throughout a 14-day monitoring campaign in a large shopping mall.

Table 2. Descriptive statistics for PM$_{2.5}$ in the renovation area (in $\mu g/m^3$).

PM$_{2.5}$—Renovation Area	
Mean	265.36
Median	213.50
Mode	N/A
Minimum	63.00
Maximum	559.00

3.3. Public Area

Figure 6 shows the levels of PM$_{2.5}$ concentration in the public area, which is also the control of this study. The results show the levels of PM$_{2.5}$ concentration to be significantly lower than in the construction and renovation area, with the peak levels being measured at only 37 $\mu g/m^3$ on day 1.

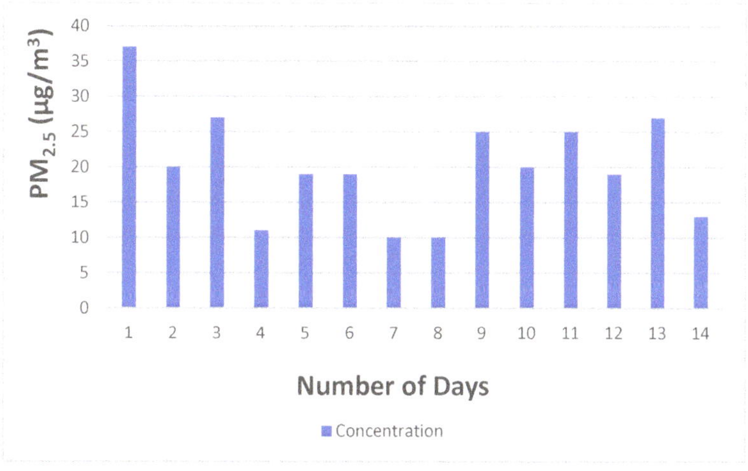

Figure 6. The levels of PM$_{2.5}$ concentration in the public area throughout a 14-day monitoring campaign in a large shopping mall.

Table 3 shows a descriptive analysis of the public area. The level of $PM_{2.5}$ concentration at the public area has a mean of 20.14 $\mu g/m^3$, which is well below the levels measured in the construction and renovation area. This shows that the processes being performed in the construction and renovation area have led to a burst of $PM_{2.5}$ in a short period of time, causing a health hazard to the on-site personnel.

Table 3. Descriptive statistics for $PM_{2.5}$ in the public area (in $\mu g/m^3$).

$PM_{2.5}$—Public Area	
Mean	20.14
Median	19.50
Mode	19.00
Min	10.00
Maximum	37.00

Figure 7 shows a cement slurry mixer commonly found in the construction industry and is used to mix cement and water into a uniform slurry, which can lower the emission of $PM_{2.5}$ in a construction environment.

Figure 7. Wet cement sand mixing for building a brick wall at a renovation site.

3.4. Correlation between Workers and PM$_{2.5}$ Concentrations

According to Table 4, the correlation coefficient (r) between point A and point B is very low (r = 0.10). Thus, there is no correlation between these two collection points. Nevertheless, the number of workers and $PM_{2.5}$ concentration in point A has a low-to-medium correlation (r = 0.38), and the number of workers and $PM_{2.5}$ concentration in point B has a medium-to-high correlation (r = 0.51). Overall, the correlation between the levels of $PM_{2.5}$ and the number of workers in the construction area (r = 0.38 to 0.51) is too low to show any relationship between them.

Table 4. Correlation coefficient (r) between $PM_{2.5}$ and workers in Point A and B of the construction area.

$PM_{2.5}$	Construction Point A	Construction Point B	Workers
Construction Point A	1.00		
Construction Point B	0.10	1.00	
Workers	0.38	0.51	1.00

As shown in Table 5, the number of workers and $PM_{2.5}$ concentration in the renovation area shows a low correlation (r = 0.11), which show there is no relationship between the levels of $PM_{2.5}$ and the number of workers in the renovation area.

Table 5. Correlation coefficient (r) between $PM_{2.5}$ and workers in the renovation area.

$PM_{2.5}$	Renovation Area	Workers
Renovation area	1.00	
Workers	0.11	1.00

As shown in Table 6, the average temperature and relative humidity during the 14 days of the monitoring campaign are 21.9 °C and 50.1%, respectively.

Table 6. Temperature and humidity during the 14-day monitoring campaign.

Day	Temperature (°C)	Humidity (%)
1	21.7	47.1
2	21.3	47.5
3	22.0	51.3
4	22.5	54.4
5	21.9	50.9
6	21.8	48.8
7	22.5	48.5
8	22.4	53.5
9	22.5	52.0
10	22.0	51.4
11	21.0	48.0
12	21.5	47.5
13	21.0	47.0
14	22.0	54.0

As shown in Table 7, the correlation coefficient between the levels of $PM_{2.5}$ concentration in the construction area A and B and the temperature is low (r = 0.23 to 0.50) and the humidity is very low (r = −0.17 to 0.24). The correlation coefficient between the levels of $PM_{2.5}$ concentration in the renovation area and the temperature is medium (r = 0.55) and the humidity is low (r = 0.17). The correlation coefficient between the levels of $PM_{2.5}$ concentration in the public area and the temperature is medium (r = 0.52) and the humidity is medium (r = −0.55). Overall, there is a low-to-medium correlation between the meteorological parameters and the levels of $PM_{2.5}$ concentrations in all areas.

Table 7. Correlation coefficient (r) between meteorological parameters and $PM_{2.5}$ concentrations in different areas.

	Temperature (°C)	Humidity (%)	Construction Area A	Construction Area B	Renovation Area	Public Area
Temperature (°C)	1					
Humidity (%)	0.722724945	1				
Construction Area A	0.233882808	0.2411858	1			
Construction Area B	−0.504839483	−0.170542272	0.095182185	1		
Renovation Area	0.548430574	0.171488869	0.62435975	−0.041693677	1	
Public Area	−0.518369199	−0.545497874	−0.437081689	0.322059792	−0.293991677	1

Hence, the use of prefabricated cement parts and structures constructed in the factory and only brought on-site for installation is recommended. In addition, optimizing work team structures and introducing robots for certain tasks will help to minimize human exposure to $PM_{2.5}$ generated by cement-sand mixing tasks. Additionally, it is recommended that there should be fewer workers for demolition work, which could reduce threats to workers' health due to poor IAQ. The shopping mall may consider optimizing the work team structure and even introduce robots to replace some manual labor.

3.5. On-Site Survey

As shown in Figure 8, 75.9% of the respondents believed that watering and spraying were the best control measures due to the relatively low cost of the humidification and dust reduction equipment and the smaller portable machines needed for this. Over 72.4% of respondents believed blower fans to be effective measures, with the advantages of relatively low cost, lightness, and flexibility of use. Furthermore, only 62.1%, and 37.9% of respondents believed water curtains and water mist blowers to be effective due to their large size and inability to be moved freely. Other measures, such as dust filters and curtain separators, were the least popular, only selected by 17.2% of the respondents. According to the survey results, preventing dust at the source is the most effective mitigation method.

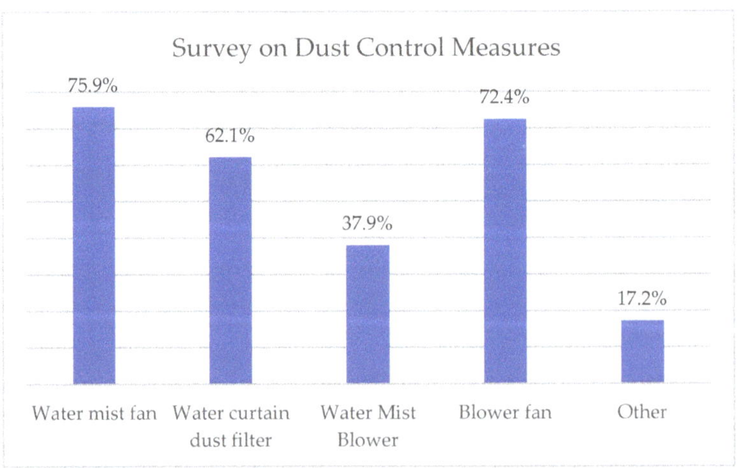

Figure 8. Response from the survey on dust control measures (multiple answers).

4. Conclusions

IAQ is very important to the health of operators and ensuring good IAQ in these environments will minimize human and economic losses. The highest level of $PM_{2.5}$ concentration recorded in this study was 559.00 μg/m³ in the renovation area, which is extremely hazardous even for short-term exposure and poses great threats to the health of the on-site personnel. The correlation between the number of workers and the levels of $PM_{2.5}$ concentrations in different areas is low-to-medium (r = 0.11 to 0.51). In addition, the correlation between the meteorological parameters and levels of $PM_{2.5}$ in different areas is low to medium (r = -0.55 to 0.55). Thus, it can be concluded that the number of workers and meteorological parameters have no direct influence on the levels of $PM_{2.5}$ concentrations, but rather the processes being performed on-site is the most important factor that influences the levels of $PM_{2.5}$ concentrations in the shopping mall. After the conclusion of this study, it has become a requirement to install an IAQ monitoring system on-site to monitor the level of real-time $PM_{2.5}$ concentration in the construction and renovation area of the shopping mall. Further studies may be conducted to monitor different pollutants presented in the indoor environment. In addition, it is recommended that the local authorities establish an occupational safety standard for the levels of $PM_{2.5}$ concentration in an indoor construction environment to protect the well-being of the on-site personnel and to serve as a reference for the construction management team.

Author Contributions: Conceptualization, T.M.T.L. and Y.W.I.C.; methodology, T.M.T.L.; software, T.M.T.L. and Y.W.I.C.; validation, T.M.T.L. and M.S.M.N.; data curation, T.M.T.L. and Y.W.I.C.; writing—original draft preparation, T.M.T.L. and Y.W.I.C.; writing—review and editing, T.M.T.L. and M.S.M.N.; supervision, T.M.T.L.; funding acquisition, T.M.T.L. All authors have read and agreed to the published version of the manuscript.

Funding: This research received no external funding.

Data Availability Statement: 3rd Party Data. Restrictions apply to availability of these data.

Conflicts of Interest: The authors declare no conflict of interest.

References

1. UN. 68% of the world population projected to live in urban areas by 2050. *United Nations News*, 16 May 2018; pp. 2–5.
2. Liu, H.; Cui, W.; Zhang, M. Exploring the causal relationship between urbanization and air pollution: Evidence from China. *Sustain. Cities Soc.* **2022**, *80*, 103783. [CrossRef]
3. Raju, S.; Siddharthan, T.; McCormack, M.C. Indoor Air Pollution and Respiratory Health. *Clin. Chest Med.* **2020**, *41*, 825–843. [CrossRef]
4. Szmigiera, M. Cities with the Highest Population Density Globally. 2021. Available online: https://www.statista.com/statistics/1237290/cities-highest-population-density/: (accessed on 1 November 2022).
5. DSEC. Demographic Statistics. 2021. Available online: https://www.dsec.gov.mo/en-US/Statistic?id=101: (accessed on 1 November 2022).
6. DSPA. Guidelines for Indoor Air Quality of General Public Places in Macau. 2022. Available online: https://www.dspa.gov.mo/pdf/GuideIndoorAir_tc.pdf: (accessed on 1 November 2022).
7. SMG. Definition of Air Quality Index. 2021. Available online: https://www.smg.gov.mo/en/subpage/784/page/347: (accessed on 1 November 2022).
8. Air Quality Guidelines. 2021. Available online: https://apps.who.int/iris/bitstream/handle/10665/345329/9789240034228-eng.pdf?sequence=1&isAllowed=y: (accessed on 1 November 2022).
9. EPA. Indoor Air Quality. 2021. Available online: https://www.epa.gov/report-environment/indoor-air-quality: (accessed on 1 November 2022).
10. Boxer, P.A. Indoor Air Quality A Psychosocial Perspective. *J. Occup. Med. Toxicol.* **1990**, *32*, 425–428. [CrossRef] [PubMed]
11. Gulgun, D.; Cemile, D. Determination of Indoor Air Quality in University Student Canteens. *Sak. Univ. J. Sci.* **2021**, *25*, 1322–1331.
12. Payus, C.; Carmen, C. Indoor air quality at shopping malls in Kota Kinabalu, Sabah (particulate matter and ozone). *Ann. Trop. Med. Public Health* **2017**, *10*, 31–35. [CrossRef]
13. Barbour, N.; Menon, N.; Mannering, F. A statistical assessment of work from-home participation during different stages of the COVID-19 pandemic. *Transp. Res. Interdiscip. Perspect.* **2021**, *11*, 100441. [CrossRef]
14. Sands China. 2021 Annual Report. 2021. Available online: https://investor.sandschina.com/system/files-encrypted/nasdaq_kms/assets/2022/03/24/17-13-14/20220324%281%29_Annual%20Report.pdf: (accessed on 1 November 2022).
15. Saini, J.; Dutta, M.; Marques, G. A comprehensive review on indoor air quality monitoring systems for enhanced public health. *Sustain. Environ. Res.* **2020**, *30*, 6. [CrossRef]
16. EPA. Indoor Air Facts No. 4 Sick Building Syndrome EPA—Air Radiat. (6609J) Res. Dev. (MD-56) **1991**, 1–4. Available online: https://www.epa.gov/sites/default/files/2014-08/documents/sick_building_factsheet.pdf (accessed on 9 March 2023).
17. Sarkingobir, Y.; Muhammadzayyanu, M.; Sarkingobir, S. Sick Building Syndrome: A Review of Related Literatures. *Int. J. Med. Biosci.* **2017**, *1*, 1.

18. Joshi, S.M. The sick building syndrome. *Indian J. Occup. Environ. Med.* **2008**, *12*, 61. [CrossRef]
19. IARC. *IARC Monographs on the Evaluation of Carcinogenic Risks to Humans*; World Health Organization International Agency for Research on Cancer: Lyon, France, 2006; Volume 88.
20. Zhang, Q.; Gangupomu, R.H.; Ramirez, D.; Zhu, Y. Measurement of ultrafine particles and other air pollutants emitted by cooking activities. *Int. J. Environ. Res. Public Health* **2010**, *7*, 1744–1759. [CrossRef]
21. Marvel, J.; Tronville, P. Ultrafine particles: A review about their health effects, presence, generation, and measurement in indoor environments. *Build. Environ.* **2022**, *216*, 108992. [CrossRef]
22. Hospodsky, D.; Qian, J.; Nazaroff, W.W.; Yamamoto, N.; Bibby, K.; Rismani-Yazdi, H.; Peccia, J. Human occupancy as a source of indoor airborne bacteria. *PLoS ONE* **2012**, *7*, 4. [CrossRef]
23. Luongo, J.C.; Fennelly, K.P.; Keen, J.A.; Zhai, Z.J.; Jones, B.W.; Miller, S.L. Role of mechanical ventilation in the airborne transmission of infectious agents in buildings. *Indoor Air* **2016**, *26*, 666–678. [CrossRef]
24. Piscitelli, P.; Miani, A.; Setti, L.; DeGennaro, G.; Rodo, X.; Artinano, B.; Vara, E.; Rancan, L.; Arias, J.; Passarini, F.; et al. The role of outdoor and indoor air quality in the spread of SARS-CoV-2: Overview and recommendations by the research group on COVID-19 and particulate matter (RESCOP commission). *Environ. Res.* **2022**, *211*, 113038. [CrossRef]
25. Power, M.C.; Kioumourtzoglou, M.A.; Hart, J.E.; Okereke, O.I.; Laden, F.; Weisskopf, M.G. The relation between past exposure to fine particulate air pollution and prevalent anxiety: Observational cohort study. *BMJ* **2015**, *350*, h1111. [CrossRef]
26. Alhorr, Y.; Arif, M.; Katafygiotou, M.; Mazroei, A.; Kaushik, A.; Elsarrag, E. Impact of indoor environmental quality on occupant well-being and comfort: A review of the literature. *Int. J. Sustain. Built Environ.* **2016**, *5*, 1–11. [CrossRef]
27. Crook, B. Legionella risk in evaporative cooling systems and underlying causes of associated breaches in health and safety compliance. *Int. J. Hyg. Environ. Health* **2020**, *224*, 113425. [CrossRef]
28. Gallego, E.; Roca, X.; Perales, J.F.; Guardino, X. Determining indoor air quality and identifying the origin of odour episodes in indoor environments. *J. Environ. Sci.* **2009**, *21*, 333–339. [CrossRef]
29. Scott, P.; Dan, C.; Paul, F.; Beth, H.; Terry, B. *Weatherization and Indoor Air Quality: Measured Impacts in Single-Family Homes under the Weatherization Assistance Program*; Oak Ridge National Laboratory: Oak Ridge, TN, USA, 2014.
30. Zahaba, M.; Sarah Fatihah Tamsi, N.; Engliman, S.; Faiz Kamaruddin, A.; Artika Hassan, N.; Ariffin, A. Indoor air quality (IAQ) in a naval ship after refit program: A time variation analysis. *IOP Conf. Ser. Earth Environ. Sci.* **2022**, *1013*, 012004. [CrossRef]
31. Otuyo, M.K.; Nadzir, M.S.M.; Latif, M.T.; Saw, L.H. In-train particulate matter (PM10 and PM2.5) concentrations: Level, source, composition, mitigation measures and health risk effect—A systematic literature review. *Indoor Built Environ.* **2022**, *32*, 460–493. [CrossRef]
32. Rahman, S.A.A.; Yatim, S.R.M.; Abdullah, A.H.; Zainuddin, N.A.; Samah, M.A.A. Exposure of Particulate Matter 2.5 (PM2.5) on Lung Function Performance of Construction Workers. *AIP Conf. Proc.* **2019**, *2124*, 020030.
33. Rosman, P.S.; Samah, M.A.A.; Yunus, K.; Hussain, M.R.M. Particulate Matter (PM2.5) At Construction Site: A Review. *Int. J. Recent Technol. Eng. (IJRTE)* **2019**, *8*, 255–259.

Article

An Empirical Analysis of the Aircraft Emissions by Operating from Scheduled Flights within the Domestic Market in Spain

Antonio Martínez Raya *, Alejandro Segura de la Cal and Rafael Eugenio González Díaz

Department of Organizational Engineering, Business Administration and Statistics,
Technical University of Madrid—Universidad Politécnica de Madrid (UPM), 28040 Madrid, Spain
* Correspondence: amartinez@cee.uned.es

Abstract: Over the past 20 years, civil aviation has substantially reduced its environmental impact to augment sustainable transportation. In Spain, the domestic market has been habitually characterized by a few enterprises providing air transport services linked to scheduled flights on domestic corridors. Because of geographic diversity and the highly concentrated population characterizing this southern European country, many of them could not be supplied by alternative transport modes in terms of both time and distance by comparison with air transportation. For air quality monitoring from 139 national corridors, this paper aims to study related aviation emissions to conduct an economic analysis in terms of positive or negative externalities. For such purposes, the study focused on these domestic routes served by the five most important Spanish airports, specifically on the number of passengers transported from 2011 to 2020. Up to 10 aircraft types representing no more than 89% of regular operations on these flyways were subsequently identified. In addition, certain engine types also were selected as representatives to evaluate their emissions, depending on the great-circle distance in each route. The research findings, though particularly conditioned by aviation peculiarities of such a domestic market, point decisively to significant dependence upon emissions in connection with the seasonality of the demand and the concentration of flights with low occupancy indices from any one of them. Results suggest that airlines would benefit from operating turboprops instead of turbojets on selected routes, especially when oil prices are high. However, it is not always easy to find a balance between uncompromising economic profitability and effective fleet availability, since nowadays air transport undertakings tend to unify their fleets by using a few aircraft families, mostly powered by jet engines, apart from regional carriers.

Keywords: environment; sustainable transportation; air quality monitoring; aviation emissions; economic analysis; externalities; economic profitability; fleet availability

Citation: Martínez Raya, A.; Segura de la Cal, A.; González Díaz, R.E. An Empirical Analysis of the Aircraft Emissions by Operating from Scheduled Flights within the Domestic Market in Spain. *Processes* **2023**, *11*, 741. https://doi.org/10.3390/pr11030741

Academic Editors: Daniele Sofia and Paolo Trucillo

Received: 1 February 2023
Revised: 20 February 2023
Accepted: 27 February 2023
Published: 2 March 2023

1. Introduction

The civil aviation sector is a fundamental part of the planning process for developing sustainable transportation in the European Union (EU), according to one of the Sustainable Development Goals (SDGs), namely Goal 11. The so-called 'European Green Deal' has set up an action plan to combat climate change within the EU. With specific regard to the transportation sector and within the scope of EU's Green Deal's objectives, specifically, the 'Fit for 55' has been included in ambitious proposals to reduce related emissions by up to 55% by 2030 from two transport modes, both maritime and air. For carbon dioxide emissions from aviation, the Emissions Trading System (EU ETS) [1] has increasingly been used since 2005 as the principal instrument for calculating the EU's total emissions through a system based on allowances (EUA) reporting from the European Union Transaction Log (EUTL). Indeed, to entail adjustments to EU climate legislation, the European Commission (hereafter the Commission) has recently amended ETS Directive [2].

Regarding greenhouse gas emissions from the European aviation sector, air transport has been facing both global warming and climate change and is considered a common priority nowadays for the entire industry within the EU framework on ETS. Over the past years, significant efforts have been made to promote the efficient use of airspace by optimizing flight operations and digitizing airline performance. For that purpose, the whole sector (manufacturers, carriers, airport infrastructure managers, and air navigation service providers) has worked towards this as their sole aim for more sustainable aviation. However, initial allocations of emission permits from the implementation of EU ETS should be continually reassessed and the necessary actions defined and revised to avoid the creation of regulatory trade barriers in the aviation sector [3]. Indeed, such a system has often been adapted to respond to the challenges of the European airline industry. More specifically, after the full inclusion of aviation in the EU ETS on 1 January 2012 [4] in accordance with those aviation activities concerned [5], there was a temporary derogation from the greenhouse gas emission allowance scheme [6], namely "Stop the Clock", which gave concerned parties time to overcome the constraints arising from the prompt implementation of ETS-related penalty systems. Then, for 2013–2020, there was an application of the system with a narrow scope allowing airlines, among other special dispensations, to use a simplified methodology to report emissions annually of less than 3000 tons of CO_2 per year on intra-community flights [7]. Lastly, regarding the emissions for the period 2021–2023, airlines concerned, according to EU ETS, must notify of their emissions by flights operated within the European Economic Area (EEA), in addition to those whose destination is any airport located either in the Swiss Confederation (hereafter, Switzerland) or the United Kingdom of Great Britain and Northern Ireland (hereafter, the United Kingdom) for the previous year every 28 February of the year following [8].

Since sprawled environments contribute to climate change directly and indirectly, due to the certain effects of transportation characteristics, among other factors [9], the air industry has often addressed the question of whether full decarbonization of aviation, using alternative energy sources, is actually practicable in the years to come before 2050. Unlike rail transport, the extensive use of aviation fuels for commercial flying based on renewable energy sources is currently not easy to achieve, as a sort of comprehensive technological solution for a partially optimized sustainable fuel is, for the time being, the only remedy available, including those based on Sustainable Aviation Fuel (SAF). In fact, using SAF in commercial airplanes is currently the most workable solution to mitigate aviation emissions by operating existing aircraft. One of the great assets of SAF is that it can be mixed with any fossil jet fuel to reduce emissions. It could even be beneficial for emissions compared to the baseline scenario where airlines only use conventional jet fuel based on simulations made with the Fleet-Level Environmental Evaluation Tool (FLEET) [10]. However, a solution relying entirely on SAF seems to be a temporary arrangement, since there are still unresolved issues regarding their availability problems and production costs. In addition, alternative energy sources, either through electric or hydrogen engines, cannot completely replace carbon-based fuels in the aviation sector. Nevertheless, promoting the research effort in sustainable alternative fuels for aviation is essential, albeit not sufficient. Building on this basis, the challenge now is how to reduce the impact of commercial flights on public transportation by using affordable and sustainable fuels and operating efficient fleets. Without restricting demand, this is unlikely in terms of emission stabilization at levels consistent with risk-averse climate targets [11]. Research in this field has become even more important as time has gone by, since the question is raised as to the feasibility of supplying domestic short-haul air routes that are sensitive to mitigating climate change through real actions towards sustainability. The present paper aims at achieving precisely that objective, thereby contributing to a knowledge of such an important issue and the search for possible solutions. On one hand, there are precisely framed sectorial matters on aviation-related issues. On the other hand, there is a rather vague global technological order on the current dependence on fossil fuels and moving toward complete decarbonization by 2050, or even earlier.

2. Conceptual Research Framework

As pointed out earlier, it has been wondered whether aviation emissions from domestic scheduled flights can be only reduced by operating more effective aircraft or, failing that, whether further types of real actions, either technical or operational, may be necessary for this matter. To that end, the research has been focused on those flights operating from the most representative airports in Spain. Hence, the study was designed under a mixed approach from a conceptual framework created to find answers to the research question. On one hand, the selection of a primary source for data collection to be properly applied to the study is necessary. On the other hand, an empirical analysis with timely and trusted data for the purpose of contributing to a better understanding of the research scope, not only in terms of emissions, but also concerning fuel consumption, thus affecting variable costs of airlines, is also necessary. From a sectorial perspective, promoting lower emissions among carriers creates a sustainable business culture and efficient mobility. At the same time, stakeholders in the air transport system have been facing several uncertainties, casting a shadow on the economic prospects of the aviation sector. Fleet diversification can be a great ally for airlines in dealing with periodical market swings, whether due to seasonal demand (i.e., tourism) or unexpected events (i.e., coronavirus). However, most airlines are subject to leasing contracts limited to a single manufacturer for reasons of efficiency and cost-effectiveness, either due to maintenance or certification. Consequently, as outlined below, there are only a few commercial aircraft types per carrier, which is often only available to one single aircraft family. Thus, during the period under review (2011–2020), and in accordance with usual classification criteria based on the most representative aircraft of European airlines from previous works [12,13], the most utilized commercial airplanes (10) of only single-aisle families, either turboprop or turbojet, in domestic corridors (139) operating in the busiest airports (5) in Spain have been selected for the study. The case of the Spanish domestic market is also a very good approximation as to what occurs in the European aviation sector in terms of emissions, as almost two-thirds of passenger air traffic correspond to France, Germany, Italy, Spain, and the United Kingdom (UK) [14]. With the approach to sustainability regarding the transport of passengers by air over regular routes, specifically focused on the Spanish domestic market, a few earlier papers have handled the issue of aviation emissions with certain routings. For instance, regarding the specific case of the Public Service Obligation (PSO) schema applied in Spain, the emissions, depending on the aircraft type of some domestic routes, such as those serving island territories [15], the EU territories with special status [16], or even intra-regional routes [17] have been discussed. This article brings a comprehensive and consistent view of the sustainability of passenger air transportation by segmenting emissions per distance band and aircraft type to reveal which corridors contribute more to the overall aviation emissions in the Spanish market.

Although the causes of climate global warming are very diverse and highly disturbing, it is widely accepted that the effect of human activities is one thereof, and even a critical one. Using energy based on fossil fuels in transport creates harmful emissions. Regarding the emission analysis of greenhouse gas (GHG), the research has only been focused on carbon dioxide emissions from the aviation sector, since this is the primary driver of global climate change, in line with scarce similar works focusing on sustainable transportation in Spanish domestic market, though without any concern about passenger transportation, but intra-national freight transport instead [18]. Because commercial aircraft emissions are affected by flight-specific variables that change continuously, it becomes necessary to develop average factors to consider the effect of these flight parameters. While these factors cannot entirely be identified from a specific flight basis, the normal methodology considers them for the purpose of developing a more robust estimation of flight emissions. As shown in Figure 1, the empirical analysis carried out in the study has been mainly supported by two primary sources, one for air traffic statistics and the other for the carbon aircraft emissions database, respectively from AENA [19] and International Civil Aviation Organization (ICAO) [20], in line with International Air Transport Association (IATA). Consequently,

Figure 2 summarizes the analysis methodology used in designing the research for the estimation of main climate-related emissions. As a result, the contribution of the paper is intended to evaluate the impact of commercial aviation on carbon dioxide (CO_2) emissions from the analysis of representative domestic flights in such a demanding transportation market as the Spanish route. The research question does arise, therefore, whether regular air passenger services on domestic routes can be always sustainable in terms of excessive pollution by scheduling turbojets, or otherwise deploying turboprops on certain regional corridors where feasible and economical.

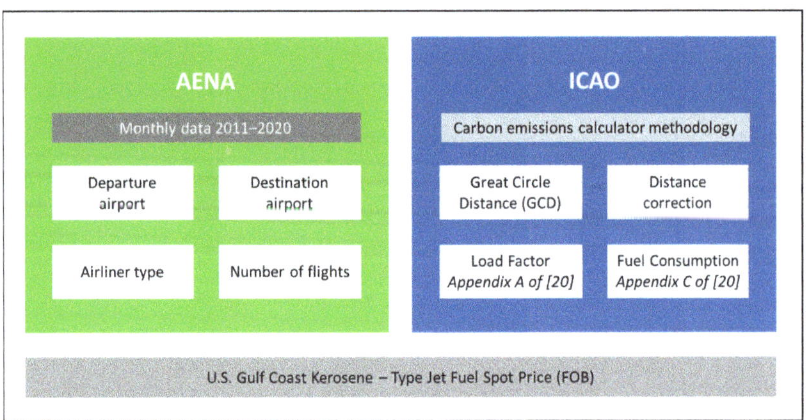

Figure 1. Block diagram for data set collection. Source: own compilation based on research sources according to Appendix A—Load Factors by Route Group from [20] (pp. 12–13), and Appendix C—ICAO Fuel Consumption Table from [20] (pp. 17–23).

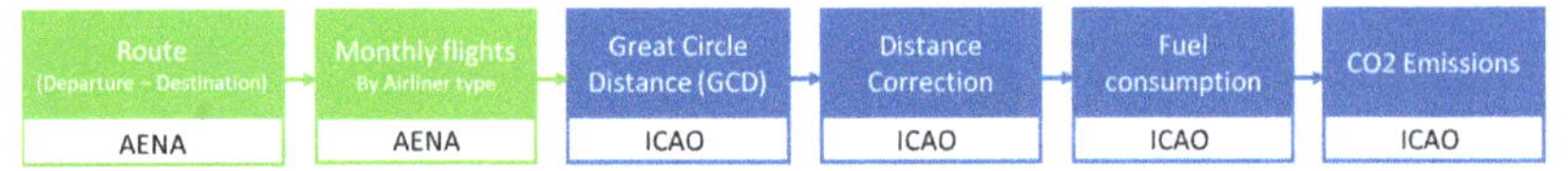

Figure 2. Flow chart research process. Source: own elaboration according to research design.

2.1. Related Considerations in Operating within Airport Network in Spain

Spain has a very dense airport network and an extensive aviation infrastructure. Due to the widespread displacement of the population in certain areas by orography, there are air facilities in almost every Spanish province. Most of them belong to a public business entity, officially named Aena SME, S.A. (hereafter, AENA), specifically 46 airports and 2 heliports. Of these, as shown in Table 1, the five most representative airports according to the number of scheduled passengers boarded by commercial airlines from 2001 to 2020 have been selected for research. Furthermore, as can be seen in Table 2, there are a couple of smaller facilities belonging to other public entities, or even semi-public bodies, at the regional level, which have not been considered in the study due to a slight impact on the research matter. This simplification has no significant influence on the result of the CO_2 emissions from the aviation sector within the Spanish domestic market, since the five airports cover almost the entire domestic network of corridors. In that regard, such consideration is sufficient to support the research approach and hence the validity of their findings, since the routes displayed in Figure 3 represent 62.5% of total aviation-related emissions. Significantly, the following map shows the set of 149 destinations from Table 2 according to 139 routes from Table 3 that have been analyzed when the scheduled passenger transport service occurs in the operative application of domestic flights during the studied ten-year period for each selected airport. Likewise, such a map shows the distribution of those destinations most representative of the study, and specifically whether or not they have a significant impact on the overall emissions. In all of them, the corridor MAD-LPA (hereafter, EC5, as shown in Table 3) makes the greatest impact on the environmental footprint from aviation activity; nine percent to be precise, as explained later in this article.

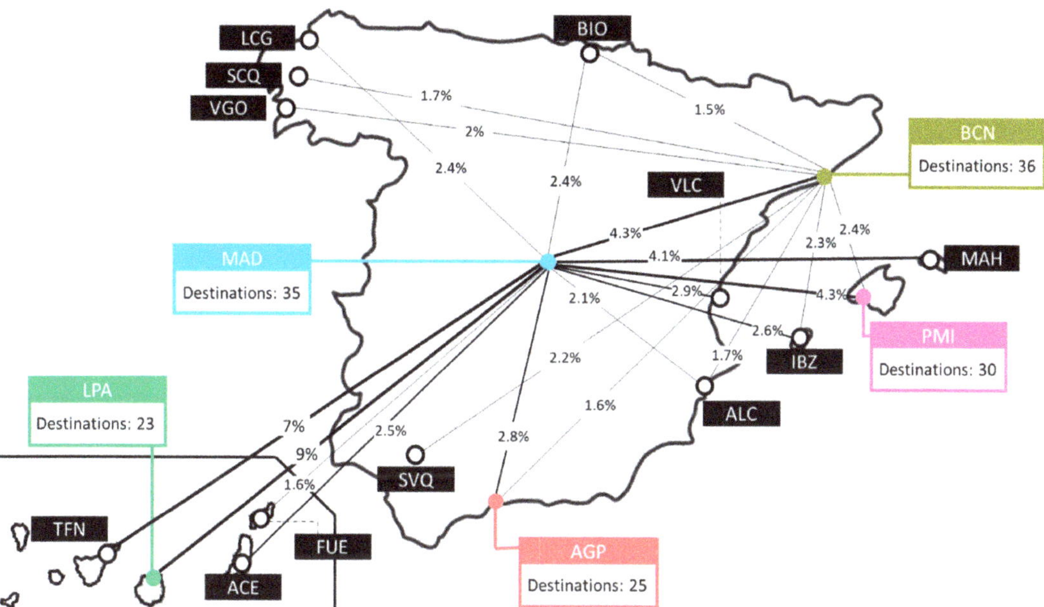

Figure 3. Overview graphic of the routes analyzed in the study whose pollutant emissions have resulted in estimated tons of carbon dioxide equivalent greater than 1.5% with regard to the five airports. Source: own calculations based on data provided from primary sources ([19,20]). Note: the use of color-coded airports, in addition to their IATA designators, has been harmonized throughout the paper, depending on each of them, for a better understanding.

Table 1. Key facts of air passenger traffic at main airports in Spain between 2004 and 2019.

IATA Code	Airport Group [a]	Airport Typology [b]	Airport Category [c]	Average Annual Passenger Traffic [d]	CAGR [e]
MAD	Major	Hub	Level 3	48,414,000	0.7
BCN	Major	Hub	Level 3	36,116,000	1.3
PMI	Major	Touristic	Level 3	23,781,000	0.6
AGP	I	Touristic	Level 3	14,284,000	0.8
LPA	Canary	Touristic	Level 3	10,747,000	0.5

Source: own calculation based on information from Aena database [19]; additional information provided by the Spanish Directorate-General for Civil Aviation (DGAC) upon request. Explanatory notes: [a,b] according to the classification provided by Aena; [c] according to the categorization under Article 3 of [21] and Article 4 of [22], Spanish air facilities have been classified as non-coordinated airports (Level 1), airports with schedules facilitated (Level 2), or coordinated airports (Level 3); [d] total amount, including both arrivals and departures at each airport; [e] Compound Annual Growth Rate (expressed as a decimal figure).

Table 2. Existing civil air facilities in Spain (as of 14 November 2022).

Airport Full Name	IATA Designator	ICAO Designator	Airport Operator	Airport Type	Ownership Type
A Coruña	LCG	LECO	Aena	General Interest	Statewide
Adolfo Suárez Madrid-Barajas	MAD	LEMD	Aena	General Interest	Statewide
Albacete	ABC	LEAB	Aena	General Interest	Statewide
Algeciras †	AEI	LEAG	Aena	General Interest	Statewide
Alicante-Elche Miguel Hernández	ALC	LEAL	Aena	General Interest	Statewide
Almería	LEI	LEAM	Aena	General Interest	Statewide
Andorra–La Seu d'Urgell	LEU	LESU	GenCat	Autonomic Interest	Regional
Asturias	OVD	LEAS	Aena	General Interest	Statewide
Badajoz	BJZ	LEBZ	Aena	General Interest	Statewide
Bilbao	BIO	LEBB	Aena	General Interest	Statewide
Burgos	RGS	LEBG	Aena	General Interest	Statewide
Castellón-Costa Azahar	CDT	LECN	Aerocas	General Interest	Statewide
Aeropuerto de Ciudad Real	CQM	LERL	CRIA	Industrial Interest	Private
Ceuta †	JCU	GECE	Aena	General Interest	Statewide
César Manrique-Lanzarote	ACE	GCRR	Aena	General Interest	Statewide
Córdoba	ODB	LEBA	Aena	General Interest	Statewide
El Hierro	VDE	GCHI	Aena	General Interest	Statewide
Federico García Lorca Granada-Jaén	GRX	LEGR	Aena	General Interest	Statewide
Fuerteventura	FUE	GCFV	Aena	General Interest	Statewide
Girona-Costa Brava	GRO	LEGE	Aena	General Interest	Statewide
Gran Canaria	LPA	GCLP	Aena	General Interest	Statewide
Huesca-Pirineos	HSK	LEHC	Aena	General Interest	Statewide
Ibiza	IBZ	LEIB	Aena	General Interest	Statewide
Internacional Región de Murcia *	RMU	LEMI	Aena	General Interest	Statewide
Jerez	XRY	LEJR	Aena	General Interest	Statewide
Josep Tarradellas Barcelona-El Prat	BCN	LEBL	Aena	General Interest	Statewide
La Gomera	GMZ	GCGM	Aena	General Interest	Statewide
La Palma	SPC	GCLA	Aena	General Interest	Statewide
León	LEN	LELN	Aena	General Interest	Statewide
Lleida-Alguaire	ILD	LEDA	GenCat	Autonomic Interest	Regional
Logroño-Agoncillo	RJL	LERJ	Aena	General Interest	Statewide
Madrid-Cuatro Vientos ‡	-	LECU	Aena	General Interest	Statewide
Melilla	MLN	GEML	Aena	General Interest	Statewide
Menorca	MAH	LEMH	Aena	General Interest	Statewide
Málaga-Costa del Sol	AGP	LEMG	Aena	General Interest	Statewide
Palma de Mallorca	PMI	LEPA	Aena	General Interest	Statewide
Pamplona	PNA	LEPP	Aena	General Interest	Statewide
Reus	REU	LERS	Aena	General Interest	Statewide
Sabadell	QSA	LELL	Aena	General Interest	Statewide
Salamanca	SLM	LESA	Aena	General Interest	Statewide
San Sebastián	EAS	LESO	Aena	General Interest	Statewide
Santiago-Rosalía de Castro	SCQ	LEST	Aena	General Interest	Statewide
Seve Ballesteros-Santander	SDR	LEXJ	Aena	General Interest	Statewide
Sevilla	SVQ	LEZL	Aena	General Interest	Statewide
Son Bonet ‡	-	LESB	Aena	General Interest	Statewide
Tenerife Norte-Ciudad de La Laguna	TFN	GCXO	Aena	General Interest	Statewide
Tenerife Sur	TFS	GCTS	Aena	General Interest	Statewide
Teruel	TEV	LETL	Plata	Industrial Interest	Semipublic
Valencia	VLC	LEVC	Aena	General Interest	Statewide
Valladolid	VLL	LEVD	Aena	General Interest	Statewide
Vigo	VGO	LEVX	Aena	General Interest	Statewide
Vitoria	VIT	LEVT	Aena	General Interest	Statewide
Zaragoza	ZAZ	LEZG	Aena	General Interest	Statewide

Source: own elaboration based on IATA and ICAO airport codes. Explanatory note: A dagger "†" denotes a heliport with scheduled flights, while a double dagger "‡" indicates an airport mainly intended for general aviation. (*) The last civil airport belonging to Aena's network, which replaced the former one, named Murcia-San Javier (IATA: MJV, ICAO: LELC), specifically intended for military use so far.

Table 3. Main domestic routes in the Spanish airport network over the study period (2011–2020).

Departure ↓\Arrival →	MAD	LPA	BCN	AGP	PMI
LCG	EC1	EC2	EC3		EC4
MAD	•	EC5	EC6	EC7	EC8
ABC			EC9		
AEI					
ALC	EC10	EC11	EC12	EC13	EC14
LEI	EC15		EC16	EC17	EC18
LEU					
OVD	EC19	EC20	EC21	EC22	EC23
BJZ	EC24		EC25		
BIO	EC26	EC27	EC28	EC29	EC30
RGS	EC31		EC32		
CDT					
CQM	EC33		EC34		
JCU				EC35	
ACE	EC36	EC37	EC38	EC39	
ODB					
VDE		EC40			
GRX	EC41	EC42	EC43	EC44	EC45
FUE	EC46	EC47	EC48	EC49	
GRO	EC50				EC51
LPA	EC5	•	EC52	EC53	EC54
HSK					
IBZ	EC55		EC56	EC57	EC58
RMU/MJV *	EC59		EC60		
XRY	EC61		EC62	EC63	EC64
BCN	EC6	EC52	•	EC65	EC66
GMZ		EC67			
SPC	EC68	EC69	EC70		
LEN	EC71		EC72	EC73	EC74
ILD					EC75
RJL	EC76				EC77
LECU **					
MLN	EC78	EC79	EC80	EC81	EC82
MAH	EC83		EC84	EC85	EC86
AGP	EC7	EC53	EC65	•	EC87
PMI	EC8	EC54	EC66	EC87	•
PNA	EC88		EC89		EC90
REU			EC91		EC92
QSA					
SLM			EC93		EC94
EAS	EC95		EC96		
SCQ	EC97	EC98	EC99	EC100	EC101
SDR	EC102	EC103	EC104	EC105	EC106
SVQ	EC107	EC108	EC109		EC110
LESB **					
TFN	EC111	EC112	EC113	EC114	EC115
TFS	EC116	EC117	EC118	EC119	EC120
TEV					
VLC	EC121	EC122	EC123	EC124	EC125
VLL			EC126	EC127	EC128
VGO	EC129	EC130	EC131		
VIT	EC132		EC133		EC134
ZAZ	EC135	EC136	EC137	EC138	EC139

Source: own codification based on IATA airport codes. Explanatory note: Black circle "•" denotes an airway whose origin and destination are the same, which applies to non-commercial flights, either for training or calibration purposes, while a blank space indicates "not applicable" according to the analysis carried out on national scheduled air transport services in Spain. (*) For the purpose of the analysis carried out within the research period, both airports have been considered as a unit, since the former one (MJV) remained operating until 15 January 2019, and the new one (RMU) has been active since 16 January 2019. (**) In the absence of the IATA code, the ICAO code is shown.

Of all possible routes identified among the air facilities identified below operating from the key airports selected in the study, only 139 corridors fulfilled the criteria for the review. From the air traffic database provided by the related trusted source [19] with regard to possible routes analyzed, properly listed, and adequately codified from Table 3, according to the criteria previously noted, the research has evolved around traffic ancillary data in order to calculate jet fuel consumption and evaluate emissions.

2.2. Related Considerations in Supplying Domestic Air Routes in Spain

On the basis of the above considerations, as summarized in Table 4, jet fuel consumption and carbon dioxide emissions have been calculated according to the most representative carrier and airliner on each corridor selected in the study.

Table 4. Air routes' standard operational and emission facts over the study period (2011–2020).

Air Route Code	Travel Distance (km) [a]	Travel Time (h)	Initial Heading Vector	Block Fuel (kTon) [b,c]	CO$_2$ Emissions (kTon) [b,c]	Emissions CO$_2$/PAX (kg) [b,c]	Most Common Airline [d]	Most Common Airliner [d]	Standard Cabin Config.	Share of Airliner [e]
EC1	506	0:54	126° (SE)	3.5	11.1	97.67	IB	319	Y144	0.31
EC2	1818	2:21	202° (SSW)	8.9	28.2	190.28	IB	320	CY180	0.51
EC3	890	1:19	101° (E)	5.3	16.7	113.35	VY	320	CY180	0.61
EC4	1017	1:28	110° (ESE)	5.0	15.9	107.31	IB	32A	CY180	0.59
EC5	1765	2:12	221° (SW)	10.2	32.1	170.70	FR	73H	Y189	0.26
EC6	484	0:52	77° (ENE)	4.0	12.5	81.04	IB	320	CY180	0.44
EC7	432	0:49	191° (S)	2.4	7.7	77.98	I2	320	CY180	0.46
EC8	548	0:55	99° (E)	3.7	11.7	80.39	FR	73H	Y189	0.52
EC9	425	0:50	51° (NE)	1.7	5.3	127.73	YW	CR2	Y50	1.00
EC10	357	0:44	315° (NW)	2.4	7.6	73.45	IB	320	CY180	0.34
EC11	1794	2:22	234° (SW)	6.1	19.1	234.80	YW	CR9	Y90	0.40
EC12	404	0:47	33° (NNE)	3.5	11.2	76.21	VY	320	CY180	0.63
EC13	392	0:53	244° (WSW)	0.9	2.7	n/a	n/a	CNJ	VIP	1.00
EC14	319	0:41	63° (ENE)	2.3	7.1	66.57	VY	320	CY180	0.33
EC15	418	0:49	346° (NNW)	2.3	7.1	95.72	YW	CRK	CY100	0.38
EC16	626	1:02	36° (NE)	3.8	11.9	92.87	VY	320	CY180	0.41
EC17	191	0:40	265° (W)	0.3	0.8	n/a	n/a	BE2	VIP	0.77
EC18	539	0:58	54° (NE)	2.6	8.2	138.74	YW	CR2	Y50	0.40
EC19	396	0:46	148° (SSE)	3.1	9.8	89.98	V7	319	Y144	0.35
EC20	1927	2:22	209° (SSW)	8.0	25.4	182.16	FR	73H	Y189	0.32
EC21	713	1:07	108° (ESE)	4.6	14.5	98.26	VY	320	CY180	0.66
EC22	776	1:17	170° (S)	4.6	14.4	137.67	V7	717	Y125	0.36
EC23	856	1:17	118° (ESE)	4.4	14.7	110.60	V7	320	CY180	0.29
EC24	331	0:43	56° (ENE)	1.3	4.1	113.77	YW	CR2	Y50	0.64
EC25	804	1:15	68° (ENE)	2.5	7.8	137.39	YW	CR9	Y90	0.40
EC26	468	0:51	190° (S)	2.8	8.8	79.11	IB	319	Y144	0.28
EC27	2041	2:35	217° (SW)	9.7	30.7	208.06	VY	320	CY180	0.39
EC28	468	0:51	117° (ESE)	3.7	11.8	80.06	VY	320	CY180	0.56
EC29	748	1:10	191° (S)	4.7	14.9	101.54	VY	320	CY180	0.60
EC30	629	1:02	130° (SE)	4.2	13.4	92.99	VY	320	CY180	0.39
EC31	488	0:54	179° (S)	1.9	6.0	133.43	YW	CR2	Y50	0.75
EC32	619	1:03	102° (ESE)	2.0	6.3	153.28	YW	CR2	Y50	1.00
EC33	185	0:34	11° (N)	0.5	1.6	n/a	n/a	CNJ	VIP	1.00
EC34	583	0:59	60° (ENE)	4.2	13.4	90.29	n/a	320	CY180	1.00
EC35	113	0:30	40° (NE)	0.1	0.3	n/a	n/a	NDE	VIP	1.00
EC36	1574	2:00	33° (NNE)	8.0	25.2	157.27	FR	73H	Y189	0.55
EC37	208	0:44	238° (WSW)	0.6	1.9	31.80	NT	AT7	Y68	0.56
EC38	1974	2:31	42° (NE)	9.8	31.0	181.56	VY	32B	CY232	0.36
EC39	1208	1:40	42° (NE)	6.6	20.8	140.95	VY	320	CY180	0.43
EC40	247	0:49	86° (E)	0.6	1.9	35.13	YW	AT7	Y68	0.63
EC41	367	0:45	3° (N)	2.5	7.9	91.93	YW	CRK	CY100	0.39
EC42	1495	1:59	230° (SW)	8.4	26.5	147.39	VY	32B	CY232	0.60
EC43	681	1:05	46° (NE)	4.6	14.7	96.17	VY	320	CY180	0.56
EC44	86	0:30	229° (SW)	0.1	0.4	n/a	n/a	BE2	VIP	0.59
EC45	627	1:03	63° (ENE)	3.8	12.1	111.62	YW	E95	C12Y108	0.55
EC46	1634	2:04	32° (NNE)	8.2	25.8	161.34	FR	73H	Y189	0.44
EC47	160	0:39	249° (WSW)	0.5	1.5	25.77	NT	AT7	Y68	0.62
EC48	2032	2:29	41° (NE)	9.6	30.2	189.21	FR	73H	Y189	0.54
EC49	1265	1:44	42° (NE)	6.2	19.7	146.16	VY	320	CY180	0.43
EC50	553	0:55	256° (WSW)	4.1	13.0	83.54	FR	73H	Y189	1.00
EC51	261	0:37	180° (S)	2.5	8.0	53.60	UX	73H	Y189	0.98
EC52	2175	2:44	42° (NE)	10.6	33.5	196.04	VY	32B	CY232	0.33

Table 4. Cont.

Air Route Code	Travel Distance (km) [a]	Travel Time (h)	Initial Heading Vector	Block Fuel (kTon) [b,c]	CO_2 Emissions (kTon) [b,c]	Emissions CO_2/PAX (kg) [b,c]	Most Common Airline [d]	Most Common Airliner [d]	Standard Cabin Config.	Share of Airliner [e]
EC53	1409	1:54	44° (NE)	7.2	22.9	157.57	VY	320	CY180	0.48
EC54	2110	2:44	48° (NE)	8.4	26.6	249.65	YW	CRK	CY100	0.28
EC55	460	0:49	295° (WNW)	3.0	9.5	75.55	FR	73H	Y189	0.36
EC56	276	0:38	12° (NNE)	2.6	8.2	58.57	YW	320	CY180	0.41
EC57	572	0:56	246° (WSW)	3.4	10.9	84.30	FR	73H	Y189	0.66
EC58	140	0:30	57° (ENE)	0.7	2.3	42.69	YW	CRK	CY100	0.25
EC59	366	0:46	323° (NW)	1.7	5.3	123.50	YW	CR2	Y50	0.79
EC60	476	1:20	31° (NNE)	0.9	2.9	n/a	n/a	CNJ	VIP	1.00
EC61	469	0:51	27° (NNE)	2.9	9.3	95.05	IB	319	Y144	0.36
EC62	867	1:15	52° (NE)	5.1	16.0	104.02	VY	73H	Y189	0.49
EC63	140	0:36	93° (E)	0.1	0.3	n/a	n/a	PA1	VIP	1.00
EC64	831	1:15	65° (ENE)	5.2	16.6	108.36	VY	320	CY180	0.39
EC65	766	1:11	230° (SW)	4.8	15.0	102.83	VY	320	CY180	0.45
EC66	202	0:33	164° (SSE)	1.9	6.1	43.04	FR	73H	Y189	0.40
EC67	180	0:41	93° (E)	0.6	2.0	51.60	NT	AT7	Y68	0.45
EC68	1847	2:23	41° (NE)	9.3	29.5	192.50	I2/IB	320	CY180	0.66
EC69	245	0:49	108° (ESE)	0.7	2.2	36.28	YW	AT7	Y68	0.60
EC70	2283	2:52	47° (NE)	10.6	33.6	227.04	VY	320	CY180	0.60
EC71	291	0:40	143° (SE)	1.3	4.2	101.43	YW	CR2	Y50	1.00
EC72	657	1:06	100° (E)	2.4	7.6	154.96	YW	CR2	Y50	0.45
EC73	664	1:06	171° (S)	2.0	6.4	156.14	YW	CR2	Y50	1.00
EC74	782	1:15	113° (ESE)	2.2	7.1	169.14	YW	CR2	Y50	0.65
EC75	305	0:41	142° (SE)	1.4	4.4	103.84	YW	CR2	Y50	0.44
EC76	242	0:37	206° (SSW)	1.1	3.6	87.00	YW	CR2	Y50	1.00
EC77	534	0:57	126° (SE)	2.0	6.4	156.52	YW	CR2	Y50	1.00
EC78	581	1:28	335° (N)	1.2	3.9	70.57	YW	AT7	Y68	0.97
EC79	1432	3:09	239° (WSW)	2.5	7.9	140.98	YW	AT7	Y68	1.00
EC80	800	1:54	319° (NW)	1.5	4.9	87.24	YW	AT7	Y68	1.00
EC81	208	0:45	138° (SE)	0.6	1.7	31.80	YW	AT7	Y68	0.82
EC82	692	1:41	45° (NE)	0.6	1.8	31.80	YW	AT7	Y68	1.00
EC83	667	1:39	279° (W)	3.5	10.9	117.65	YW	CRK	CY100	0.32
EC84	241	0:48	312° (NW)	2.1	6.7	52.41	VY	320	CY180	0.50
EC85	840	1:59	248° (WSW)	5.1	16.2	109.31	VY	320	CY180	1.00
EC86	132	0:35	255° (WSW)	0.7	2.1	42.72	YW	CR9	Y90	0.28
EC87	710	1:05	61° (ENE)	4.5	14.2	91.91	UY	73H	Y109	0.41
EC88	299	0:40	213° (SSW)	1.8	5.6	78.74	YW	CRK	CY100	0.44
EC89	349	0:44	117° (ESE)	1.7	5.3	118.58	YW	CR2	Y50	0.79
EC90	513	0:56	133° (SE)	1.4	4.5	80.83	YW	AT7	Y68	1.00
EC91	78	0:25	77° (ENE)	0.7	2.2	22.70	n/a	320	CY180	0.78
EC92	222	0:34	143° (SE)	2.3	7.2	46.38	FR	73H	Y189	0.97
EC93	638	1:05	84° (E)	2.0	6.2	153.58	YW	CR2	Y50	0.98
EC94	718	1:10	100° (E)	1.7	5.5	161.71	YW	CR2	Y50	1.00
EC95	350	1:01	205° (SSW)	1.1	3.6	50.13	YW	AT7	Y68	0.55
EC96	392	0:46	124° (SE)	2.6	8.4	89.48	VY	319	Y144	0.77
EC97	484	0:52	122° (ESE)	3.7	11.6	81.04	IB	320	CY180	0.44
EC98	1775	2:13	203° (SSW)	7.9	24.9	171.54	FR	73H	Y189	0.86
EC99	885	1:16	98° (E)	5.2	16.5	105.33	FR	73H	Y189	0.42
EC100	768	1:09	153° (SSE)	4.8	15.1	96.49	FR	73H	Y189	0.54
EC101	1005	1:24	108° (ESE)	5.3	16.8	114.58	FR	73H	Y189	0.47
EC102	327	0:43	176° (S)	2.0	6.4	112.26	YW	CR2	Y50	0.28
EC103	2008	2:27	215° (SW)	9.0	28.3	187.93	FR	73H	Y189	0.97
EC104	540	0:54	114° (ESE)	3.5	11.1	79.95	FR	73H	Y189	0.51
EC105	752	1:08	185° (S)	4.6	14.7	95.39	FR	73H	Y189	1.00
EC106	696	1:04	126° (SE)	4.3	13.7	91.14	FR	73H	Y189	0.71
EC107	396	0:46	30° (NNE)	2.8	8.9	75.72	I2	320	CY180	0.62
EC108	1377	1:51	223° (SW)	7.1	22.6	154.92	I2	320	CY180	0.42
EC109	810	1:14	55° (NE)	5.0	15.7	106.55	I2	320	CY180	0.48
EC110	789	1:10	70° (ENE)	4.9	15.4	97.96	FR	73H	Y189	0.53
EC111	1771	2:12	38° (NE)	10.5	33.2	171.06	UX	73H	Y189	0.27
EC112	112	0:33	123° (ESE)	0.5	1.5	19.61	NT	AT7	Y68	0.45
EC113	2195	2:39	45° (NE)	10.5	33.1	200.74	FR	73H	Y189	0.40
EC114	1434	1:51	48° (NE)	7.4	23.3	147.14	UX	73H	Y189	0.41
EC115	2140	2:46	50° (NE)	7.0	22.0	252.24	YW	CRK	CY100	0.67
EC116	1824	2:16	37° (NE)	8.9	28.3	174.78	FR	73H	Y189	0.39
EC117	117	0:34	96° (E)	0.5	1.5	20.22	NT	AT7	Y68	0.28

Table 4. *Cont.*

Air Route Code	Travel Distance (km) [a]	Travel Time (h)	Initial Heading Vector	Block Fuel (kTon) [b,c]	CO2 Emissions (kTon) [b,c]	Emissions CO2/PAX (kg) [b,c]	Most Common Airline [d]	Most Common Airliner [d]	Standard Cabin Config.	Share of Airliner [e]
EC118	2246	2:42	44° (NE)	10.0	31.6	204.26	FR	73H	Y189	0.62
EC119	1483	1:54	47° (NE)	7.3	23.0	150.69	FR	73H	Y189	0.71
EC120	2189	2:45	49° (NE)	10.5	33.1	223.69	VY	320	CY180	1.00
EC121	286	0:40	294° (WNW)	1.5	4.8	75.57	YW	CRK	CY100	0.37
EC122	1880	2:19	231° (SW)	7.4	23.3	178.80	FR	73H	Y189	0.59
EC123	296	0:41	46° (NE)	3.2	10.1	n/a	n/a	763	C18Y246	0.25
EC124	471	0:50	230° (SW)	1.8	5.6	76.26	FR	73H	Y189	0.33
EC125	277	0:38	88° (E)	2.0	6.4	56.06	FR	73H	Y189	0.28
EC126	580	0:57	92° (E)	3.5	10.9	84.74	FR	73H	Y189	0.60
EC127	559	0:55	177° (S)	4.1	13.0	83.82	FR	73H	Y189	1.00
EC128	685	1:03	108° (ESE)	2.9	9.3	83.82	FR	73H	Y189	0.66
EC129	465	0:52	113° (ESE)	3.3	10.4	96.00	UX	E95	C12Y108	0.37
EC130	1701	2:16	203° (SSW)	5.6	17.5	213.49	YW	CRK	CY100	0.51
EC131	896	1:19	93° (E)	5.3	16.8	113.79	VY	320	CY180	0.53
EC132	274	0:37	195° (SSW)	2.9	9.0	n/a	n/a	752	Y216	0.40
EC133	435	0:48	112° (ESE)	3.2	10.1	n/a	n/a	73Y	FREIGHT	0.39
EC134	589	0:57	127° (SE)	4.5	14.3	91.82	UX	73H	Y189	1.00
EC135	249	0:41	239° (WSW)	0.6	2.0	n/a	n/a	CNJ	VIP	1.00
EC136	2005	2:27	225° (SW)	9.2	29.2	187.58	UX	73H	Y189	1.00
EC137	264	0:37	98° (E)	2.4	7.6	n/a	n/a	734	Y150	0.63
EC138	629	1:00	209° (SSW)	3.3	10.3	87.78	UX	73H	Y189	0.80
EC139	397	0:47	125° (SE)	3.0	9.6	90.90	UX	E95	C12Y108	0.46

Source: own calculation based on information provided by ICAO [20]. Explanatory note: [a] an orthodromic distance on cruise flight; [b] standard assumptions: Jet A1 density as of 0.804 kg/L, CO_2 emissions as of 3.16 kg of CO_2 per kg of Jet A1; [c] on route assumptions: maximum payload, no wind, one aircraft per leg under technical specifications by International Standard Atmosphere (ISA), Joint Aviation Requirements (JAR), EU Aviation Safety Agency (EASA), passengers (PAX); [d] IATA designator; [e] Emissions share per aircraft of each route regarding total routes thereof (expressed as a decimal figure). Not applicable (n/a) when scheduled flights (i.e., general aviation) have not been identified through analysis based on related air traffic data.

3. Results

As pointed out previously, the research analysis has been focused on the empirical evaluation of partial greenhouse gas emissions from the aviation sector in Spain by estimating fuel consumption due to commercial domestic corridors for scheduled passenger air transport services according to the distribution from Table 5. Therefore, significant analysis work has been carried out to evaluate carbon dioxide emissions from commercial aviation activities on major national routes. To that end, after gathering information sourced from the trusted database [19], primary screening of data records was made by identifying relevant passenger flights within the Spanish domestic market, thereby rejecting air services likely operated by charter carriers, as well as those with little significance when compared with the total number of 25,000 flights covered by the preliminary analysis. For instance, those origin-destination-aircraft combinations below 10 flights in the 10-year period studied were removed from the analysis. This has thus made it possible to simplify the subsequent calculation up to 2500 flights. Then, based on the five major airports introduced earlier and the corresponding 139 routes finally identified for the study, the following considerations have been considered throughout the 10-year analysis period. Firstly, the Passenger Load Factor (PLF) is estimated at 82.3% [20] (p. 12). Secondly, as shown in Table 6, the label 'Distance Segment' is designed to break down the total of 'corridors', and any subtotals, in an attempt to identify the existing different distance segments to said emissions relating to the research matter. While aviation routes are typically categorized by flight time, short-haul flights (0–3 h in duration), medium-haul flights (3–6 h in duration), and long-haul flights (6–12 h in duration), the research has identified three types of segments for classifying the corridors from Table 4, as ranked in Table 6. Thirdly, most representative airplanes (see Appendix A Table A1) have been identified from the corridors previously selected for the emissions analysis. Similarly, as the fourth point, most representative airlines (see Appendix A Table A2) have been identified for the study. Fifthly, similar to above, most representative engine families have been cleared for the study (see Appendix A Table A3). Sixthly, and finally, only narrow-body airplanes have been considered for the analysis of 139 routes, since only intra-country passenger transport services, without distinction for engine families, have been analyzed.

Table 5. Air corridor distribution on the travel distance [1].

Distance Segment (km)	Percentage (%)	Emission Share (%)	Average (km)
Short Route: d ≤ 500	67.31	39.36	296
Medium Route: 501 ≤ d ≤ 1000	19.96	24.57	680
Long Route: d ≥ 1001	12.73	36.07	1799
Average Distance Traveled (km)			564

Source: own calculation in connection with the 2500 routes previously defined for the study. [1] Considered as orthodromic distance on cruise flight.

Table 6. Air corridor distribution on route type [1].

Route Type	Number of Routes	Percentage
Short	58	41.73%
Medium	47	33.81%
Long	34	24.46%

Source: own calculation in connection with the 139 routes previously defined for the study. [1] According to the classification provided in Table 5.

In addition to the aforementioned considerations, the emissions estimation has been subjected to calculation methodology provided by ICAO [20] that is affected by flight-specific variables, which in turn change continuously. According to such guidelines, it has been necessary to develop average factors to consider the effect of various flight parameters. While the factors cannot be captured on a specific flight basis, the methodology considers them to develop a more robust estimation of flight emissions.

Each aircraft has been mapped with one of the 312 equivalent airplane types to calculate fuel consumption per trip based on the driving distance between the airports involved in the routes selected. Based on the traffic and operational data collected by ICAO over several past years, this methodology has been applied according to both PLF and the corresponding weight ratio of freight and passengers, better known as a pax-to-freight factor (PFF), for estimating the total fuel (TF) burned for the passengers carried on each route selected. Then, the average fuel consumption for the journey, weighted by the departure frequency of each equivalent aircraft type, was calculated by dividing the total number of equivalent economy class passengers (Y-seats). This calculates, therefore, an average fuel consumption per such class passenger. Lastly, the result must be multiplied by 3.16 to obtain the carbon dioxide footprint attributed to passengers on each route selected.

Hence, the equation for calculating the CO_2 emissions per passenger (E_p):

$$E_p = 3.16 * (TF * PFF)/(Y\text{-seats} * PLF), \tag{1}$$

where variables are as follows according to ICAO [20] (p. 6):

- TF = Weighted average fuel consumed on all flights between the departure airport and arrival airport. The weighting factor is thus the ratio between the number of departures for each equivalent aircraft type and the total number of departures;
- PPF = Ratio calculated from the ICAO statistical database based on the number of passengers and the tonnage of mail and cargo transported on particular routes;
- Y-seats = Total number of economy class equivalent seats available on all flights serving the given pair of airports;
- PLF = Ratio calculated from the ICAO statistical database based on the number of passengers transported and the number of seats available on particular routes;

- 3.16 = Constant representing the number of tons of CO_2 produced by burning one ton of aviation fuel.

3.1. Technical and Operational Issues

In addition to the equation above (1), the estimation of carbon dioxide emissions in serving the routes concerned for the analysis period had to consider elements relating to aircraft performance and operating limitations, even though this problem is sometimes exacerbated by a lack of open access databases with full flight records. Looking at air operation when airplanes fly between two points, the pathway used is the distance runs on the great-circle distance (GCD). However, due to temporary limitations of air traffic flow management on certain air routes, as well as temporal and spatial variation of air space and possible restrictions (e.g., night flying restrictions), the performance of both airports and aircraft involved in operating every one of the flights on routes selected for the research could have been deviated from the optimal route to fly, better known as 'business trajectory', which is indeed a function of the required time of arrival at the airport. Delayed flights, mainly due to air traffic jams, increase the time of flight and fuel consumption. Airport congestion during the approach to landing is a further cause of leaving airplanes in a holding pattern, which can even result in a total diversion, resulting in significant unnecessary emissions. To confront these uncertainties of calculating emissions, specific considerations relating to technique as well as operation on the 139 routes have been adopted. At the technical level, based on the most common aircraft per route, differentiating between turboprop and turbojet, emissions have been calculated considering the standard engines of each representative aircraft, as expressed in ICAO [20] (p. 9). Regarding operational aspects, the travel distance of each route has been considered without take-off and the landing cycle, while a GCD correction factor (+50 km, +100 km, +125 km) has been made according to guidelines provided by ICAO [20] (pp. 7–8). As shown below, results vary based on the specific airplane model and seasonal pattern, including thus, if relevant, the impact of both technical and operational factors in the emissions calculated.

3.1.1. Aircraft Performance

In line with each representative aircraft, Figure 4 is a graphical view of performances on all the routes selected, showing how they relate to various distance trips identified.

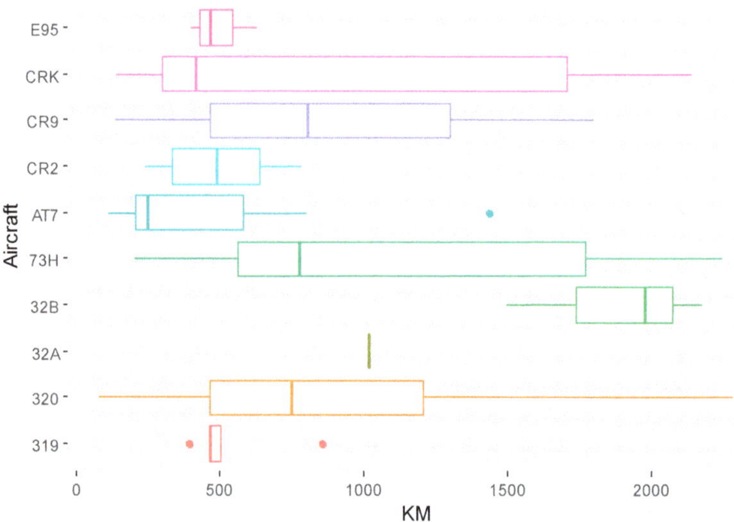

Figure 4. Overall fleet performance. Source: own elaboration from data published by [19].

3.1.2. Airport Performance

As shown in Figure 5, an analysis of the scheduled seasons of air transport services operating in the five airports has been provided to identify the seasonal trends of regular flights performed on 139 domestic routes, thus calculating their respective emissions.

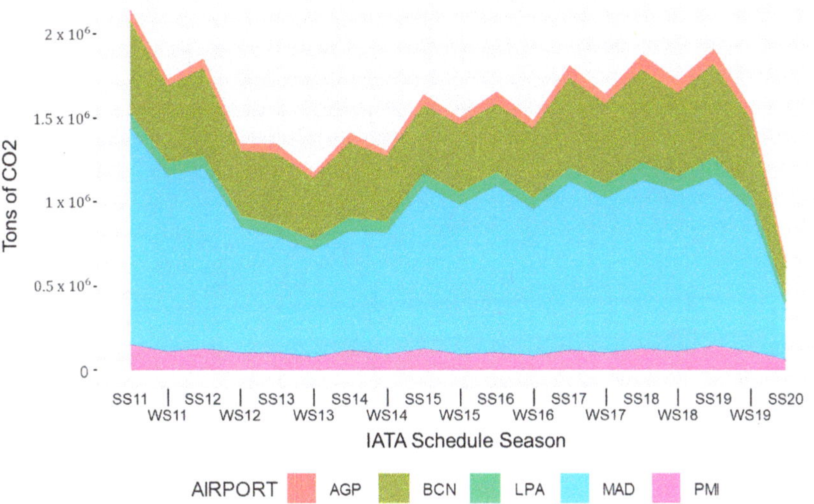

Figure 5. Cumulative carbon dioxide emissions in tons (T) of carbon dioxide equivalent associated with scheduled seasons between 2001 and 2020. Source: own calculation based on data published by [19] according to guidelines provided by [20]. Note: although the IATA Summer schedule begins on the last Sunday of March and ends on the last Saturday of October, and the IATA Winter schedule begins on the last Sunday of October and ends on the last Saturday of March, the calculation has been made by considering the same semester period in every year for simplification purposes.

3.2. Emissions Analysis Disaggregated by Airport Concerned

In order to analyze carbon dioxide emissions over the years under the set of 139 routes operating in the airport network identified as representative of the domestic market in Spain, collected flight data has been compiled to explain its monthly evolution, as summarized in Figure 6. These indicate emissions have been estimated on the basis of charging the quantity of fuel consumed monthly per route to the departure airport instead of the destination airport, as with the results shown in Figure 5. In other words, where emissions are associated with a certain corridor whose flights have been operating between two airports (e.g., A and B), the calculation is obliged to count each flight up to two legs, that is, from A to B, and then B to A. However, in this example, to simplify the approach for calculating emissions, the estimation is allocated to airport A. Under this scenario, as can be seen in Figure 6, aside from the effect of travel restrictions due to Coronavirus disease 2019 (hereafter, COVID-19) on mobility, and hence on aviation activity, there is apparently a pronounced relationship between what national demand dropped and what supply has shrunk (see, for example, the period 2013–2014).

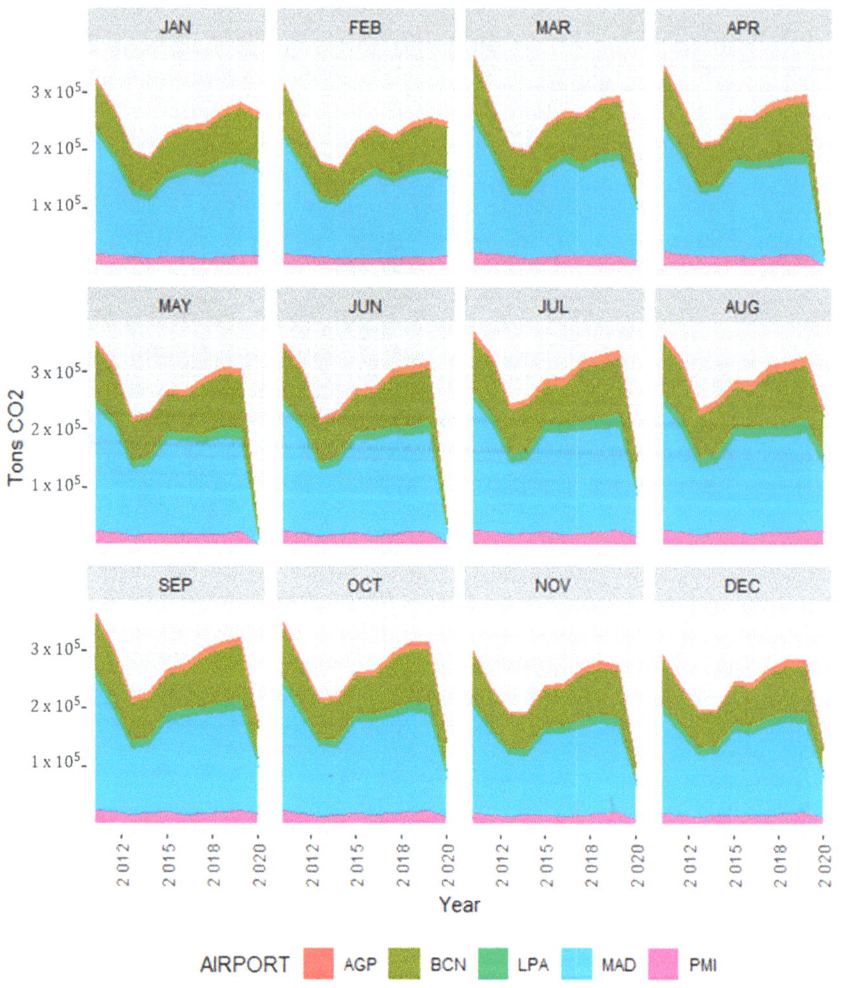

Figure 6. Monthly carbon dioxide emissions (disaggregated by each airport) between 2001 and 2020. Source: own calculation based on data published by [19] according to guidelines provided by [20].

3.2.1. Related Aspects Concerning Domestic Routes Operating in the MAD Airport

As noted earlier in Table 1, the Adolfo Suárez Madrid-Barajas airport (IATA: MAD, ICAO: LEMD) concentrates most of the regular flights in the Spanish domestic market. As can be seen in Figure 7a, there are some routes, such as EC83, EC102, and EC24, with very high fuel consumption per passenger, and thus a larger impact on specific emissions. Where total consumption is concerned, though, these routes will have scarcely any effect in terms of emissions from flights operated at such an airport. Indeed, as shown in Figure 7b, only two routes, EC6 and EC8, are significant to the total emissions chargeable to the airport in the specified analysis period. Besides no long-distance segment of both routes (484 and 548 km, respectively), two main factors have a direct impact on emission levels thereof. Both aspects are not connected, but they are weakening the capabilities for reducing emissions per passenger from domestic flights, in comparison with other related corridors, even shorter routes (e.g., EC24). On one hand, the intensive use of turbojets instead of turboprops on paths up to 550 km range, such as EC6 and EC8, and on the other hand, the overall capacity in terms of supplied seats has been especially intense on both routes because they have had a robust demand over the last 10 years analyzed, even after taking account of unfavorable seasonality. Specifically, the EC6 is the legendary air shuttle in Spain, which has been habitually operated by Iberia (IATA: IB, ICAO: IBE), whose offer of flights is primarily targeted at frequent travelers, and mainly for business reasons. Furthermore, other airlines have been operating on the route, such as Vueling (IATA: VY, ICAO: VLG) and Air Europa (IATA: UX, ICAO: AEA). All of them are usually operated using single-aisle jets, either Airbus A320 family or Boeing B737-800 series, which have recently been upgraded by retrofitting scimitar winglets on existing aircraft to cut fuel, or even renewed with A320neo family and B737 MAX series, respectively, and offer enhanced performance in terms of fuel consumption thanks to the improved design of both engines and aerodynamic shapes. Additionally, it should be noted that, within the potential market in Spain of domestic corridors, there are about 30 million passengers in the internal market of air transport services, of which about 2.5 million are concentrated on the EC6, according to the figures of 2019, i.e., over 200,000 passengers carried on such a route monthly before people went into the lockdown in March 2020 because of the COVID-19 outbreak. That and, above all, the importance of this corridor in the overall aviation emissions of domestic flights in Spain is shown.

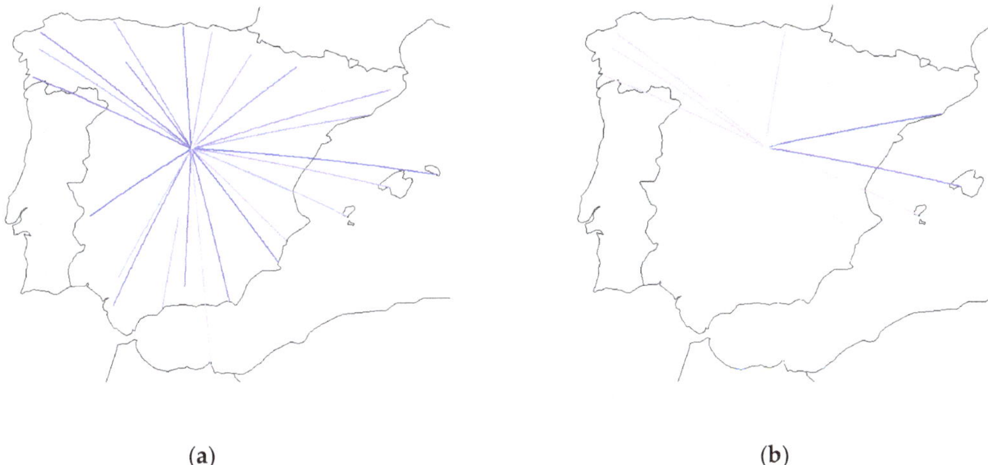

(**a**)   (**b**)

Figure 7. Carbon dioxide emissions from main domestic corridors operating in MAD airport, where a rough sketch means more emissions than the thinner one as follows: (**a**) Partial emission per passenger and per route (arithmetic average: 72.58, standard deviation: 13.76); (**b**) Total emission of routes concerned. Source: own elaboration from the data provided by [19,20].

3.2.2. Related Aspects Concerning Domestic Routes Operating in the LPA Airport

The Gran Canaria Airport (IATA: LPA, ICAO: GCLP), as can be seen in Table 1, has habitually held the fifth position in terms of passengers carried within the Spanish domestic market of regular air transport services. Moreover, the airport has a two-fold role: on the one hand, the major destination for flights from the Spanish mainland are always operated with turbojets, either narrow-body or wide-body aircraft; and on the other hand, this airport is headquarters to the main regional airline operating in the Canary Islands, named Binter Canarias (IATA: NT, ICAO: IBB), whose fleet has traditionally been comprised of short-haul regional airliners, mainly turbofans, such as the ATR 72 (IATA: AT7, ICAO: AT72), and more recently of medium-range jet airliners, the newest Embraer E2 (IATA: 295, ICAO: E295). Furthermore, both aircraft types have been designed with tight specifications that are tuned for reduced emissions per passenger carried, even with low levels of PLF. As a result, while the related flights into the Spanish mainland present a high carbon dioxide emission level, inter-island flights in such Spanish archipelago have proved to be the lowest level of the whole study in the specific years examined. Further to be noted is the seasonal character of the demand for air services such as from the main airport of the Canary Islands to mainland Spain, whereas the related inter-island air traffic market in the period being analyzed has become more stable than the supply of flights out of the archipelago, partly owing to generated extra demand for regional routes from subsidizing air transport and Public Service Obligation (PSO) implementation [23]. Moreover, six of the 23 routes so far imposed in Spain as essential transportation services under such EU schema have been linked to LPA, and two of them have been often operating as a type 3 PSO, that is, a closed route with both exclusivity and compensation [24]. Nevertheless, the existence of affordable regular flights from the Canary Islands, belonging to the EU's outermost regions group, helps to ensure the mobility of their inhabitants. Indeed, the specific matter resulting from such a geographical situation must not be misunderstood. As noted in the pooled results shown below, respectively Figure 8a,b, the highest emissions per passenger carried appear in those corridors to mainland destinations, and concerning absolute values for all of the corridors selected, the EC5 has been intensively operated by twin-engine turbojets, some of them even wide-body, such as the A330 family (IATA: 330, ICAO: A330) due to a greater haul capacity for cargo and mail.

(**a**)   (**b**)

Figure 8. Carbon dioxide emissions from main domestic corridors operating in LPA airport, where a rough sketch means more emissions than the thinner one as follows: (**a**) Partial emission per passenger and route (arithmetic average: 112.43, standard deviation: 65.99); (**b**) Total emission of routes concerned. Source: own elaboration from the data provided by [19,20].

3.2.3. Related Aspects Concerning Domestic Routes Operating in the BCN Airport

As can be seen from Table 1 above, the Josep Tarradellas Barcelona-El Prat Airport (IATA: BCN, ICAO: LEBL) is the second facility belonging to Aena's network in terms of passenger traffic and aircraft operations. The analysis of domestic flights operated from BCN over the period under study has been summarized in Figure 9. As with the case of MAD, it shows that related routes linking to BCN have had a high impact on the total aviation emissions of the Spanish aviation market, despite comparatively little supply in comparison with the total operations, the majority of which were medium-haul flights to travel destinations in other European countries, and in a few cases, even long-haul flights over the Atlantic Ocean. Interestingly, though being operated at almost full load on national corridors, some of them with a trip distance up to 500 km, carriers still mostly operated turbojets, possibly indicating the limited availability of regional turboprop airliners based especially to such an airport, even because of the absence of eco-friendly planes for regional routes in related airlines' fleets. For instance, scheduling on short routes, such as EC9, EC12, EC16, and EC25, an AT7 can be used instead of operating small turbojets, such as CJR200 (IATA: CR2, ICAO: CRJ2) and CRJ900 (IATA: CR9, ICAO: CRJ9), which can lead to a reduction of up to 50% less carbon dioxide emissions in similar circumstances [15]. In contrast, regional turbofans in scheduled air carriers are very scarce in Spain and limited to a few airlines. In the absence of any aircraft model of Bombardier Q Series, for example, the DHC-8-400 (IATA: DH4, ICAO: DH8D), the turboprops existing in Spain for commercial purposes belong to the ATR aircraft family, and the vast majority of them are AT7s, habitually operated by Air Nostrum (IATA: YW, ICAO: ANE) in addition to those operated on the Canary Islands' market mainly by NT, and to a much lesser extent, by Canaryfly (IATA: PM, ICAO: CNF). Nonetheless, in the case at hand, because of increased requirements of payload, a small aircraft operated on short-haul routes has sometimes been replaced by a large aircraft. It is also worth noting the fact of considering fleet availability as a key factor for airlines when scheduling their offer of flights. In such a context, Figure 9a shows a similar impact of specific emissions on each route selected, while Figure 9b highlights significant emissions on domestic routes with high demand, particularly those linking to three major cities in Spain, such as EC6, EC65, and EC109.

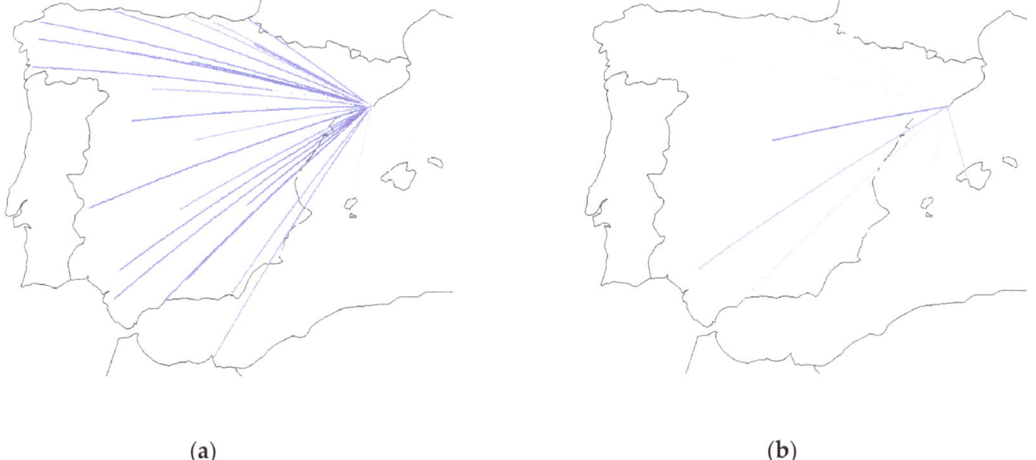

(a)   (b)

Figure 9. Carbon dioxide emissions from main domestic corridors operating in BCN airport, where a rough sketch means more emissions than the thinner one as follows: (**a**) Partial emission per passenger and route (arithmetic average: 79.39, standard deviation: 25.64); (**b**) Total emission of routes concerned. Source: own elaboration from the data provided by [19,20].

3.2.4. Related Aspects Concerning Domestic Routes Operating in the AGP Airport

The Málaga-Costa del Sol airport (IATA: AGP, ICAO: LEMG), as noted in Table 1, is the most important air transport infrastructure situated in Southern Spain, linking to several destinations (mostly European) whose flights have habitually been scheduled by low-cost carriers instead of legacy airlines, that has an enormous presence of international travelers. The airport has witnessed a substantial year-on-year increase in demand as a direct result of recent facility expansion through both terminal building and runway, contributing towards the 6% and 18% increase in traffic according to the growth rate estimation, respectively [25]. The significant improvement of such infrastructure has led to a remarkable increase in airport capacity, not just in terms of passengers, but also in terms of freight carried. Regarding related domestic routes, as can be seen in Figure 10, involves identifying what impacts the overall emissions have and how to explain them. In that sense, Figure 10a shows significant specific emissions per passenger on some air corridors into Northern Spain and Central Spain, such as EC22, EC29, EC73, and EC105. By contrast, despite the EC07 playing the role of a feeder route for both the UX and IB hubs in MAD, the route does not have significant specific emissions since carriers have been flying some of their various regional aircraft just 432 km to MAD for shuttle services according to demand, in particular by using aircraft with extremely low engine emissions figures, such as AT7 and Embraer E195LR (IATA: E95, ICAO: E195). It should also be noted that unit emissions for EC85 lead to a high rate in comparison with the rest of the routes selected in the study. At the same time, the calculated emissions for EC65, in absolute terms, have been revealed to be of critical importance within the overall estimation of emissions with regard to those domestic flights departing from AGP, as shown in Figure 10b. Commercial airlines operating such a corridor have not only operated twinjets, but also all flights under consideration were scheduled by six-abreast cabins belonging both to VY and Ryanair (IATA: FR, ICAO: RYR). In fact, these carriers today operate solely narrow-body aircraft on short-haul routes within Europe (and therefore includes domestic routes in Spain), specifically the Airbus A320 family in VY and Boeing B737-800 series in FR. Interestingly, this was not the case for EC81, since related flights have always been operated with turboprops, mostly AT7, due to the existing approach and landing procedures in Melilla (IATA: MLN, ICAO: GEML). That also made them remarkable for low emissions.

(a)(b)

Figure 10. Carbon dioxide emissions from main domestic corridors operating in AGP airport, where a rough sketch means more emissions than the thinner one as follows: (**a**) Partial emission per passenger and route (arithmetic average: 75.69, standard deviation: 22.37); (**b**) Total emission of routes concerned. Source: own elaboration from the data provided by [19,20].

3.2.5. Related Aspects Concerning Domestic Routes Operating in the PMI Airport

The Palma de Mallorca airport (IATA: PMI, ICAO: LEPA), as shown in Table 1, has been identified as Spain's third leading air facility in terms of the number of passengers and air operations over the study period. Because of the singular nature of the Balearic Islands, which stands as a highly popular tourist destination in Europe, such an airport has been characterized for attracting several air carriers whose total number of seats offered has been growing steadily in the last few years, with the exception of restrictions placed on movement around Europe due to coronavirus lockdowns between the years 2020 and 2021. According to the last expected growth at the airports in the Balearic Islands, PMI has recently led to an increase of 11% in scheduled seats (31.2 million) [26]. Indeed, for the present year, once travel restrictions eased and then completely lifted, the three airports of such a Spanish archipelago have undergone unprecedented growth. All of these factors have created an ample offer of flights at PMI that have not only been operating towards many major European city destinations scheduled and charter services, but also into medium-sized cities, and even sometimes outlying towns to major cities across Europe. All of this means that travelers interested in flying into PMI usually have access to the low prices and wide choices that are available in the truly competitive EU single aviation market. The outcome is an enormous representative offer of the common European air transport sector with great pulling power, which includes both low-cost and legacy airlines. Nonetheless, this great expansion of direct flights into such an attractive tourist destination situated in the Mediterranean Sea has not always ensured that the long-term level of capacity would be better adjusted to meet the level of demand, either in terms of optimizing traffic density or in the range of the destination from tourist source countries. As shown in Figure 11a, in the case of those national departure airports that serve PMI, widespread high specific emission values apart from some routes between islands, respectively EC58 and EC86, have been observed. Unlike domestic routes into mainland Spain, some of them are even related to seasonal flights from small airports, and a very seasonal demand on both corridors, whose usual carriers (YW and UX) have traditionally operated turboprop aircraft (AT7), thus achieving a low emission performance [21], has not been noted. By contrast, as can be seen in Figure 11b, the route (EC8) with the greater choice and the largest number of seats offered has the highest emission levels.

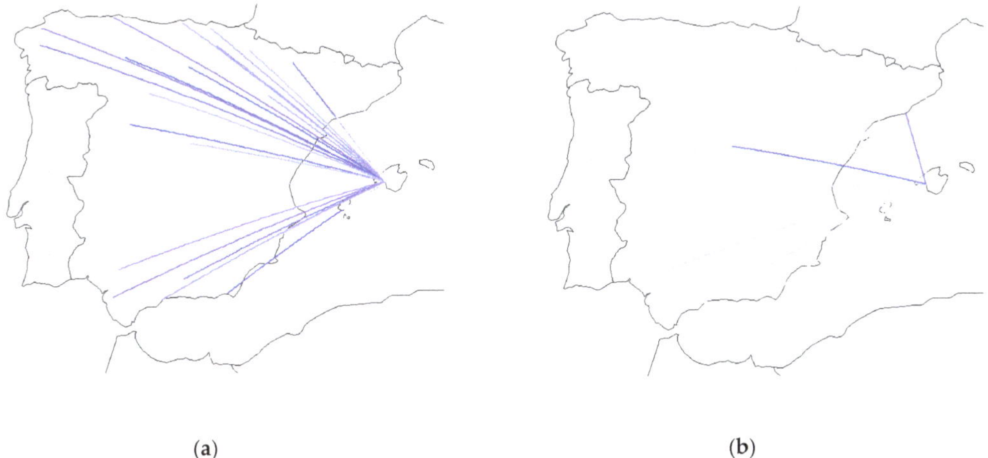

(**a**) (**b**)

Figure 11. Carbon dioxide emissions from main domestic corridors operating in PMI airport, where a rough sketch means more emissions than the thinner one as follows: (**a**) Partial emission per passenger and route (arithmetic average: 76.18, standard deviation: 30.21); (**b**) Total emission of routes concerned. Source: own elaboration from the data provided by [19,20].

3.3. Overall Findings Based on Key Performance Indicators

In order to make a comparative evaluation of direct impacts resulting from aviation emissions in operational terms, as can be seen below, proper ratios of the specific emissions to air according to the ten selected aircraft types have been calculated and presented.

3.3.1. Benchmarking of Air Pollution

Regarding carbon dioxide emissions on the basis of the distance flown through Spanish airspace in domestic routes only, twin-turboprop aircraft appear to be most appropriate in scheduling corridors up to 550 km, as shown in Figures 12 and 13, respectively.

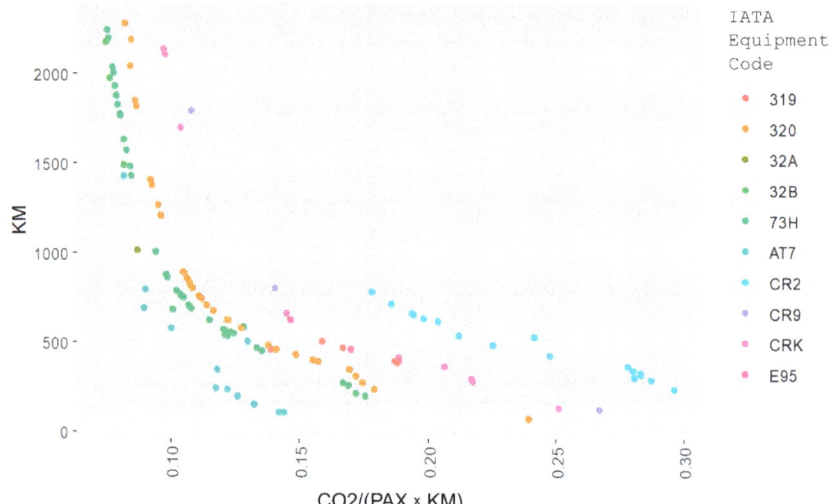

Figure 12. The relative ratio of carbon dioxide emissions in kilograms (kg) per passenger (PAX) and kilometer (km) according to the distance traveled in kilometers (km) by each aircraft selected. Source: own elaboration based on data compiled from [19] and weighted from [20].

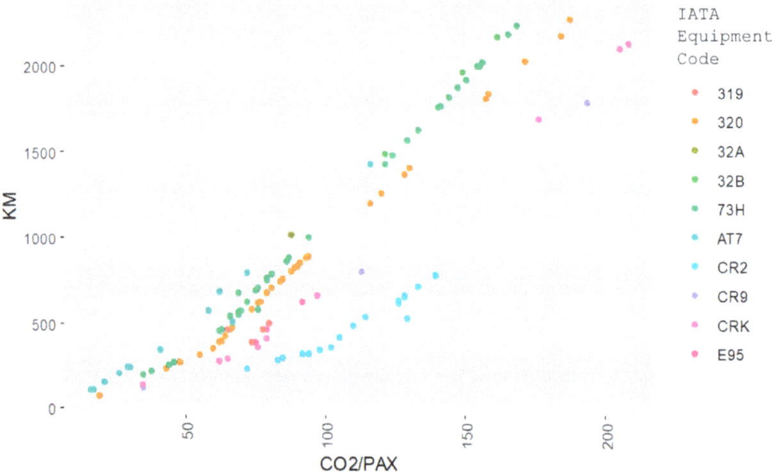

Figure 13. The relative ratio of carbon dioxide emissions in kilograms (kg) per passenger (PAX) according to the distance traveled in kilometers (km) by each aircraft selected. Source: own elaboration based on data compiled from [19] and weighted from [20].

3.3.2. Benchmarking of Fuel Consumption

Concerning fuel consumption (jet A1) caused by related aviation activity from operating those 139 domestic routes over 10 years between 2011 and 2020, its evolution has been calculated according to the U.S. Gulf Coast Kerosene-Type Jet Fuel Spot Price in terms of Free on Board (FOB) [27]. As presented below in Figures 14 and 15, respectively, there appears to be a trend toward increasing emissions during periods of low fuel prices.

Figure 14. Comparative performance of carbon dioxide emissions and fuel prices. Source: own elaboration based on data compiled from [19] and weighted from [20] according to [28,29].

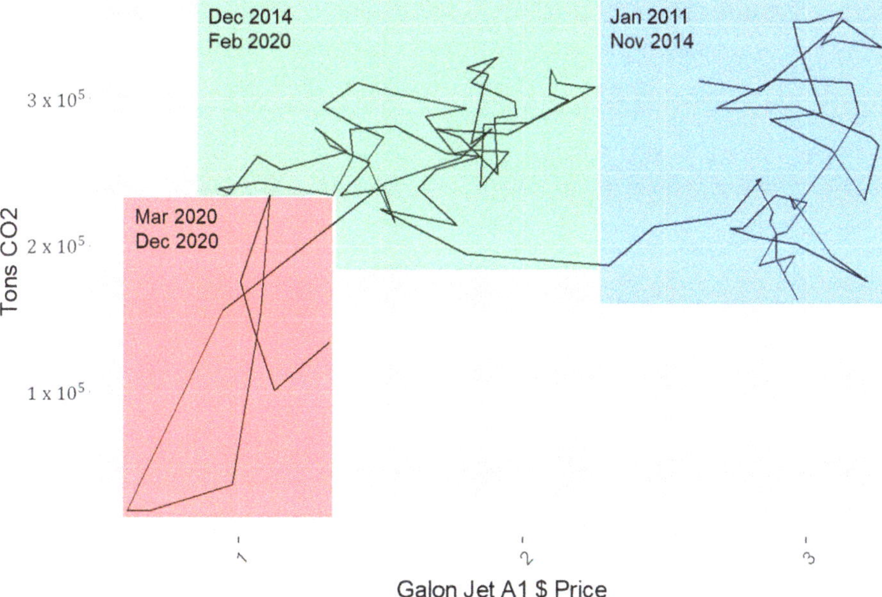

Figure 15. Trend analysis of carbon dioxide emissions and fuel prices. Source: own elaboration based on data compiled from [19] and weighted from [20] according to [28,29].

4. Discussion

According to the results obtained, and taking into consideration other previously published works on greenhouse gas emissions from commercial aviation, the findings are now discussed. In most EU member states, over the years, the public transportation sector has been enhancing steadily, with most reporting an increase in the number of routes in operation. Regarding collective modes of transportation until 2020, due to the outbreak of COVID-19, the number of passengers carried by air over regular routes and on regular schedules had experienced vibrant growth in the European aviation market. However, the air transport sector has recently been hampered by structural difficulties that make it more vulnerable to other transport modes. This is because the air sector is both highly cost-intensive and highly dependent on the global aviation leasing market, as well as fuel costs and wage bills due in part to the existence of minimum fuel requirements and high professional qualifications, respectively. Results from the study show there is a wide dynamic performance of air activity and, consequently, jet fuel consumption. Apparently, there are a number of factors that contribute to this fact, such as the aircraft's operational availability, adequacy of aircraft and crew for operational purposes, and the impact of the offered routes on airline profitability. While many airlines analyzed in this study have different types of cabins and seat configurations on various aircraft in their fleets, it is particularly important to consider their business strategies in which fleet modernization and renewal programs are carried out, not only for enhancing efficiency and profitability in transport activities, but also the promotion of sustainable transport. This statement is also precisely in line with what some previous papers pointed out on such a matter. For instance, regarding carbon emissions of European commercial air traffic, it was found that the main driver for efficiency in emissions is the modernization of the aircraft fleet [12]. The other essential element for more sustainable aviation is related to the use of alternative fuels, including exploring opportunities for alternative raw materials, such as renewable jet fuel by using advanced fermentation (AF) fuel from perennial grasses [30], since scaling up the SAF production is vital to enable an effective aviation's low-carbon transition.

In the research, only domestic routes have been examined, sometimes of seasonal nature, others even operated on some of the 23 routes under the Public Service Obligation (PSO) scheme so far imposed in Spain [21], the majority being feeder air routes, since legacy airlines are often looking to fly more short-haul domestic routes with regional aircraft with up to 100 seats for more efficient operation under the hub and spoke model. In contrast, low-cost carriers usually fly more point-to-point routes with medium-haul airliners for more profitable operations. Of the 139 corridors analyzed in the study, the highest fuel consumption ratio per kilometer has precisely been found in those serving as feeders to hubs from smaller airports due to the lack of international destinations at the source. Moreover, it is because they have been operated several times with small regional turbojets, such as CR2 or CR9, that emissions have been estimated by exceeding 0.20 kg of CO2 per passenger and kilometer. This concerns in particular the following routes: EC76, EC71, EC24, EC102, EC75, EC89, EC60, EC86, EC58, EC9, EC77, EC91, EC32, EC121, EC88, EC18, EC41, and EC93. Although the research results have been calculated according to the most representative airliner scheduled on each route over the study period, the previous corridors have been operated by using regional airplanes with a low seating capacity that are most suitable for thin route service, as the air traffic density is flat. Furthermore, outcomes of emission monitoring have highlighted the highly varying importance of scheduling airplanes on thin routes over the past years until the beginning of recent trends toward fleet harmonization with the entry of more fuel-efficient aircraft for sustainable transportation. Examples of this are, among others, the EC83, EC72, and EC15, whose scheduled flights had been mostly operated with up to six, three, and five different airplane types, respectively, and then it became necessary to introduce a unique regional jet model, specifically the CRK. These routes account for the most significant cases in facing the challenge of estimation methods used to calculate related emissions. This underlines the need for identifying the most representative airplane on each route at the statistical level.

5. Conclusions

As already stated in the opening of the paper, the principal aim of the research has been to carry out the empirical analysis of commercial airplane emissions on some short-haul flights previously selected for the study in terms of efficient performance and sustainable operation from a representative domestic market in the EU aviation sector. Specifically, this study underscores the influence that various aircraft have, in accordance with their nominal engine performances, on the outdoor air quality of such transport activities concerning environmental pollution from carbon dioxide emissions only. All of that is enveloped by the domestic market of passenger transport in Spain, particularly those air corridors operating from the five main airports, as shown in Table 1. In light of the research findings, this study is meaningful in that it does not simply analyze the emissions before and after the cruise flight as part of commercial transport services, but examines changes in fuel behavior depending on whether the airplane is provided in a situation where the aircraft is fully aware of the system.

Regarding overall consumption, it is important to highlight the routes with a more significant emission impact in the set of 139 routes selected; for instance, those with emissions above 2.2 million tons of carbon dioxide in the 10 years analyzed, such as EC6, EC111, and EC5, and in addition to that, 1.39 million tons emitted in the period (EC8). Despite that EC6 was the corridor most intensively operated in the study period, specifically 178,577 flights, the highest consumption has been detected in the corridors between Madrid and the Canary Islands. This is particularly significant in the case of EC111 and EC116, whose emissions are 2.84 million and 3.30 million of tons CO2 respectively, because of the large number of seats required all year round on such a main air corridor. Globally, as previously noted, five airports have been considered to become the most representative of the Spanish aviation market from the study, with MAD and BCN as the key facilities representing the 84.6% of total emissions. The remaining 15.4% is shared among the other main airports, namely PMI, LPA, and AGP (6.8%, 5%, and 3.6%, respectively).

Concerning consumption per passenger and kilometer, the majority of the 139 routes assessed are highly dependent on the airplane used for each of them, particularly in the case of regional aircraft types, such as CR2 and CRK, with 50 and 100 standard seats, respectively. Consumption is highly dependent on the distance traveled and, on the airliner, either turboprop or turbojet, operating flights scheduled on a given route. In line with some previous works, both in general matters [31] and in specific cases, such as in Georgia [32] and New Zealand [33], this research reinforces the importance of aircraft type used, load factors, and seat configuration to this issue. Such a clear case is that of the former Spanish air carrier named Spanair (IATA: JK, ICAO: JKK), which ceased operations on 28 January 2012, operating airplanes with high fuel costs per seat-mile (or km), such as those belonging to the McDonnell Douglas MD-80 series on routes EC6 and EC8, up to 148 and 212 flights a month in 2011, respectively. Another example can be found in those corridors traditionally operated by CR2, where a CRK has been introduced, thus leading to a significant emission reduction over the analysis period; specifically, from an average of 21,892 tons of carbon dioxide equivalent per month in May 2011 to an average of 398 tons per month in 2020. This was despite the non-variation impact on the supply of flights over the period, mainly on thin routes, such as EC121, EC72, and EC32, as well as a couple of PSO corridors, such as EC24 and EC25. Of particular note is the case of the introduction of CRK at the expense of CR9 in the Spanish domestic market, which had been used intensively on domestic routes selected until 2018, then gradually removed from the active fleet in 2019, and finally withdrawn from the corridors analyzed in April 2020. This has been caused by identifying the less efficient airliners for use on regional routes and even on thin routes, which can be gradually replaced with more efficient and environmentally friendly aircraft. The main example of this is provided by EC102, which was intensively scheduled with CR2, and later passed to be operated with 319 and CRK. This assessment indicates that the implementation of fleet renewal programs has allowed airliners to slash emissions beyond the levels mandated in the EU ETS rules on aviation. That answers the research question as to whether such investment efforts would do sustainable short-haul routes with a traveled distance of up to 500 km, and the answer to that question is an unequivocal yes when mainly operating turboprops. In any event, despite short-haul routes usually being considered as being shorter than 1100–1500 km, only 23 of 139 routes have a greater distance than 1500 km. Therefore, the limited number of medium-haul flights reinforces the idea that most domestic routes can be operated with turboprops.

6. Research Limitations and Future Directions

Despite the limited number of previous works on the matter of aircraft emissions caused by short-haul flights operated by regular airlines in Europe's single aviation market, in particular very scarce in the case of the Spanish domestic market, the findings of the present paper seem to suggest the opportunity to locate any common patterns among case studies from various geographical areas. However, there are several limitations to the research issue, and the results may not apply to the whole of the wider global aviation market in passenger transport, since fleets are often not homogeneous across the world, so a benchmark is difficult to establish. Owing to the limitations of the study, the research has evolved around two primary sources. On the one hand, related air traffic data were collected from the national airport services operator [19]. On the other hand, standard estimations of fuel consumption and emissions for each type of aircraft were calculated according to guidelines provided by the international organization of reference in the field of aviation [20]. However, all calculations have been based on the average values from aggregated data concerning cruise flights during the sampling period on the 139 routes selected. Thus, it has not been possible to consider a harmonized approach using an adequate intake estimation model based on flight data recorded by the airport operators in Spain, including serial number and registration of the aircraft, engine type, number of passengers carried, and emission factors. The reason for such a research limitation is due to the existence of operational data belonging to the business domain of both airport managers and air navigation service providers being specifically declared as confidential by either party, such as airlines or airliner manufacturers. Despite these limitations, the study has estimated the aggregated emissions per route, thus comparing the situation of each one according to the most representative aircraft operated over the research period. This has made it possible to analyze the difference in emissions per distance, thus highlighting the environmental impact depending on each aircraft type, both in absolute and relative terms (per passenger and kilometer).

Even though findings from the research are confined to the passenger transport services provided by scheduled airlines operating in the Spanish domestic market, it can be assumed that similar results could be achieved by future works on similar cases in the EU single aviation market. To deepen the knowledge of the research issue, it may be interesting to study the behavior of intercontinental air corridors, not only by transport capacity but also by the passenger, hence allowing the assessment of the actual use of the aircraft and its consumption. In that sense, it can be interesting to examine the wage costs of aircraft crews, when available, in order to make a full cost breakdown, thus highlighting the impact of fuel on the total operating costs of the airlines' operating corridors selected. The research issue also addresses the challenge of sustainable and efficient jet fuel procurement in the aviation sector, which is of great interest to both carriers and public authorities. Hence, it would be of great interest for further studies to focus on the economic viability of alternative fuels for intensive aviation purposes on scheduled flights, since economic analysis is potentially useful in identifying and clarifying the issues involved in making public policies relating to the environmental impact of air transportation. Another environmental challenge with regard to which the aviation sector has lately faced is the intensive use of general aviation for private chartered flights. Given that the present study focuses on carbon emissions from regular commercial flights into and out of Spain, charter flights have not been accounted for in the base of air quality monitoring estimation. Perhaps future research should shed some light on such a particular matter, as business aviation is usually operated with small-engine airplanes, mainly turbojets that often burn excessive fuel per passenger carried. Additionally, it would be interesting to look at how certain airlines are currently managing to evolve their fleet with the worst jet CO_2 emissions to eco-efficient, sustainable clean-sheet aircraft, such as the Airbus A320neo family (32D, 32A, 32B), as it would allow to determinate how this contributes to the optimization of the use of more efficient airplanes, and thereby more sustainable scheduled flights.

Although the research approach has only been focusing on carbon dioxide (CO_2) emissions from the domestic aviation sector in Spain, it would be interesting to analyze other outdoor air contaminants, such as nitrogen oxides (NOx), hydrocarbons (HC), carbon monoxide (CO), and sulfur gases, as well as soot and metal particles. Future studies on the matter might be carried out in a manner similar to some earlier works about air quality monitoring tools, both with low-cost multi-channel monitors under specific laboratory conditions for indoor environments [34,35] and with low-cost sensors for outdoor environments [36,37]. This could serve as the basis for planning experiments to accurately estimate emissions depending on each model engine if abundant funding from external sources were available, due to the high costs of this kind of testing. Nevertheless, the variety of flight conditions in operating passenger transport services is large and some aspects of aircraft engine emissions may need to be reviewed to understand the performance of the aircraft as a whole.

Author Contributions: Conceptualization, A.M.R., A.S.d.l.C. and R.E.G.D.; methodology, A.M.R. and A.S.d.l.C.; software, A.S.d.l.C.; validation, A.M.R. and A.S.d.l.C.; formal analysis, A.M.R. and A.S.d.l.C.; investigation, A.M.R. and A.S.d.l.C.; resources, A.M.R. and A.S.d.l.C.; data curation, A.M.R. and A.S.d.l.C.; writing—original draft preparation, A.M.R. and A.S.d.l.C.; writing—review and editing, A.M.R.; visualization, A.S.d.l.C.; supervision, R.E.G.D.; project administration, A.M.R. All authors have read and agreed to the published version of the manuscript.

Funding: This research received no external funding.

Institutional Review Board Statement: Not applicable.

Informed Consent Statement: Not applicable.

Data Availability Statement: Data is available on request due to restrictions. The data presented in this study are available on request from the corresponding author. The primary traffic data source is publicly available from the Statistical Site property of Aena SME, S.A., as described in references.

Conflicts of Interest: The authors declare no conflict of interest.

Appendix A

Table A1. List of Commercial Airplanes Considered in the Research.

IATA Designator	ICAO Designator	Aircraft Type	Manufacturer (as of 14 October 2022)
AT7	AT72	Aerospatiale-Alenia ATR 72	Avions de Transport Régional GIE
319/32D	A319	Airbus A319 ceo/sharklets	Airbus SAS
320/32A	A320	Airbus A320 ceo/sharklets	Airbus SAS
321/32B	A321	Airbus A321 ceo/sharklets	Airbus SAS
738/73H	B738	Boeing 737-800/winglets	The Boeing Company
712	B712	Boeing 717-200	The Boeing Company
CR2	CRJ2	Canadair Regional Jet (CRJ) 200	MHI RJ Aviation ULC
CR9	CRJ9	Canadair Regional Jet (CRJ) 900	MHI RJ Aviation ULC
CRK	CRJX	Canadair Regional Jet (CRJ) 1000	MHI RJ Aviation ULC
E95	E195	Embraer 195	Embraer S.A.
295	E295	Embraer 195-E2	Embraer S.A.
M81/M82/M83/M87	MD81/MD82/MD83/MD87	McDonnell Douglas MD-80/90	The Boeing Company

Table A2. List of Scheduled Airlines Considered in the Research.

IATA Designator	ICAO Designator	Trade Name	Air Operator's Certificate (AOC) Issued by DGAC (as of 14 October 2022)
UX	AEA	Air Europa	ES.AOC.004
X5	OVA	Air Europa Express	ES.AOC.020
YW	ANE	Air Nostrum	ES.AOC.002
NT	IBB	Binter Canarias	ES.AOC.011
PM	CNF	Canaryfly	ES.AOC.100
IB	IBE	Iberia	ES.AOC.001
I2	IBS	Iberia Express	ES.AOC.117
formerly JK	formerly JKK	Spanair	withdrawn
V7	VOE	Volotea	ES.AOC.115
VY	VLG	Vueling	ES.AOC.060

Table A3. List of Representative Engines Considered in the Research.

Propulsion	Aircraft	Engines
Turboprop	ATR 72-500 (72-212A)	2 × Pratt & Whitney Canada PW100
Turboprop	ATR 72-600 (72-212A)	2 × Pratt & Whitney Canada PW100
Turbofan	Airbus A320-271N	2 × CFM Leap-1A/PW1100G-JM
Turbofan	Airbus A321-271N	2 × General Electric CF34-8C5A1
Turbofan	Airbus A319-100	2 × General Electric CF34-8C5
Turbofan	Airbus A320-214	2 × CFM56-5B4/P
Turbofan	Airbus A321-231 (WL)	2 × IAE V2533-A5
Turboprop	ATR 72-500 (72-212A)	2 × Pratt & Whitney Canada PW100
Turboprop	ATR 72-600 (72-212A)	2 × Pratt & Whitney Canada PW100
Turbofan	Boeing 737-8AS (WL)	2 × CFM56-7B
Turbofan	Boeing 737-85P (WL)	2 × CFM56-7
Turbofan	Boeing 717-2BL/-23S/-2CM	2 × BMW RR BR715
Turbofan	Bombardier CRJ1000 (CL-600-2E25)	2 × General Electric CF34-8C5A1
Turbofan	Bombardier CRJ900 (CL-600-2D24)	2 × General Electric CF34-8C5
Turbofan	Bombardier CRJ200 (CL-600-2B19)	2 × General Electric CF34-3B1
Turbofan	Embraer E195-E2 (ERJ 190-400 STD)	2 × Pratt & Whitney Canada PW1900G
Turbofan	Embraer ERJ-195LR (ERJ 190-200 LR)	2 × General Electric CF34-10E
Turbofan	McDonnell Douglas MD-80/90	2 × PW JT8D-219

References

1. European Union. Directive 2003/87/EC of the European Parliament and of the Council of 13 October 2003 establishing a scheme for greenhouse gas emission allowance trading within the Community and amending Council Directive 96/61/EC. *Off. J. Eur. Union* **2003**, *L 275/32*, 32003L0087.
2. European Union. Proposal for a Directive of the European Parliament and of the Council amending Directive 2003/87/EC as regards aviation's contribution to the Union's economy-wide emission reduction target and appropriately implementing a global market-based measure. *Off. J. Eur. Union* **2021**, *COM/2021/552*, 52021PC0552.
3. Kopsch, F. Aviation and the EU Emissions Trading Scheme—Lessons learned from previous emissions trading schemes. *Energy Policy* **2012**, *49*, 770–773. [CrossRef]
4. European Union. Directive 2008/101/EC of the European Parliament and of the Council of 19 November 2008 amending Directive 2003/87/EC so as to include aviation activities in the scheme for greenhouse gas emission allowance trading within the Community. *Off. J. Eur. Union* **2009**, *L 8/3*, 32008L0101.
5. European Union. Commission Decision 2009/450/EC of 8 June 2009 on the detailed interpretation of the aviation activities listed in Annex I to Directive 2003/87/EC of the European Parliament and of the Council. *Off. J. Eur. Union* **2009**, *L 149/69*, 32009D0450.
6. European Union. Decision No 377/2013/EU of the European Parliament and of the Council of 24 April 2013 derogating temporarily from Directive 2003/87/EC establishing a scheme for greenhouse gas emission allowance trading within the Community. *Off. J. Eur. Union* **2013**, *L 113/1*, 32013D0377.
7. European Union. Regulation (EU) No 421/2014 of the European Parliament and of the Council of 16 April 2014 amending Directive 2003/87/EC establishing a scheme for greenhouse gas emission allowance trading within the Community, in view of the implementation by 2020 of an international agreement applying a single global market-based measure to international aviation emissions. *Off. J. Eur. Union* **2014**, *L 129/1*, 32014R0421.

8.	European Union. Regulation (EU) 2017/2392 of the European Parliament and of the Council of 13 December 2017 amending Directive 2003/87/EC to continue the current limitations of scope for aviation activities and to prepare to implement a global market-based measure from 2021. *Off. J. Eur. Union* **2017**, *L 350/7*, 32017R2392.

9.	Feng, Q.; Gauthier, P. Untangling Urban Sprawl and Climate Change: A Review of the Literature on Physical Planning and Transportation Drivers. *Atmosphere* **2021**, *12*, 547. [CrossRef]

10.	Jain, S.; Chao, H.; Mane, M.; Crossley, W.; DeLaurentis, D. Estimating the Reduction in Future Fleet-Level CO_2 Emissions from Sustainable Aviation Fuel. *Front. Energy Res.* **2021**, *9*, 771705. [CrossRef]

11.	Macintosh, A.; Wallace, L. International aviation emissions to 2025: Can emissions be stabilized without restricting demand? *Energy Policy* **2009**, *37*, 264–273. [CrossRef]

12.	Amizadeh, F.; Alonso, G.; Benito, A.; Morales-Alonso, G. Analysis of the recent evolution of commercial air traffic CO_2 emissions and fleet utilization in the six largest national markets of the European Union. *J. Air Transp. Manag.* **2016**, *55*, 9–19. [CrossRef]

13.	Alonso, G.; Benito, A.; Lonza, L.; Kousoulidou, M. Investigations on the distribution of air transport traffic and CO_2 emissions within the European Union. *J. Air Transp. Manag.* **2014**, *36*, 85–93. [CrossRef]

14.	Martínez Raya, A.; González-Sánchez, V.M. Some Considerations about Value Creation in Regard to Entrepreneurship and Innovation from Public Service Obligations on Scheduled Air Transport. In *Analyzing the Relationship between Innovation, Value Creation, and Entrepreneurship*, 1st ed.; Galindo-Martín, M., Mendez-Picazo, M., Castaño-Martínez, M., Eds.; IGI Global: Hershey, PA, USA, 2020; pp. 289–306. [CrossRef]

15.	Martínez Raya, A.; González-Sánchez, V.M. Efficiency and Sustainability of Regional Aviation on Insular Territories of the European Union: A Case Study of Public Service Obligations on Scheduled Air Routes among the Balearic Islands. *Sustainability* **2021**, *13*, 3949. [CrossRef]

16.	Martínez Raya, A. Sustainable and efficient transportation in special territories of the European Union through public service obligations: The case of scheduled air services serving the Spanish city of Melilla. *World Rev. Intermodal Transp. Res.* **2021**, *10*, 361–377. [CrossRef]

17.	Martínez Raya, A.; González-Sánchez, V.M. Efficiency and sustainability of public service obligations on scheduled air services between Almeria and Seville. *Econ. Res. Ekon. Istraž.* **2020**, 2751–2768. [CrossRef]

18.	Llano, C.; Pérez-Balsalobre, S.; Pérez-García, J. Greenhouse Gas Emissions from Intra-National Freight Transport: Measurement and Scenarios for Greater Sustainability in Spain. *Sustainability* **2018**, *10*, 2467. [CrossRef]

19.	Aena Air Traffic Statistics. Statistical Data for the Period 2011–2020. Available online: https://www.aena.es/es/estadisticas/inicio.html (accessed on 14 October 2022).

20.	ICAO. *Carbon Emissions Calculator. Guidelines Methodology Manual 2018, Version 11*; ICAO: Montreal, QC, Canada, 2018.

21.	European Union. Council Regulation (EEC) No 95/93 of 18 January 1993 on common rules for the allocation of slots at community airports. *Off. J. Eur. Union* **1993**, *L 14/1*, 31993R0095.

22.	The Kingdom of Spain. Royal Decree 20/2014, of 17 January, supplementing the legal regime on the allocation of slots at Spanish airports. *Off. State Gaz.* **2014**, *16*, 2748–2761, BOE-A-2014-508.

23.	Martín-Cejas, R.R. The Influence of the Resident Subsidy on Regional Carrier Economies and the Environment in the Canary Interisland Air Traffic Network. *Tour. Hosp.* **2022**, *3*, 558–572. [CrossRef]

24.	Martínez Raya, A.; González-Sánchez, V.M. Tender Management Relating to Imposition of Public Service Obligations on Scheduled Air Routes: An Approach Involving Digital Transformation of Procurement Procedures in Spain. *Sustainability* **2020**, *12*, 5322. [CrossRef]

25.	Eugenio-Martin, J.L. Estimating the Tourism Demand Impact of Public Infrastructure Investment: The Case of Malaga Airport Expansion. *Tour. Econ.* **2016**, *22*, 254–268. [CrossRef]

26.	Aena Press Release of 23 March 2022. Available online: https://www.aena.es/doc/pressdetail/220323-aena-temporada-verano-2022-eng.pdf (accessed on 14 October 2022).

27.	IndexMundi. Commodity Prices Indices. Available online: https://www.indexmundi.com/commodities/?commodity=jet-fuel&months=180 (accessed on 14 October 2022).

28.	IATA. Jet Fuel Price Monitor. Statistical Data for the Period 2011–2020. Available online: https://www.iata.org/en/publications/economics/fuel-monitor/ (accessed on 14 October 2022).

29.	US Energy Information Administration. Spot Prices for Crude Oil and Petroleum Product. Available online: https://www.eia.gov/dnav/pet/hist/eer_epjk_pf4_rgc_dpgD.htm (accessed on 14 October 2022).

30.	Winchester, N.; Malina, R.; Staples, M.D.; Barrett, S.R.H. The impact of advanced biofuels on aviation emissions and operations in the U.S. *Energy Econ.* **2015**, *49*, 482–491. [CrossRef]

31.	Miyoshi, C.; Mason, K.J. The Carbon Emissions of Selected Airlines and Aircraft Types in Three Geographic Markets. *J. Air Transp. Manag.* **2009**, *15*, 138–147. [CrossRef]

32.	Tokuslu, A. Estimation of Aircraft Emissions at Georgian International Airport. *Energy* **2020**, *206*, 118219. [CrossRef]

33.	Tarr, A.P.; Smith, I.J.; Rodger, C.J. Carbon dioxide emissions from international air transport of people and freight: New Zealand as a case study. *Environ. Res. Commun.* **2022**, *4*, 075012. [CrossRef]

34.	Baldelli, A. Evaluation of a low-cost multi-channel monitor for indoor air quality through a novel, low-cost, and reproducible platform. *Meas. Sensors* **2021**, *17*, 100059. [CrossRef]

35. Saini, J.; Dutta, M.; Marques, G. A comprehensive review on indoor air quality monitoring systems for enhanced public health. *Sustain. Environ. Res.* **2020**, *30*, 6. [CrossRef]
36. Clements, A.L.; Griswold, W.G.; RS, A.; Johnston, J.E.; Herting, M.M.; Thorson, J.; Collier-Oxandale, A.; Hannigan, M. Low-Cost Air Quality Monitoring Tools: From Research to Practice (A Workshop Summary). *Sensors* **2017**, *17*, 2478. [CrossRef]
37. Karagulian, F.; Barbiere, M.; Kotsev, A.; Spinelle, L.; Gerboles, M.; Lagler, F.; Redon, N.; Crunaire, S.; Borowiak, A. Review of the Performance of Low-Cost Sensors for Air Quality Monitoring. *Atmosphere* **2019**, *10*, 506. [CrossRef]

Article

The Impact of Coronavirus Disease of 2019 (COVID-19) Lockdown Restrictions on the Criteria Pollutants

Puneet Verma [1,*], Sohil Sisodiya [2,*], Sachin Kumar Banait [3], Subhankar Chowdhury [3], Gaurav Dwivedi [3] and Ali Zare [4]

1   School of Earth and Atmospheric Sciences, Queensland University of Technology, Brisbane City, QLD 4000, Australia
2   Department of Civil Engineering, University Departments, Rajasthan Technical University, Kota 324010, Rajasthan, India
3   Energy Centre, Maulana Azad National Institute of Technology, Bhopal 462003, India
4   School of Engineering, Waurn Ponds Campus, Building KE 75 Pigdons Road, Deakin University, Waurn Ponds, Victoria, VIC 3216, Australia
*   Correspondence: puneet.verma@connect.qut.edu.au (P.V.); sohilsisodiya@gmail.com (S.S.)

Abstract: Air pollution is accountable for various long-term and short-term respiratory diseases and even deaths. Air pollution is normally associated with a decreasing life expectancy. Governments have been implementing strategies to improve air quality. However, natural events have always played an important role in the concentration of air pollutants. In Australia, the lockdown period followed the Black Summer of 2019–2020 and coincided with the season of prescribed burns. This paper investigates the changes in the concentration of criteria pollutants such as particulate matter, nitrogen dioxide, ozone, and sulphur dioxide. The air quality data for the lockdown period in 2020 was compared with the pre-lockdown period in 2020 and with corresponding periods of previous years from 2016 to 2019. The results were also compared with the post-lockdown scenario of 2020 and 2021 to understand how the concentration levels changed due to behavioural changes and a lack of background events. The results revealed that the COVID-19 restrictions had some impact on the concentration of pollutants; however, the location of monitoring stations played an important role.

Keywords: air quality; PM_{10}; $PM_{2.5}$; NO_2; COVID

Citation: Verma, P.; Sisodiya, S.; Banait, S.K.; Chowdhury, S.; Dwivedi, G.; Zare, A. The Impact of Coronavirus Disease of 2019 (COVID-19) Lockdown Restrictions on the Criteria Pollutants. *Processes* **2023**, *11*, 296. https://doi.org/10.3390/pr11010296

Academic Editors: Daniele Sofia, Paolo Trucillo and Avelino Núñez-Delgado

Received: 14 November 2022
Revised: 11 January 2023
Accepted: 12 January 2023
Published: 16 January 2023

1. Introduction

According to the World Health Organization's (WHO's) report, seven million deaths occur every year because of exposure to air pollution [1]. Air pollution is a cause of one in every eight deaths globally [2]. Unfortunately, worldwide nine out of ten people are exposed to a high level of air pollutants [1]. The severity of the effects of air pollution on human health is immense and there is an immediate need to curb this problem, and effective management of ambient air pollution can lead to a substantial reduction of pollutant concentrations [3]. However, the unprecedented outbreak of coronavirus disease of 2019 (COVID-19) has made a remarkable breakthrough in unresolved air pollution management [4].

The WHO declared COVID-19 a global pandemic on 12 March 2020 [5–7]. All affected countries enforced lockdowns and preventive measures to stop the virus's spread [8,9]. However, the situation was a little different in Australia. From 23 March 2020, the Queensland Department of Health announced that some businesses would be required to close or limit their operation [10]. On 5 June 2020, the Premier announced that the Queensland Tourism and Accommodation Industry COVID-19 Safe Plan had been approved, with the plan applying to thousands of tourism and accommodation businesses. This made Queensland's situation unique as there were some COVID-19-related restrictions, but no complete lockdown was enforced.

Pyrogenic events such as bushfires are a major environmental issue as they constitute up to 40% of carbon emissions every year [11] and worsen air quality around the world [12]. In general, the air quality levels in Brisbane and the rest of Queensland remain below the national air quality standards but particulate matter emissions have been an important air quality issue due to pyrogenic events such as bushfires, dust storms, and prescribed burning [13]. Due to restrictions on anthropogenic activities, the air quality around the world reportedly improved. Nitrogen dioxide (NO_2) concentrations in the lockdown periods of Wuhan, Delhi, New York, and Rome decreased by 65%, 69%, 56%, and 57%, respectively; $PM_{2.5}$ (levels of particles less than 2.5 μm) decreased by 49%, 69%, 53%, and 68%, respectively [14]. Similarly, a reduction of 24% in PM_{10} (particles less than 10 μm) concentrations was reported in Taiwan [4].

However, not all cities experienced declines in air pollution [15]. The impact of COVID-19 restrictions on air quality levels was not uniform across Australia. Some media publications reported from satellite imagery that NO_2 concentration levels decreased in Sydney but not in Perth and Melbourne [16]. As noted in the above paragraph, in most studies investigating changes in air quality, significant improvement was observed. However, in Australia, the lockdown period followed the Black Summer of 2019–2020 and coincided with the season of prescribed burns [17]. This paper investigates the changes in the concentration of criteria pollutants such as $PM_{2.5}$, PM_{10}, nitrogen dioxide, ozone, and sulphur dioxide during the lockdown period of 2020 compared to the pre-lockdown period of 2020, the preceding years from 2016 to 2019, and 2021.

2. Materials and Methods

2.1. Study Area

This study focuses on Brisbane City, which is situated in the Queensland state of Australia. Brisbane has coordinates of 27°28′2.39″S 153°01′24.00″E, and is the third largest city on the southeastern coast of Queensland. The city is surrounded by hills in the northwest [18]. Brisbane has the second-largest area in Australia, at 15,842 km². Brisbane's total population is 2,568,927 as of the 2021 Australian Census [19].

The air quality monitoring data were retrieved from the Queensland government's website (https://www.qld.gov.au, last accessed on 15 December 2022). The Queensland government's Department of Environment and Science maintains a range of air quality stations across the state. The locations of the monitoring stations used in this study are presented in Figure 1.

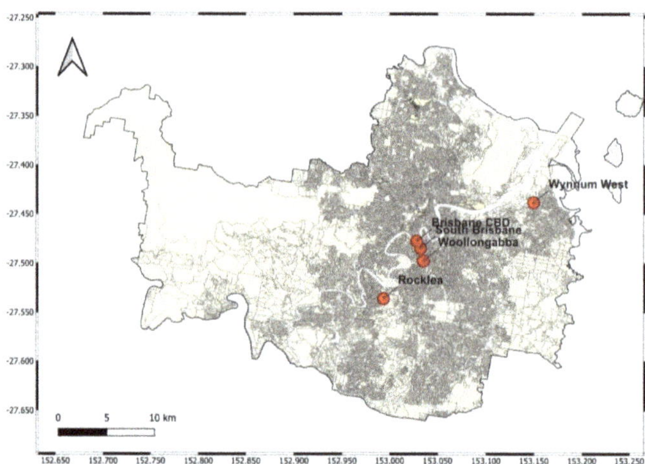

Figure 1. Study area boundary and location of monitoring stations.

We chose Rocklea monitoring station to collect most of the data on various air quality parameters such as PM_{10}, $PM_{2.5}$, NO_2, and ozone. The SO_2 monitoring data was collected from Springwood monitoring station. Meteorological parameters such as wind speed, wind direction, relative humidity, and ambient temperature were also collected from Rocklea monitoring station. Located approximately 500 metres southeast of the railway line in the grounds of the Oxley Common, the Rocklea monitoring station is surrounded by open parkland with light industry and residential uses surrounding the site. The site is well-placed to represent the background air quality for Brisbane and often used to establish the background concentrations in the air quality dispersion modelling assessments for different industrial emission sources in Brisbane. To investigate the impact of restricted emissions and behavioural changes, we analysed the $PM_{2.5}$, PM_{10}, and NO_2 data from other monitoring stations such as Brisbane Central Business Districts (Brisbane CBD), Woolloongabba, and South Brisbane.

2.2. Analysis Methodology

The raw data with hourly averages was downloaded and cleaned for any error values due to instrument malfunction such as -9999 and -1111. The error values were replaced with empty data cells. The plots were generated using the OpenAir package in the R programming environment. To analyse the impact of COVID-19 restrictions, the pollutant concentration data were segregated into periods: pre-lockdown (1 January 2020 to 19 March 2020), lockdown period (20 March 2020 to 5 June 2020), and post-lockdown period (6 June 2020 to 31 December 2020). The different periods were again compared with the same periods for each of the last four years (2016 to 2019) and for the year 2021. The comparison of pollutant concentrations for one year does not give a very clear idea of the variation in the emissions of the pollutants as the variations in the pollutants are very much dependent on the meteorological parameters [20], hence the meteorology of the other years was also compared.

3. Results and Discussion

3.1. Analysis of Meteorological Parameters

The changes in the meteorological parameters are presented in Figures 2 and 3. The wind speed and wind direction are presented in the form on wind-rose plots for different monitoring periods of 2020 in Figure 2. In general, the prevalent winds were from the northeast and southwest quadrants. There was a lower frequency of winds from the northeast quarter, which could be due to seasonal variation during the lockdown period and early parts of the post-lockdown period.

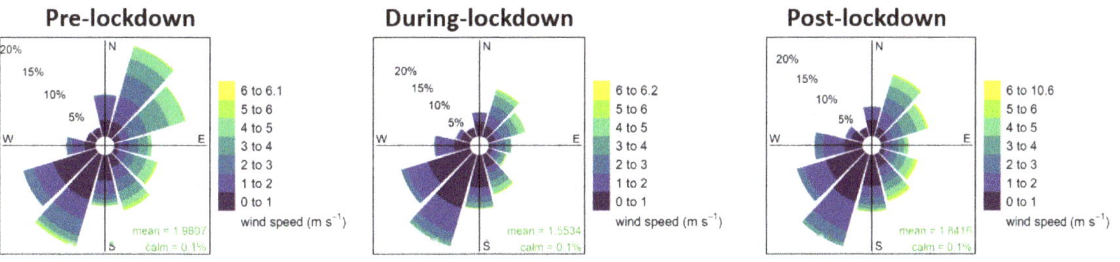

Figure 2. Wind-rose plots for different monitoring periods of the year 2020.

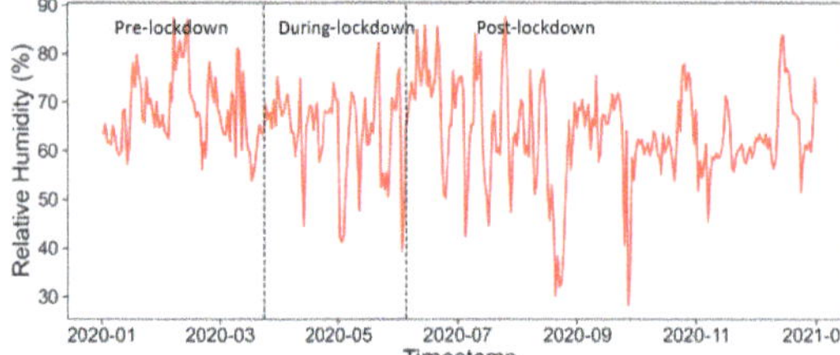
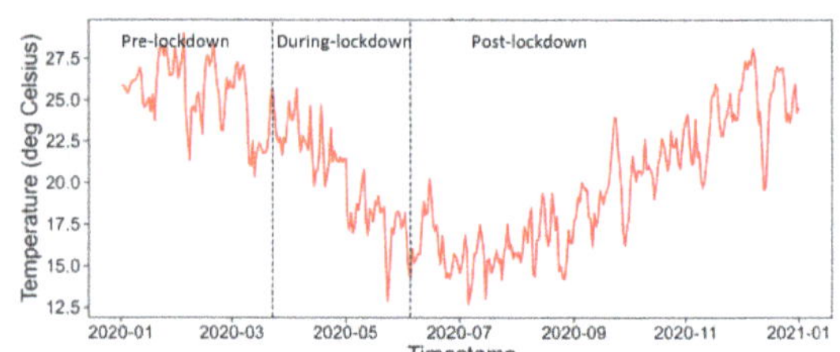

Figure 3. The variation in daily average relative humidity and temperature for different monitoring periods of the year 2020.

Figure 3 illustrates the daily average variation in ambient temperature and relative humidity. During the pre-lockdown period, temperatures were higher compared to during lockdown and the post-lockdown period, such that the highest temperature in the post-lockdown period was 0.1 °C lower than the lowest temperature in the pre-lockdown period. The average temperature of the study periods was found to follow a monotonically decreasing trend, having the highest average of 25.4 °C during the pre-lockdown period. The summer weather for Brisbane starts from December and winter starts from June [21].

A similar trend was observed for wind speed. With an average wind speed of 1.5 m/s, in the during-lockdown period, more frequent winds were from the south and southwest. According to the Bureau of Meteorology, Australia, Brisbane falls under the category of 'light winds' because the maximum daily average wind speed was 4 m/s, even though Brisbane comes under a tropical cyclone risk area [22]. Nonetheless, hourly wind speeds of up to 6.3 m/s were observed in Brisbane during the pre-lockdown and during-lockdown periods, which were mostly southerly winds. For the sudden increment of wind speed, we can consider Brisbane's topography, as well as the pressure-gradient force, the vertical exchange of horizontal momentum in the boundary layer, and the Coriolis force as some of the factors affecting it [23]. In terms of wind speed, the post-lockdown period was relatively calm, and slow and calm wind speeds are due to northerly winds. The statistical comparison of the different meteorological parameters for the during-lockdown period for the years 2016 to 2021 are shown in Table S1. There were no substantial differences in the meteorological variables among years during the lockdown period.

3.2. Analysis of Pollutants for 2020 Events

This section explains the differences in the 24 h average concentrations for the different pollutants in 2020. Table S2 in the Supplementary Materials summarises the changes in different statistical parameters for 24 h concentrations of all pollutants for 2020.

3.2.1. Particulate Matter

The time-series 24 h average PM_{10} and $PM_{2.5}$ concentrations are presented in Figure 4. The average 24 h PM_{10} concentration at Rocklea dropped by 15.8% during the lockdown period compared to the pre-lockdown concentration. Similarly, the 24 h PM_{10} concentration dropped by 8.2%, 3.1%, and 3.1% in Brisbane-CBD, South-Brisbane, and Woolloongabba, respectively. The average 24 h PM_{10} concentration witnessed a reduction of 8.3% during the lockdown period. The maximum 24 h concentration in the pre-lockdown period was 48 $\mu g/m^3$ recorded on 20 February 2020. However, this concentration was the only peak recorded in contrast to the general trend. This could be due to some regional event that resulted in the elevated concentration. Similarly, the during-lockdown period also recorded two minor peaks of 32 and 23 $\mu g/m^3$. The overall air quality levels did not change significantly on removing the peaks due to background events such as dust storms. The average 24 h PM_{10} concentration dropped by 7.5% during the lockdown period compared to the pre-lockdown period on removing the peaks due to background events. Hence, the reduction in the PM_{10} concentration could be due to the restricted transportation traffic and industrial emissions. Similar observations were made in other metropolitan cities around the world [24,25]. In addition to anthropogenic activities, the role of meteorology cannot be ignored. For example, there was a higher frequency of winds from the northeastern quadrant in 2020 during the lockdown period, as shown in Figure S1 (Supplementary Materials). Similarly, there were some notable differences in the frequency of the counts by wind direction for other years, as shown in Figures S2 and S3.

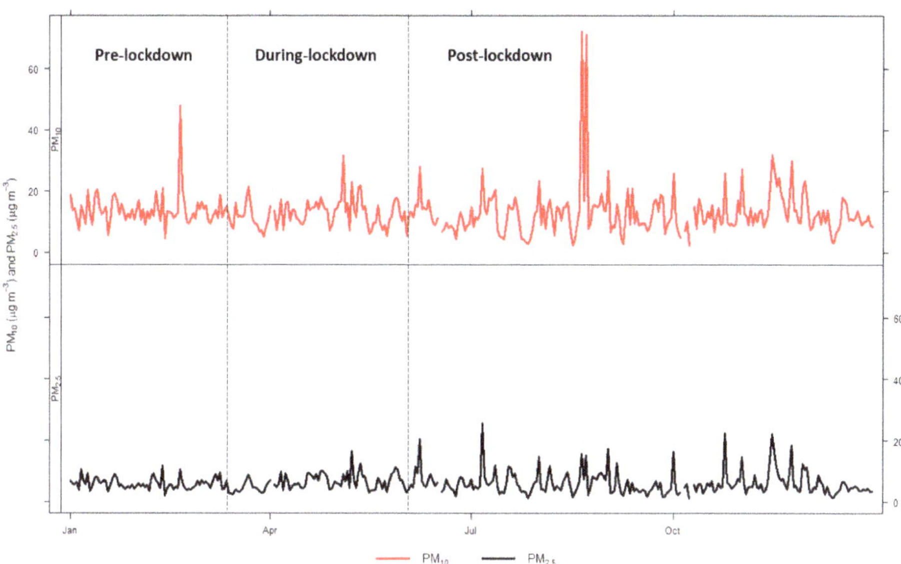

Figure 4. The variation in 24 h PM$_{10}$ and PM$_{2.5}$ concentrations at Rocklea monitoring station for different periods.

In comparison to the pre-lockdown period, the during-lockdown period witnessed an increase of 10% of the average 24 h PM$_{2.5}$ concentration. The 24 h concentration during the post-lockdown period dropped marginally by 0.6% compared to during the lockdown period. The post-lockdown period also observed some peaks such as one of 25.6 µg/m^3 on 7 June 2020, which could have resulted from pyrogenic emissions or other regional events. Corresponding peaks were also observed in the PM$_{10}$ plot. Hence, lockdown had little to no effect on the PM$_{2.5}$ concentration in the Rocklea area. This could be due to several reasons such as regional events, and the fact that vehicle movement was not as prominent as in the rest of Brisbane due to its proximity to light industry. The lockdown restrictions imposed in Brisbane were not like the curfew-like conditions in the rest of the world. Essential industries were allowed to operate as usual. Further, some peaks were noticed in the PM$_{2.5}$ concentration levels during the lockdown period, which could be due to a regional event and pyrogenic emission episodes [26–28].

Therefore, comparison of different periods at other monitoring stations gives a good indication of the impact of restrictions on PM$_{2.5}$ and PM$_{10}$ concentration levels, as shown in Figure 5. The 24 h PM$_{2.5}$ concentration levels dropped by 13.5%, 12.3%, and 5.4% in Brisbane-CBD, South-Brisbane, and Woolloongabba, respectively. In general, Rocklea monitoring station recorded the lowest 24 h PM$_{10}$ concentration and the highest PM$_{2.5}$ concentration compared to the other monitoring stations. However, for the post-lockdown period, Rocklea monitoring station recorded the lowest 24 h PM$_{2.5}$ concentration compared to the other stations.

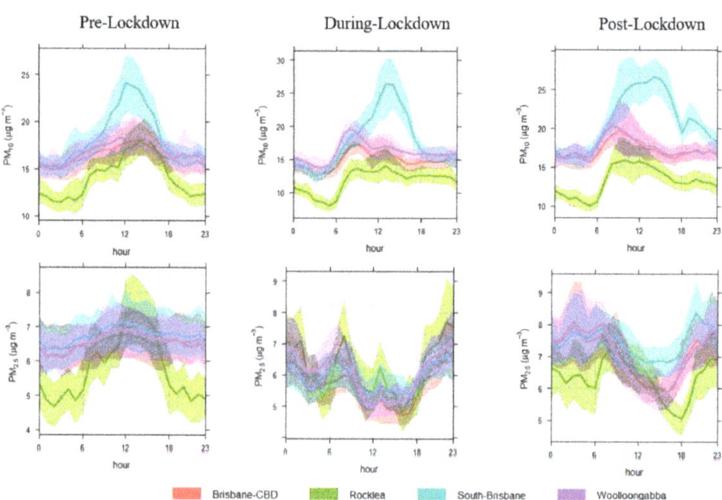

Figure 5. Hourly diurnal profile of PM_{10} and $PM_{2.5}$ concentrations at different monitoring stations during various periods.

3.2.2. Nitrogen Dioxide

Anthropogenic emissions from sources such as the combustion of fossil fuels (coal, gas, and oil), especially fuel used in vehicles and industrial activities, are mainly responsible for increasing level of nitric oxides (90 to 95%) and nitrogen dioxides (5 to 10%) [29–34]. In the atmosphere, hydroperoxyl radicals (HO_2) and organic peroxy radicals (RO_2) react with nitric oxide (NO) to convert it to NO_2; thus, increases in NO emissions lead to increases in NO_2 concentrations [35]. In Brisbane, the lockdown conditions were not similar to the rest of world. They did not restrict local transportation to ensure social distancing, but for maintaining social distancing during travel people were using personal vehicles. Exhaust emissions from personal vehicles could be a major source of nitrogen oxide (NO) that is converted to nitrogen dioxide by atmospheric chemistry. Also, during the post-lockdown period there were 10%, 12%, and 24% increases in the use of private vehicles in January 2021, July 2021, and January 2022, respectively, which could be responsible for the increased levels of NO_2 during the post-lockdown period compared to pre-lockdown conditions [36].

The 24 h average NO_2 concentration level and diurnal trends at Rocklea monitoring station are presented in Figure 6. As observed, the 24 h average concentration shows an increasing trend during the lockdown period. The 24 h average NO_2 concentration increased by 39% during the lockdown period compared to the pre-lockdown period. To further investigate the impact of restrictions, we analysed the changes in NO and NOx concentration levels at Rocklea monitoring station. The diurnal trends of NO, NO_2, and NOx are presented in Supplementary Materials Figure S4. The NO levels increased by 32% during lockdown compared to pre-lockdown levels. The NO levels were relatively similar during the post-lockdown period, i.e., they increased marginally by 0.3% compared to during the lockdown period. This means vehicular movement was higher during lockdown and the post-lockdown period. This is well-supported by public surveys in Australia [36]. Figure 7 shows a comparison of diurnal patterns of NO_2 concentration levels at different monitoring stations. There is a clear similarity in the diurnal profile of NO_2 in the during- and post-lockdown periods. The NO_2 concentration levels at South Brisbane and Woolloongabba monitoring stations were substantially higher than at Rocklea monitoring station. This could be due to the placement of the monitoring stations, as they are close to a major highway. A similar trend was observed in the $PM_{2.5}$ and PM_{10} emissions, as discussed in Section 3.2.1.

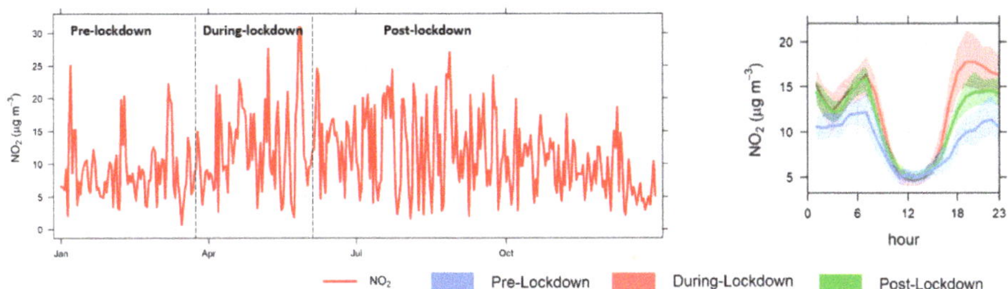

Figure 6. The variation in 24 h concentrations and hourly diurnal profile of NO_2 at Rocklea monitoring station for different periods.

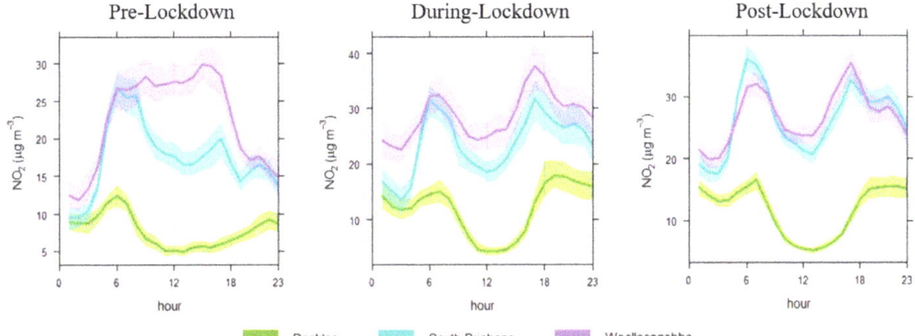

Figure 7. Hourly diurnal profile of NO_2 concentrations at different monitoring stations during various periods.

However, the post-lockdown period observed a decreasing trend as the average 24 h concentration dropped by 10% compared to the during-lockdown period. The NO_2 to NO_x ratio increased by 2% during the lockdown period compared to the pre-lockdown period. On the other hand, the NO_2 to NO_x ratio decreased by 6% during the post-lockdown period. The lower share of NO_2 levels for different periods in NO_x could be related to NO_2 changes due to seasonal variation and the increments in NO concentration levels could be related to more vehicles on the roads [36]. A similar pattern was also observed in Sydney by Duc et al. [37]. The authors mentioned that the variation in NO_2 concentration could be due to the photochemical reaction change involving nitrogen oxides, carbon monoxide, volatile organic compounds, and ozone from Australian summer months (December to February) to Australian autumn months (March to May) and winter months (June to August), which results in the reduction of ambient temperature. As the solar insolation decreased during Australian winters, there was a reduction in photochemical activities, and ozone was rapidly consumed by nitrogen monoxide (NO) and nitrogen dioxide produced [38].

The increasing trend of NO_2 levels is also evident in the diurnal pattern which gives the finer details of the average change in the NO_2 concentration levels at each hour of the day. This type of analysis has been used by several authors to assess the impact of different ambient conditions on the pollutant concentration. The NO_2 concentration levels were typically higher in the morning and then dropped during the mid-day before peaking in the evening at around 18:00 h. This increment in the NO_2 concentration levels was more dominant during the lockdown period.

3.2.3. Ozone

The presence of nitrogen oxides (NO_x), volatile organic compounds (VOCs), and solar irradiation are the cause of the formation of the secondary pollutant ozone [33]. The composition of photochemically aged air advected into a region (urban or rural) and mixed with local emissions has long been recognised as a key control on the production rates of ozone and other photochemical species. For O_3 production, nitrogen dioxide serves as a precursor [24]. The oxidation of NO by O_3, and combustion processes such as industrial boilers, vehicles, and ships are responsible for the direct emission of NO_2, even if there was a restriction related to the pandemic on public transit [25,38,39]. The 24 h average ozone concentration levels and the hourly diurnal profile are presented in Figure 8.

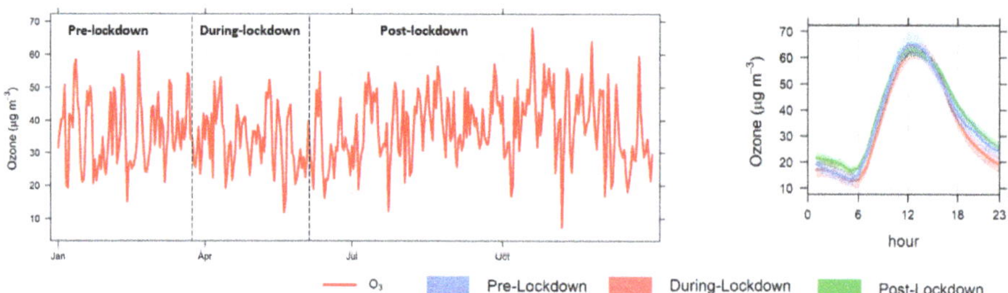

Figure 8. The variation in 24 h concentrations and hourly diurnal profile of ozone at Rocklea monitoring station for different periods.

As observed in Figure 8, the maximum daily concentration in the pre-lockdown period was 60.8 µg/m^3 in addition to some of the other peaks. In contrast, the maximum daily concentration during lockdown and post-lockdown was observed to be 53.1 and 68.1 µg/m^3. The average 24 h concentration dropped by 7.2% during the lockdown period compared to the pre-lockdown period and then increased by 12.4% during the post-lockdown period. On the other hand, the reduction in the median 24 h ozone levels was relatively lower (4.4%).

Similar trends were observed in the diurnal profile (right side of Figure 8). Peak ozone was observed during 10:00 to 16:00 h with the minimum concentration in the morning during 06:00 to 08:00 h. This agrees with the local diurnal temperature profiles and the typical NOx titration from local motor-vehicle-related emissions in the morning. The decreasing trend of ozone levels during the lockdown period could also be attributed to the seasonal variation as higher ozone levels occur in the warmer months (October to March) and peak ozone is usually recorded in January. In addition, the ambient emission concentrations also change seasonally as the solid-fuel heating and temperature-dependent emissions change due to weather changes [40]. Some evidence comes from a similar study conducted on ten major world cities from February to April. Sydney and Perth were included in this study and a comparison was made of February to April 2020 and 2019. The authors observed a reduction in NO_2, CO, and PM in February, March, and April but an increase in O_3 in February and April [37,41]. It is not an unexpected observation that median O_3 levels increased during the lockdown period. Mostly in volatile organic compounds (VOC), a limited regime of ozone formation occurs and, because of reduced levels of NO_x, the photochemistry reaction rate tends to increase and ozone levels also increase [42].

3.2.4. Sulphur Dioxide

The variation in the 24 h average concentration and the hourly diurnal profile are presented in the Figure 9. The pre-lockdown period witnessed a series of high 24 h SO_2 concentration levels during 19 January to 24 January 2020 when the concentration ranged between 10 and 37 µg/m^3. Similarly, several peaks were observed during lockdown and post-lockdown, such as a 24 h average concentration of 18 µg/m^3 on 30 April and 22 µg/m^3 on 31 October 2020. The average 24 h concentration during the lockdown period decreased by 18% whereas a negligible increment of 0.5% was observed during the post-lockdown period. As observed in the diurnal profile of the SO_2 concentration levels (Figure 6, right hand side plot), the SO_2 concentration levels peaked between 16:00 and 19:00 h, which could be due to vehicular movement. It could be inferred that SO_2 levels dropped during the lockdown period due to less traffic activity. The registered peak could be due to the regional wildfire events [43].

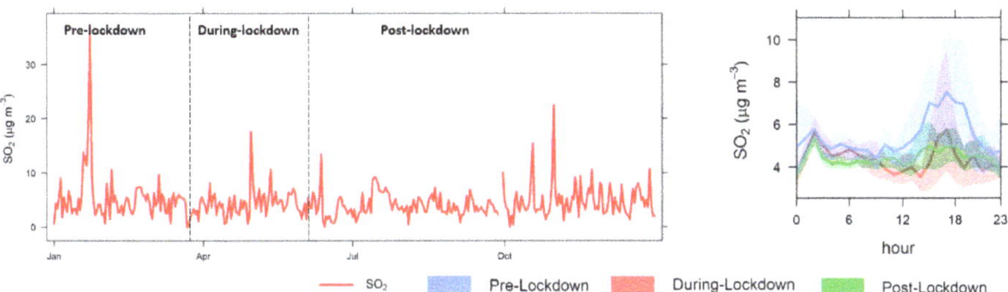

Figure 9. The variation in 24 h concentrations and hourly diurnal profile of SO_2 at Rocklea monitoring station for different periods.

3.3. Annual Trend Analysis of the Pollutants for 2016 to 2019 and 2021

The comparison of pollutant concentrations for one year does not give a very clear idea of the variation in the emissions of the pollutants as they are very much dependent on meteorological parameters [20]. As discussed above in Section 3.2, the changes in the criteria pollutants' concentration, especially NO_2 and O_3, were dependent on the seasonal variation between pre-lockdown and lockdown periods. Thus, comparison of the same period of different years gives more confidence to investigate if lockdown restrictions really affected the changes in the pollutants' concentrations. The comparison of the 24 h average concentration for different pollutants is presented in Figures S5 to S9 in the Supplementary Materials.

Figure 10 presents the comparison of PM_{10}, $PM_{2.5}$, and PM Ratio ($PM_{2.5}/PM_{10}$) for different periods from 2016 to 2021. The extreme high concentrations (24 h level of greater than 150 $\mu g/m^3$ for PM_{10} and greater than 50 $\mu g/m^3$ for $PM_{2.5}$) were excluded for better representation of these plots. As observed, the median values for 24 h PM_{10} concentrations during the pre-lockdown period remained similar for 2018 and 2019 but varied for other years. The post-lockdown period for all years observed greater variability in the median 24 h PM_{10} concentration levels. The year 2019 recorded the highest median value of 18.4 $\mu g/m^3$ whereas other years ranged between 10.4 to 15 $\mu g/m^3$. The higher median 24 h PM_{10} concentration for 2019 could be attributed to the wildfires during the summer of 2019–2020. This is the reason that PM_{10} levels were higher during the overlapping pre-lockdown period in 2020.

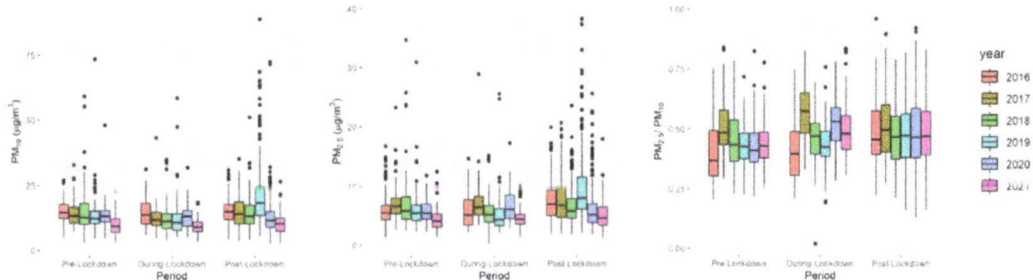

Figure 10. Comparison of PM_{10}, $PM_{2.5}$, and PM Ratio for different periods from 2016 to 2021.

In contrast to PM_{10} observations, the $PM_{2.5}$ trends were higher. The literature research has shown that $PM_{2.5}$ particles represent a main pollutant emitted from wildfire smoke constituting approximately 90% of the total particle mass [44–47]. A higher 24 h average on certain days in 2020 was observed (as shown in Figure S6). The PM ratio (ratio of $PM_{2.5}$

to PM_{10}) was higher during the lockdown period in 2020 compared to other years. This shows that during the lockdown period in 2020 $PM_{2.5}$ was impacted by background events of which the impact of the COVID-19 lockdown restrictions dominated. However, this was not the case for rest of Australia. The greater metropolitan region in New South Wales state observed a 12% reduction in $PM_{2.5}$ during the lockdown period in 2020 compared to 2019 [37].

As observed in Figure 11, the median of the 24 h NO_2 concentration levels during the lockdown period of 2020 was 14% higher than the average of the corresponding median values for 2016 to 2019. This was mainly due to the high-than-average NO_2 concentration during 2017 and 2018. Duc et al. [37] attributed the higher NO_2 concentrations during similar periods of 2017 and 2018 to higher temperatures. However, this was not the case in Brisbane as the median and mean temperature were not significantly different for all years except for the pre- and during-lockdown periods of 2021, as shown in Figure 12.

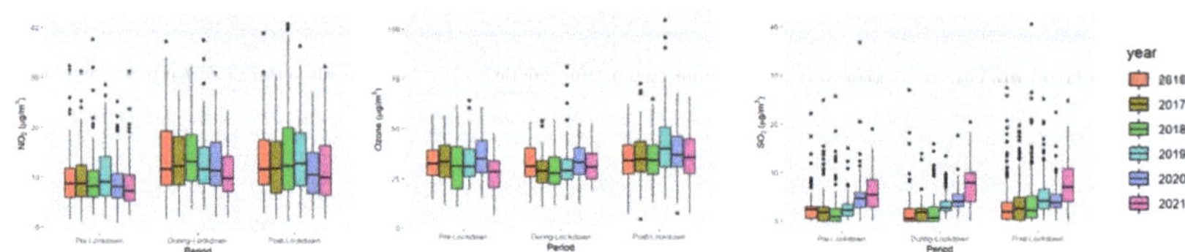

Figure 11. Comparison of NO_2, ozone, and SO_2 for different periods from 2016 to 2021.

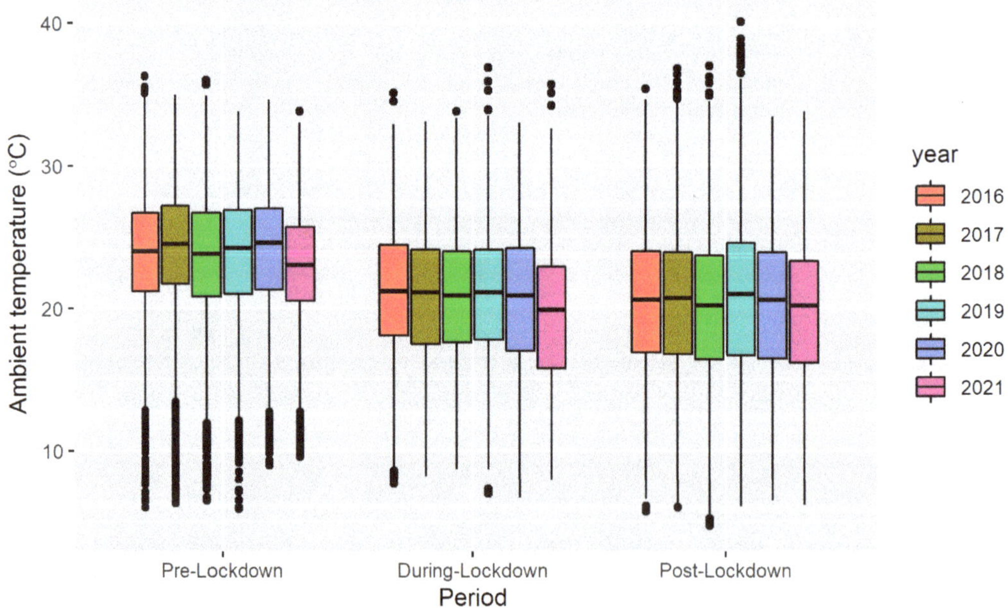

Figure 12. Comparison of ambient temperature for different periods from 2016 to 2021.

On the other hand, the ozone concentrations were higher in the pre-lockdown and lockdown periods of 2020 compared to the similar periods in the preceding years. However, the post-lockdown period in 2020 and 2021 recorded lower median ozone concentration levels compared to 2016 to 2019 levels. This inconsistency in the trend of ozone levels between 2019 and 2020 at different sites could be due to different meteorological conditions, as well as the photochemistry mechanism. The ozone levels were particularly higher in the pre-lockdown period of 2020 and post-lockdown period of 2019 which overlapped with the pyrogenic events and heatwave episodes of 2019–2020 in southeast Australia [48,49]. In contrast to the other pollutants, SO_2 concentration levels were highest in 2021.

4. Conclusions

Pyrogenic emissions such as bushfires, prescribed burns, and agricultural burning along with the COVID-19-related restrictions and post-lockdown behavioural changes have resulted in episodes of substantial changes (positive or negative) in air quality parameters. The various criteria pollutants for different periods in 2020 were analysed and their comparison with the corresponding periods in 2016 to 2019 and 2021 was conducted. The following conclusions can be drawn from this study:

1. The air quality levels during the lockdown period improved; however, some pollutant concentrations were largely impacted by background events.
2. Meteorology also played an important role in the changes in pollutant concentration across different periods. However, across the different years, meteorological conditions were only marginal different. For example, there was a higher frequency of winds from the northeastern quadrant in 2020 during the lockdown period.
3. The average 24 h PM_{10} concentration dropped by 3.1% to 15.8% during the lockdown period compared to the pre-lockdown concentration levels at different monitoring stations.
4. The lockdown period had no impact on the $PM_{2.5}$ concentration levels at Rocklea monitoring station. In comparison to the pre-lockdown period, the during-lockdown period witnessed an increase of 10% in the average 24 h $PM_{2.5}$ concentration. The 24 h concentration during the post-lockdown period dropped marginally by 0.6% compared to during the lockdown period. The $PM_{2.5}$ could be largely impacted by background events such as dust storms and bushfires. The 24 h $PM_{2.5}$ concentration levels dropped by 13.5%, 12.3%, and 5.4% in Brisbane-CBD, South-Brisbane, and Woolloongabba, respectively.
5. The 24 h average NO_2 concentration increased by 39% during the lockdown period compared to the pre-lockdown period. The median of the 24 h NO_2 concentration levels during the lockdown period of 2020 was 14% higher than the average of the corresponding median values for 2016 to 2019. This was mainly due to the higher-than-average NO_2 concentration during 2017 and 2018.
6. The average 24 h SO_2 concentration during the lockdown period decreased by 18%, whereas a negligible increment of 0.5% was observed during the post-lockdown period. The diurnal profile of the SO_2 concentration levels showed an incremental trend in the evening which could be due to vehicular movement.

The implication from COVID-19-related restrictions showed that the improvements in the concentration of criteria pollutants was small but measurable. Natural activities such as bushfires impacted the concentration substantially and contributed to elevated background levels. Hence, due to the location of monitoring stations and contextual natural events, the transition from combustion vehicles to electric or hybrid vehicles may not be enough for reducing pollutant concentrations in the future and biogenic sources such as dust and bushfires also play a critical role in air quality management in urban areas.

Supplementary Materials: https://www.mdpi.com/article/10.3390/pr11010296/s1, Figure S1: Windrose plots for pre-lockdown period for different years, Figure S2: Windrose plots for during-lockdown period for different years, Figure S3: Windrose plots for post-lockdown period for different years, Figure S4: Diurnal trends of NO, NO2 and NOx concentration levels at Rocklea monitoring station for year 2020, Figure S5: Calendar plot for Average 24-h PM10 concentration, Figure S6: Calendar plot for Average 24-h PM2.5 concentration, Figure S7: Calendar plot for Average 24-h NO2 concentration, Figure S8: Calendar plot for Average 24-h Ozone concentration, Figure S9: Calendar plot for Average 24-h SO2 concentration; Table S1. Data Attributes for Meteorological Parameters, Table S2. Statistical Data Attributes for pollutant parameters.

Author Contributions: Conceptualization, S.C., P.V. and A.Z.; Data curation, S.S., S.K.B. and P.V.; Formal analysis, S.C., S.S., S.K.B. and A.Z.; Methodology, S.C.; Project administration, G.D. and P.V.; Resources, G.D.; Software, S.C.; Supervision, G.D. and P.V.; Writing—original draft, S.C.; Writing—review & editing, G.D., A.Z. and P.V. All authors have read and agreed to the published version of the manuscript.

Funding: This research received no external funding.

Institutional Review Board Statement: Not applicable.

Informed Consent Statement: Not applicable.

Data Availability Statement: The data for this study can be accessed from the Queensland Government website (https://www.data.qld.gov.au, last accessed on 15 December 2022).

Acknowledgments: The authors duly acknowledge the Queensland Government Department of Environment and Science for maintaining the air quality monitoring stations and Don Neale providing all of the datasets used in this study.

Conflicts of Interest: The authors declare no conflict of interest.

References

1. Tiwari, A.; Williams, I. *Air Pollution: Measurement, Modelling and Mitigation, Fourth Edition*, 4th ed.; CRC Press: Boca Raton, FL, USA, 2018; Volume 13, 33487-2742; ISBN 978-1138503663.
2. Whitacre, P. Air Pollution Accounts for 1 in 8 Deaths Worldwide, According to New WHO Estimates. *WHO News Release*, 25 March 2014.
3. GBD MAPS Working Group. Burden of Disease Attributable to Major Air Pollution Sources in India—Health Effects Institute. 2018. Available online: https://www.healtheffects.org/publication/gbd-air-pollution-india (accessed on 1 May 2021).
4. Wong, Y.J.; Shiu, H.-Y.; Chang, J.H.-H.; Ooi, M.C.G.; Li, H.-H.; Homma, R.; Shimizu, Y.; Chiueh, P.-T.; Maneechot, L.; Nik Sulaiman, N.M. Spatiotemporal Impact of COVID-19 on Taiwan Air Quality in the Absence of a Lockdown: Influence of Urban Public Transportation Use and Meteorological Conditions. *J. Clean. Prod.* **2022**, *365*, 132893. [CrossRef] [PubMed]
5. Gautam, A.S.; Dilwaliya, N.K.; Srivastava, A.; Kumar, S.; Bauddh, K.; Siingh, D.; Shah, M.A.; Singh, K.; Gautam, S. Temporary Reduction in Air Pollution Due to Anthropogenic Activity Switch-off during COVID-19 Lockdown in Northern Parts of India. *Environ. Dev. Sustain.* **2021**, *23*, 8774–8797. [CrossRef] [PubMed]
6. Jain, S.; Sharma, T. Social and Travel Lockdown Impact Considering Coronavirus Disease (COVID-19) on Air Quality in Megacities of India: Present Benefits, Future Challenges and Way Forward. *Aerosol Air Qual. Res.* **2020**, *20*, 1222–1236. [CrossRef]
7. Hu, M.; Chen, Z.; Cui, H.; Wang, T.; Zhang, C.; Yun, K. Air Pollution and Critical Air Pollutant Assessment during and after COVID-19 Lockdowns: Evidence from Pandemic Hotspots in China, the Republic of Korea, Japan, and India. *Atmos. Pollut. Res.* **2021**, *12*, 316–329. [CrossRef] [PubMed]
8. Mor, S.; Kumar, S.; Singh, T.; Dogra, S.; Pandey, V.; Ravindra, K. Impact of COVID-19 Lockdown on Air Quality in Chandigarh, India: Understanding the Emission Sources during Controlled Anthropogenic Activities. *Chemosphere* **2021**, *263*, 127978. [CrossRef]
9. Gouda, K.C.; Singh, P.; Benke, M.; Kumari, R.; Agnihotri, G.; Hungund, K.M. Assessment of Air Pollution Status during COVID-19 Lockdown (March–May 2020) over Bangalore City in India. *Environ. Monit. Assess.* **2021**, *193*, 395. [CrossRef]
10. Queensland Government Business Closures and Restrictions. Available online: https://statements.qld.gov.au/statements/89582 (accessed on 15 October 2022).
11. Van Der Werf, G.R.; Randerson, J.T.; Collatz, G.J.; Giglio, L.; Kasibhatla, P.S.; Arellano, A.F.; Olsen, S.C.; Kasischke, E.S. Continental-Scale Partitioning of Fire Emissions During the 1997 to 2001 El Niño/La Niña Period. *Science* **2004**, *303*, 73–76. [CrossRef]
12. Castagna, J.; Senatore, A.; Bencardino, M.; D'Amore, F.; Sprovieri, F.; Pirrone, N.; Mendicino, G. Multiscale Assessment of the Impact on Air Quality of an Intense Wildfire Season in Southern Italy. *Sci. Total Environ.* **2021**, *761*, 143271. [CrossRef]
13. Queensland Government State of the Environment Report. Key Messages. 2021. Available online: https://www.stateoftheenvironment.des.qld.gov.au/pollution/air-quality (accessed on 6 May 2022).
14. Deng, M.; Lai, G.; Li, Q.; Li, W.; Pan, Y.; Li, K. Impact Analysis of COVID-19 Pandemic Control Measures on Nighttime Light and Air Quality in Cities. *Remote Sens. Appl.* **2022**, *27*, 100806. [CrossRef]
15. Rodríguez-Urrego, D.; Rodríguez-Urrego, L. Air Quality during the COVID-19: PM2.5 Analysis in the 50 Most Polluted Capital Cities in the World. *Environ. Pollut.* **2020**, *266*, 115042. [CrossRef]
16. Sánchez-García, E.; Leon, J. These 5 Images Show How Air Pollution Changed over Australia's Major Cities before and after Lockdown. 2020. Available online: https://theconversation.com/these-5-images-show-how-air-pollution-changed-over-australias-major-cities-before-and-after-lockdown-136723 (accessed on 6 December 2022).

17. Brisbane City Council. Planned Burns. 2023. Available online: https://www.brisbane.qld.gov.au/community-and-safety/community-safety/disasters-and-emergencies/types-of-disasters/busfires/plannedburns#:~{}:text=Most%20planned%20burns%20in%20Brisbane,day%20for%20each%20planned%20burn (accessed on 6 December 2022).

18. Latitude. GPS Coordinates of Brisbane, Australia. Available online: https://latitude.to/articles-by-country/au/australia/452/brisbane (accessed on 30 May 2021).

19. Australian Bureau of Statistics Regional Population. Statistics about the Population for Australia's Capital Cities and Regions. Available online: https://www.abs.gov.au/statistics/people/population/regional-population/latest-release (accessed on 15 October 2022).

20. Elminir, H.K. Dependence of Urban Air Pollutants on Meteorology. *Sci. Total Environ.* **2005**, *350*, 225–237. [CrossRef]

21. Tourism Australia. Weather in Australia. Available online: https://www.australia.com/en-in/facts-and-planning/weather-in-australia.html (accessed on 9 May 2021).

22. Marine Knowledge Centre Waves. Reference Material. Available online: http://www.bom.gov.au/marine/knowledge-centre/reference/wind.shtml (accessed on 9 May 2021).

23. Wu, J.; Zha, J.; Zhao, D.; Yang, Q. Changes in Terrestrial Near-Surface Wind Speed and Their Possible Causes: An Overview. *Clim. Dyn.* **2018**, *51*, 2039–2078. [CrossRef]

24. Xu, K.; Cui, K.; Young, L.-H.; Hsieh, Y.-K.; Wang, Y.-F.; Zhang, J.; Wan, S. Impact of the COVID-19 Event on Air Quality in Central China. *Aerosol Air Qual. Res.* **2020**, *20*, 915–929. [CrossRef]

25. Collivignarelli, M.C.; Abbà, A.; Bertanza, G.; Pedrazzani, R.; Ricciardi, P.; Carnevale Miino, M. Lockdown for CoViD-2019 in Milan: What Are the Effects on Air Quality? *Sci. Total Environ.* **2020**, *732*, 139280. [CrossRef]

26. Van Oldenborgh, G.J.; Krikken, F.; Lewis, S.; Leach, N.J.; Lehner, F.; Saunders, K.R.; Van Weele, M.; Haustein, K.; Li, S.; Wallom, D.; et al. Attribution of the Australian Bushfire Risk to Anthropogenic Climate Change. *Nat. Hazards Earth Syst. Sci.* **2021**, *21*, 941–960. [CrossRef]

27. Rahman, R.; Zafarullah, H. Impact of Climate Change on Human Health: Adaptation Challenges in Queensland, Australia. *Clim. Res.* **2020**, *80*, 59–72. [CrossRef]

28. Qi, F.; Zhou, Y.; Feng, S. Strengthening Destinations' Resilience from Bushfires—A Study of Eastern Australia. *J. Manag. Sustain.* **2021**, *11*, 43. [CrossRef]

29. Shah, V.; Jacob, D.J.; Li, K.; Silvern, R.F.; Zhai, S.; Liu, M.; Lin, J.; Zhang, Q. Effect of Changing NOx Lifetime on the Seasonality and Long-Term Trends of Satellite-Observed Tropospheric NO2 Columns over China. *Atmos. Chem. Phys.* **2020**, *20*, 1483–1495. [CrossRef]

30. Queensland Government. Nitrogen Oxides. 2013. Available online: https://www.qld.gov.au/environment/management/monitoring/air/air-pollution/pollutants/nitrogen-oxides (accessed on 29 December 2022).

31. United States Environmental Protection Agency. Basic Information about NO2. 2022. Available online: https://www.epa.gov/no2-pollution/basic-information-about-no2#What%20is%20NO2 (accessed on 29 December 2022).

32. Jarvis, D.J.; Adamkiewicz, G.; Heroux, M.-E.; Rapp, R.; Kelly, F.J. Nitrogen dioxide. 2010. Available online: https://www.ncbi.nlm.nih.gov/books/NBK138707/ (accessed on 29 December 2022).

33. Davis, M.L.; Cornwell, D.A. *Introduction to Environmental Engineering*; McGraw-Hill: New York, NY, USA, 2008.

34. Goldberg, D.L.; Anenberg, S.C.; Griffin, D.; McLinden, C.A.; Lu, Z.; Streets, D.G. Disentangling the Impact of the COVID-19 Lockdowns on Urban NO2 From Natural Variability. *Geophys. Res. Lett.* **2020**, *47*, e2020GL089269. [CrossRef]

35. Stockwell, W.R.; Lawson, C.V.; Saunders, E.; Goliff, W.S. A Review of Tropospheric Atmospheric Chemistry and Gas-Phase Chemical Mechanisms for Air Quality Modeling. *Atmosphere* **2012**, *3*, 1–32. [CrossRef]

36. Transurban. Urban Mobility Trends from COVID-19. 2022. Available online: https://www.transurban.com/content/dam/transurban-pdfs/03/Mobility-Trends-Report-1H22.pdf (accessed on 29 December 2022).

37. Duc, H.; Salter, D.; Azzi, M.; Jiang, N.; Warren, L.; Watt, S.; Riley, M.; White, S.; Trieu, T.; Tzu-Chi Chang, L.; et al. The Effect of Lockdown Period during the COVID-19 Pandemic on Air Quality in Sydney Region, Australia. *Int. J. Environ. Res. Public Health* **2021**, *18*, 3528. [CrossRef] [PubMed]

38. Zhang, C.; Stevenson, D. Characteristic Changes of Ozone and Its Precursors in London during COVID-19 Lockdown and the Ozone Surge Reason Analysis. *Atmos. Environ.* **2022**, *273*, 118980. [CrossRef] [PubMed]

39. Pour-Biazar, A.; McNider, R.T.; Roselle, S.J.; Suggs, R.; Jedlovec, G.; Byun, D.W.; Kim, S.; Lin, C.J.; Ho, T.C.; Haines, S.; et al. Correcting Photolysis Rates on the Basis of Satellite Observed Clouds. *J. Geophys. Res. Atmos.* **2007**, *112*, 1–17. [CrossRef]

40. Riley, M.L.; Watt, S.; Jiang, N. Tropospheric Ozone Measurements at a Rural Town in New South Wales, Australia. *Atmos. Environ.* **2022**, *281*, 119143. [CrossRef]

41. Habibi, H.; Awal, R.; Fares, A.; Ghahremannejad, M. COVID-19 and the Improvement of the Global Air Quality: The Bright Side of a Pandemic. *Atmosphere* **2020**, *11*, 1279. [CrossRef]

42. Wang, Y.; Wen, Y.; Wang, Y.; Zhang, S.; Zhang, K.M.; Zheng, H.; Xing, J.; Wu, Y.; Hao, J. Four-Month Changes in Air Quality during and after the COVID-19 Lockdown in Six Megacities in China. *Environ. Sci. Technol. Lett.* **2020**, *7*, 802–808. [CrossRef]

43. Ditto, J.C.; He, M.; Hass-Mitchell, T.N.; Moussa, S.G.; Hayden, K.; Li, S.-M.; Liggio, J.; Leithead, A.; Lee, P.; Wheeler, M.J.; et al. Atmospheric Evolution of Emissions from a Boreal Forest Fire: The Formation of Highly Functionalized Oxygen-, Nitrogen-, and Sulfur-Containing Organic Compounds. *Atmos. Chem. Phys.* **2021**, *21*, 255–267. [CrossRef]

44. Mahato, S.; Pal, S.; Ghosh, K.G. Effect of Lockdown amid COVID-19 Pandemic on Air Quality of the Megacity Delhi, India. *Sci. Total Environ.* **2020**, *730*, 139086. [CrossRef]
45. Vicente, A.; Alves, C.; Monteiro, C.; Nunes, T.; Mirante, F.; Evtyugina, M.; Cerqueira, M.; Pio, C. Measurement of Trace Gases and Organic Compounds in the Smoke Plume from a Wildfire in Penedono (Central Portugal). *Atmos. Environ.* **2011**, *45*, 5172–5182. [CrossRef]
46. Vicente, A.; Alves, C.; Monteiro, C.; Nunes, T.; Mirante, F.; Cerqueira, M.; Calvo, A.; Pio, C. Organic Speciation of Aerosols from Wildfires in Central Portugal during Summer 2009. *Atmos. Environ.* **2012**, *57*, 186–196. [CrossRef]
47. Vicente, A.; Alves, C.; Calvo, A.I.; Fernandes, A.P.; Nunes, T.; Monteiro, C.; Almeida, S.M.; Pio, C. Emission Factors and Detailed Chemical Composition of Smoke Particles from the 2010 Wildfire Season. *Atmos. Environ.* **2013**, *71*, 295–303. [CrossRef]
48. Bureau of Meteorology. Special Climate Statement 73—Extreme Heat and Fire Weather in December 2019 and January 2020. 2020. Available online: http://www.bom.gov.au/climate/current/statements/scs73.pdf (accessed on 5 December 2022).
49. Bureau of Meteorology. Annual Climate Statement 2020. 2020. Available online: http://www.bom.gov.au/climate/current/annual/aus/2020/ (accessed on 5 December 2022).

processes

Article

Comparison Process of Blood Heavy Metals Absorption Linked to Measured Air Quality Data in Areas with High and Low Environmental Impact

Nicoletta Lotrecchiano [1,2], Luigi Montano [3,4,*], Ian Marc Bonapace [5], Tenore Giancarlo [6], Paolo Trucillo [7,8] and Daniele Sofia [1,2,8,*]

1   DIIN-Department of Industrial Engineering, University of Salerno, Via Giovanni Paolo II, 132, 84084 Fisciano, Italy; nlotrecchiano@unisa.it
2   Research Department, Sense Square Srl, 84084 Salerno, Italy
3   Andrology Unit and Service of Lifestyle Medicine in UroAndrology, Local Health Authority (ASL) Salerno, Coordination Unit of the Network for Environmental and Reproductive Health (EcoFoodFertility Project), Oliveto Citra Hospital, 84020 Oliveto Citra, Italy
4   Ph.D. Program in Evolutionary Biology and Ecology, Department of Biology, University of Rome Tor Vergata, 00133 Rome, Italy
5   Department of Biotechnology and Life Sciences, University of Insubria, 21100 Varese, Italy; ian.bonapace@uninsubria.it
6   NutraPharmaLabs, Department of Pharmacy, University of Naples Federico II, Via Domenico Montesano 49, 80131 Naples, Italy; giancarlo.tenore@unina.it
7   Department of Chemical, Material and Industrial Production Engineering, University of Naples Federico II, Piazzale V. Tecchio, 80, 80125 Naples, Italy; paolo.trucillo@unina.it
8   IODO Srl, Corso Vittorio Emanuele, 143, 84123 Salerno, Italy
*   Correspondence: l.montano@aslsalerno.it (L.M.); dsofia@unisa.it (D.S.)

Citation: Lotrecchiano, N.; Montano, L.; Bonapace, I.M.; Giancarlo, T.; Trucillo, P.; Sofia, D. Comparison Process of Blood Heavy Metals Absorption Linked to Measured Air Quality Data in Areas with High and Low Environmental Impact. *Processes* 2022, 10, 1409. https://doi.org/10.3390/pr10071409

Academic Editor: Olivier Sire

Received: 28 June 2022
Accepted: 18 July 2022
Published: 19 July 2022

Publisher's Note: MDPI stays neutral with regard to jurisdictional claims in published maps and institutional affiliations.

Abstract: Air pollution is a problem shared by the entire world population, and researchers have highlighted its adverse effects on human health in recent years. The object of this paper was the relationship between the pollutants' concentrations measured in the air and the quantity of pollutant itself inhaled by the human body. The area chosen for the study has a high environmental impact given the significant presence on the territory of polluting activities. The Acerra area (HI) has a waste-to-energy plant and numerous industries to which polluting emissions are attributed. This area has always been the subject of study as the numbers of cancer patients are high. A survey on male patients to evaluate the heavy metals concentrations in the blood was conducted in the two areas and then linked to its values aero-dispersed. Using the air quality data measured by the monitoring networks in two zones, one with high environmental impact (HI) and one with low environmental impact (LI), the chronicle daily intake (CDI) of pollutants inhaled by a single person was calculated. The pollutants considered in this study are PM10 and four heavy metals (As, Cd, Ni, Pb) constituting the typical particulates of the areas concerned. The CDI values calculated for the two zones are significantly higher in the HI zone following the seasonal pollution trend.

Keywords: PM10; chronicle daily intake; pollution absorption; pollution damage; heavy metals; ecofoodfertility; land of fires

1. Introduction

The containment of pollution and its harmful effects on human health is an emergence of public health worldwide. The World Health Organization (WHO) estimates that about a quarter of the diseases occurring today are caused by prolonged exposure to environmental risks/pollutants [1]. In particular, air pollution is among the top ten global health risk factors that can lead to premature mortality [2]. Epidemiological cohort studies, mainly conducted in the United States and Europe, have shown that long-term exposure to PM2.5 (particles with an aerodynamic diameter less than 2.5 µm) is associated with increased mortality from respiratory diseases, cardiovascular and lung cancer [2–7]. For this reason, in environmental epidemiology, there has been a progressive increase in the development and application of human biomonitoring as a tool for individual exposure assessment and susceptibility and the association between pollutants and early damage. The search for biological indicators in tissues or body fluids has proven reliable for assessing exposure based on environmental measures. It represents a fundamental approach to the characterization and management of health risks. Risk assessment is better defined by direct measurement of biomarkers of exposure and effect in body tissues and fluids versus extrapolation of pollutant concentration data in soil, air, and water, reflecting actual individual exposure to specific pollutants. The direct measure of exposure and the effect of biomarkers in tissues and body fluids is much more relevant to the risk assessment than the extrapolation from the concentration of pollutants in soil, air, and water, reflecting the real individual exposure to that particular pollutant.

Air quality analysis is essential to determine the pollutants' concentration and the impact on the surrounding environment [8]. The effects of airborne pollutants purged into the atmosphere over years of human activities are evident on land, water, food, and human health [9]. According to the literature and the most capable worldwide monitoring systems, the major pollutants present in the atmosphere are gas and particulate matter (PM). NO_x, SO_x, O_3, H_2S, and volatile organic compounds (VOCs) are the most detected gases. Atmospheric particulate matter is a set of solid and liquid particles with various physical, chemical, geometric, and morphological characteristics [10]. It is considered a good indicator of air quality [11]. The sources can be natural (soil erosion, marine spray, volcanoes, forest fires, pollen dispersion, etc.) or anthropogenic (industries, heating, vehicular traffic, and combustion processes). They are generally classified according to their size: particles with a diameter less than 10 μm are identified with the abbreviation of PM10, those with a diameter less than 2.5 μm as PM2.5, and those with a diameter less than 1 μm as PM1. In particular, PM10 (thoracic fraction) can pass through the nose and reach the throat and trachea. Smaller particles such as PM2.5 represent the inhalable fraction and can penetrate even deeper into the lungs, while particles with a smaller diameter can reach the pulmonary alveoli. Particulate matter, considered a good indicator of air quality [11], is made up of a set of solid particles of different nature, chemical composition, and size; it varies from city to city according to the degree of development of the urban center and the presence of industries, fuels used and type of domestic heating. The atmospheric particulate remains in the air from 5 to 16 h [12] and can be transported long distances. Atmospheric phenomena such as wind and rain can help dilute and reduce PM concentration in the air. Particulates emitted into the atmosphere directly from the source refers to the primary type, while the chemical–physical transformations of other substances are the secondary type. The major components of atmospheric particulate matter are sulfate, nitrate, ammonia, sodium chloride, carbon, and mineral powders. Numerous chemical substances, such as polycyclic aromatic hydrocarbons (PAHs) and metals (such as lead, nickel, cadmium, arsenic, vanadium, and chromium), can adhere to the surface of fine dust causing effects on the health of the exposed population. PM causes several health effects, including many respiratory-related ailments [13,14]. The International Agency for Research on Cancer (IARC) has classified air pollution in Group 1, which is characterized by carcinogenic substances for humans. The World Health Organization has defined guidelines to establish concentration limits to control airborne pollutants. Each country has implemented these guidelines with national directives: in Italy, the pollutants' concentrations are regulated by Legislative Decree 155/2010. This decree limits all pollutants' annual, daily, and hourly concentrations. Among the heavy metals included in the legislation, we find arsenic (As), cadmium (Cd), nickel (Ni), and lead (Pb), and their annual limits are 6 ng/m^3, 5 ng/m^3, 20 ng/m^3, 0.5 μg/m^3, respectively. The toxicity of these metals on soil, water, and human health are well known and extensively studied in the scientific literature [15].

Airborne heavy metals can enter the body through the skin, by inhalation, or ingestion of contaminated water or food. Toxicity derives from the replacement of elements ordinarily present in the body (blood, urine, and hair) with heavy metals and the alteration of the normal functions of the various organs. In most cases, people are not exposed to quantities of heavy metals that would trigger symptoms or require a test. Most chronic or acute exposures have an occupational origin, particularly in the case of manufacturing industries that use metals or produce batteries containing cadmium, lead, and mercury, or pesticides, containing arsenic or may originate in places with high levels of environmental pollution [16]. Cd is very mobile in the soil and is absorbed by microorganisms that absorb organic matter from the soil. It has been suggested that exposure to Cd doses has adverse effects on both human male and female reproduction and affects pregnancy or its outcome. Exposure to Cd affects human male reproductive organs/system and deteriorates spermatogenesis and semen quality, especially sperm motility and hormonal synthesis/release [17]. The Pb accumulation can negatively affect the soil pH and absorption capacity [18]. Heavy metals can also affect the human reproductive system: cadmium and lead act on the reproductive system, damaging it and limiting its correct functionality [19]. Cadmium and lead are present, albeit in minimal quantities, in all foodstuffs that are ingested and contribute to increasing the organism's toxicity. Therefore, these heavy metals are introduced into the human body by inhalation due to the fraction present in the atmosphere and by ingestion due to their presence in food, thus increasing the risk associated with them [17]. For example, high levels of cadmium can adversely affect kidney functions, causing decompensation due to its malfunction [20]. Arsenic negatively affects the digestive, urinary, cardiovascular, and DNA methylation systems [21]. Generally, the need for nickel is completely satisfied by the diet, but an excess can accumulate in the liver, kidneys, bones, and aorta. Excessive nickel intake has also been associated with an increased carcinogenic risk, such as lung and prostate cancers, heart attack, and stroke [22].

To verify the effects of environmental pollution on human health, it is necessary to have detailed information on the concentrations of the major airborne pollutants. National monitoring networks are available to measure the pollutants' concentrations in the air. Given their characteristics, the technologies implemented by these entities have a low spatial and temporal resolution. Recently, sensors for low-cost air quality monitoring based on IoT (Internet of Things) technology have been developed. This has allowed the implementation of air quality monitoring networks with high spatial and temporal resolution [23] that enable the detection of particulate matter of different sizes (PM10, PM2.5) and the measurement of the concentration of gases in the atmosphere (SO_2, NO_2, CO, O_3, VOC), reliably protecting data integrity [24]. The new type of intelligent measuring device could be easily installed in many parts of the city following an optimization in the choice of places [25]. The available air quality data are the basis for developing mathematical models for data spatialization [26] and forecasting [27]. Furthermore, dispersion models support experimental measures to define the behavior of pollutants in the atmosphere and verify their long-term effects in case of accidental occurrence [28]. Monitoring the number of dispersed pollutants and the implementation of long-term strategies to favor their reduction is necessary to safeguard the health [29–31] of resident populations and travelers [32]. Outdoor air quality can be reflected inside buildings, highlighting the importance of monitoring pollution levels to implement filtering and ventilation systems. This is especially important for home settings or other locations frequented by vulnerable individuals, for example, schools and elderly care centers [33–35]. Comparing the data from environmental pollution and exposure to chemical compounds with the actual contamination values found in the organism can verify their consistency and deterministic association. This will also help to translate the degree of exposure from the environment to the organism as a fundamental step in understanding the environment/health relationship. Measuring these substances and the resulting biological effect and establishing a close dose–effect relationship between them represents the right path to sharpen biological risk assessment.

The Environmental Protection Agency defines exposure as the contact of an organism (humans in the case of health risk assessment) with a chemical or physical agent [36]. Exposure involves any interaction, internal (e.g., respiratory tract) or external (e.g., skin) of the human body, with a contaminant present in the atmosphere. This phenomenon requires that, at the same time, there is the presence of a human subject and high levels of pollutants in a particular time and space [37]. It is quantitatively expressed by the duration of the contact and the relevant pollutant concentration and can be assessed using air quality data provided by air monitoring devices [38]. As defined by Asante and Duah, chronic daily intake (CDI) measures long-term (chronic) exposures. It is based on the number of events assumed to occur within an assumed lifetime for potential receptors. The intake value defines the amount of a chemical absorbed during the exposure. In contrast, the absorbed chemical quantity into the bloodstream during exposure is represented by the dose and is calculated by considering physiological parameters. In general, the physiological parameters are not readily available. Intake and dose are considered the same (i.e., a 100 percent bloodstream absorption from contact is assumed) [39]. In this perspective, also considering the study participants' characteristics, the CDI was calculated for each metal compared with the blood levels. In this work, air quality data measured by the regional monitoring network were used to calculate the absorption of inhaled pollutants in areas with high and low environmental impact in the Campania region (Southern Italy). The so-called "Land of Fires" in the northern part of the Naples metropolitan area is characterized by a multiplicity of pollution sources (illegal disposal of urban, toxic, and industrial wastes, dumping practices, traffic, and intensive agriculture). It is considered an area of high environmental impact. In comparison, the area with low environmental impact is Oliveto Citra in the Province of Salerno [40]. Several serious concerns for general health, including reproductive ones, in "Land of Fires" were reported. Bio-monitoring studies indicated that trace toxics elements (e.g., Cr, Cd, Ni, Pb, Al) feature prominently in this area [41,42], including dioxins [43]. In response to such serious concerns, the EcoFoodFertility initiative [44] has been launched to better understand the environmental impact of toxicants on healthy humans [42,45]. It is a multicenter, multidisciplinary research connecting human lifestyle and dietary habits to the environmental consequences of exposure to toxicants. In the frame of this project, the present paper aims to link the heavy metals in human blood compared with the pollutant concentrations that are aero-dispersed. The measured PM10 and heavy metals concentration was used to calculate the chronicle daily intake (CDI) over a year. Using the average population characteristics, the CDI equation parameters were set. The reference year for the measurements was 2016. The calculated CDI values for PM10 and the heavy metals were compared considering the different impact areas (HI and LI).

2. Materials and Methods

2.1. Participants

The biomonitoring study was carried out in accordance with the Code of Ethics of the World Medical Association [46], approved by the Ethical Committee of the Local Health Authority Campania Sud-Salerno (Committee code 43/2015/06). The recruitment of all participants within the fertile age (20–40 years old) across the two cities of the Campania Region (Acerra in the Province of Naples and Oliveto Citra in the Province of Salerno, as explained in Section 2.3, are representative of high and low environmental impact areas) was conducted during 2016. Enrolment criteria were as follows: residence for at least 10 years in the study area, no known chronic diseases (diabetes or other systemic diseases), no varicocele, no prostatitis, and other factors that could affect semen quality (such as fever, medications, exposure to X rays, etc.), no reported history of drug abuse and no known occupational exposures to toxic chemicals. The sample's representativeness was defined in relation to the numerous biomonitoring in these two areas, which have led to the carrying out of various studies [47–51]. Data were collected by questionnaire and physical examination, including the urogenital evaluation (testis volume and transrectal prostate evaluation). Informed consent was obtained from all males before sample collection. The mean age of participants from the low impact area was 30 years, and the Body mass index (BMI) was 24.6, while for the participants living in the high impact area, the age and the BMI values were 28 and 25, respectively. Upon enrolment, a code number (1, 2, 3 ... n) was assigned to each volunteer by the recruiting andrologist (the recruiter). Each code number was uploaded into a computer database along with personal and clinical information (e.g., age, BMI, area of residence). The examining andrologist (the evaluator, different from the recruiter) performed semen quality evaluation, only having access to the code number assigned to each sample. The participants' characteristics are reported in Table 1.

Table 1. Participants' average characteristics.

	LI Area	HI Area
Number of male participants	35	40
Age, years	30	28
Weight, kg	78	77
Height, m	1.77	1.76
BMI	24.6	25

2.2. Heavy Metals Procedure to Assess the Blood Concentration

Blood samples were collected from each subject by veni-puncture. Aliquots (500 µL) of whole blood or semen contained in metal-free, no EDTA-containing tubes were placed onto a ceramic evaporating dish and heated at 90 °C for 1 hr. After cooling, 1 mL HNO_3 68% (Sigma Aldrich—Darmstadt, Germany) and 250 µL $HClO_4$ (Sigma Aldrich, Darmstadt, Germany) were added, and the mixture was heated at 150 °C until carbonization. Samples were placed in a muffle oven at 550 °C for 5 hrs, and the resulting ashes were dissolved in 10 mL 1% HNO_3. The elemental analyses of whole blood of trace elements were carried out by Inductively Coupled Plasma-Optical Emission Spectrometer (ICP-OES) by an iCAP™ 7000 Series ICP-OES (Thermo Fisher Scientific, Waltham, MA, USA). To prevent interference with calibration solutions, a matrix modifier was added to 100 µg/L yttrium solution (Merck KGaA, Darmstadt, Germany). Elemental contents were calculated using standard curves, and the final concentrations were expressed as mg/L. A stock solution (ICP multi-element standard solution Certipur®, 1000 mg/L, Merck KGaA, Darmstadt, Germany) was used to prepare the standard solutions. The limits of detection (LOD) and quantification (LOQ) were calculated as the blank signal plus three or ten times its standard deviation, respectively [42].

2.3. Air Quality

The areas considered in this study are two southern Italian cities of middle to small dimensions, which can be regarded as high and low impact areas. The High Environmental Impact area (HI) is Acerra, in the Metropolitan City of Naples, Italy, where a waste treatment plant has burnt 600,000 tons yearly since 2009. The study by Senior and Mazza underlined that cancer-related deaths in that area are much higher than the Italian average [52]. Moreover, the city is densely populated and is surrounded by a fast road that is highly trafficked. As a reference for the Low Environmental Impact area (LI), the city of Oliveto Citra (in the Southeast of the Salerno province, Italy) was chosen (Figure 1). The city, located in an area of relatively low human density, is characterized by a small industrial area. Due to the peculiarity of the areas investigated, the particulate matter PM10 (aero-dispersed particles with diameter < 10 µm) concentration was considered for analysis in 2016. The average PM10 particulate concentration data for 2016 were taken from the daily validated database of the Regional Environmental Protection Agency for the Campania Region (ARPAC). In the HI area, the ARPAC regional air quality monitoring network consisted of 3 measuring devices: Acerra Scuola (AS), Acerra Capasso (AC), and Acerra Industriale, classified as suburban, industrial, and urban traffic stations. The average of the three measuring devices was considered for the comparison with the LI area. In the LI area, there is no ARPAC air quality monitoring station. However, due to the similarity of both the orographic and the urban and extra-urban context, the population density, and the number of industrial activities, the nearby ARPAC monitoring station (OL) located in Battipaglia (province of Salerno) was considered. It can therefore be assumed that the

PM10 concentrations detected in Battipaglia are similar to those recorded in Oliveto Citra. To calculate the concentrations of heavy metals, the percentage compositions with respect to PM10 measured by ARPAC for the reference year (2016) were used. The concentrations obtained are reported in the Section 3.2 for each metal for the HI and LI zones.

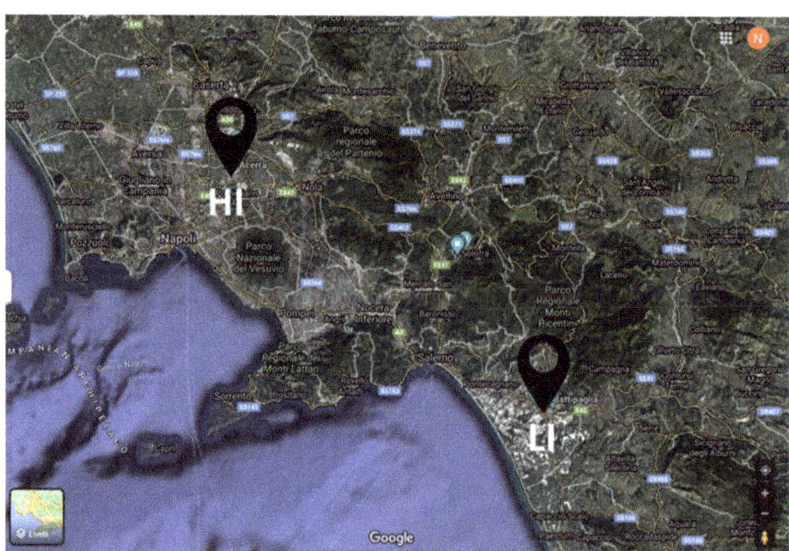

Figure 1. Localization of the high and low environmental impact areas, Acerra (HI) and Battipaglia (LI).

The measurement devices installed by ARPAC consist of buildings of approximately 6 m² which contain all the instruments suitable for controlling the parameters relating to air quality in accordance with the regulatory provisions of Legislative Decree 155/2010. The particulate matter measurement method is defined by the UNI EN 12341: 2014 standard "Ambient air-Reference gravimetric method for determining the mass concentration of suspended particulate PM10 or PM2.5". The measuring principle is gravimetry or absorption by β radiation. The concentration values used in this study are obtained gravimetrically. The reference method for the determination of PM10 particulate material is based on the collection of the PM10 fraction on a special filter and subsequent determination of its mass by gravimetric method in the laboratory after the conditioning filter has been carried out under controlled temperature conditions (20 °C ± 1 °C) and humidity (50 ± 5%). The same is carried out for PM2.5. In addition to the reference, there are equivalent methods for measuring PM10 (for example, automatic instrumentation that exploits the principle of absorption of β radiation by the sampled dust).

2.4. Pollutant Adsorption

To compare the air quality pollution with human health, the pollutant adsorption has been evaluated by computing the chronicle daily intake (CDI, [µg/kg]). This calculation has been performed according to Equation (1), developed by US EPA [53], and considers both the particle-bunds metals and the inhalation rate effects. The equation is reported as follows:

$$CDI = (C_i \times IR \times ET \times EF \times ED)/(BW \times AT) \tag{1}$$

where:

C_i is the contaminant concentration measured, [µg/m³];

IR is the inhalation rate [m³/h];

BW is the body weight [kg];
ET is the exposure time [h/day];
EF is the exposure frequency [days/year];
ED is the exposure duration [years]
AT is the averaging time [days].

Considering the population characteristics of the whole population tested, reference values for the inhalation rate for adult males (workers) were adopted according to ICRP and set at 1.5 m^3/h [54]. Likewise, the measured PM10 values were adopted for Ci. The exposure time (ET) and the exposure frequency (EF) were set equal to 24 h/day and 260 days/year, respectively. The exposure duration (ED) was considered equal to 24 years based on the US EPA [55]. Body weight (BW) was calculated as the average of the population tested, equal to 77 kg. Moreover, according to the US EPA [56], the averaging time (AT) was considered equal to 8760 days for adults for non-carcinogens, while for carcinogens, it was considered equal to 25,550 days. The CDI was calculated to assess the monthly people's exposure to account for the pollutant amount retained until the test. The contaminant concentration used to evaluate the adsorption was the PM10, while the metals (As, Cd, Ni, and Pb) that are part of PM10 were defined by the legislative decree 55/2010. Since the metal measurements in PM10 are not directly available from the 2016 ARPAC data, their corresponding presence in PM10 was assessed using the measurements available for 2019. This implies the assumption that the considered metals fraction in PM10 has been unchanged over the years.

3. Results and Discussion

3.1. Heavy Metals Blood Concentrations

The blood heavy metals average concentrations are reported in Figure 2. As explained in Section 2.2, the measurements refer to 2016, and the metals values reported in Figure 2 are the average in the period considered. The metal concentrations are higher in the patients belonging to the HI area. The average metals concentration has been calculated considering the whole patient set and adjusted for the age, BMI and attitude to smoking. The corresponding average value and the 95% confidence interval (CI) are reported in Table 2.

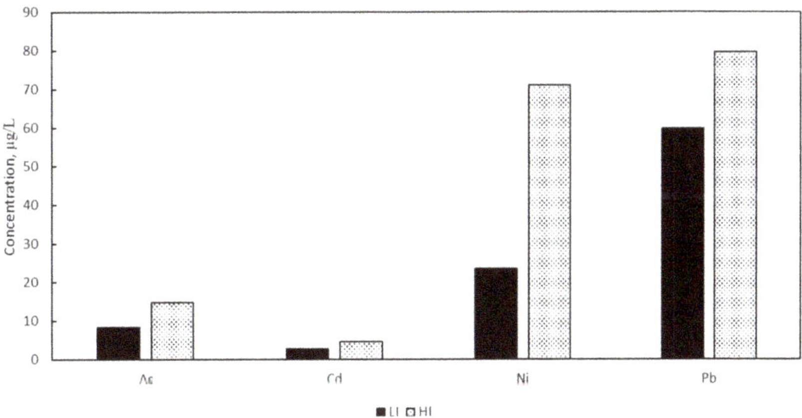

Figure 2. Heavy metals average concentrations measured in the patient blood of the LI and HI area, considering age, BMI, and attitude to smoking.

Table 2. Heavy metals metrics.

	Average Concentration, µg/L		95% CI	
	LI Area	**HI Area**	**LI Area**	**HI Area**
Arsenic	8	15	6–12	9–23
Cadmiums	2.6	4.5	1.4–5.1	2.1–9.4
Nickel	23.5	71	8–68.4	26.1–193
Lead	59.7	79.6	39.3–90.7	46.6–135.9

The reference limit values for the heavy metals have been obtained from the ISTITSAN Report [57]. According to the reference values, all the investigated heavy metals for the HI area exceeded the maximum reference concentration. In particular, nickel was 32% higher than the threshold, cadmium was 1.27%, and arsenic and lead were 1% below the threshold. The Ni and Cd average values also exceeded the maximum reference concentration in the LI area.

3.2. Air Quality

Analyzing the air quality data during 2016, there is a difference in the PM10 annual average between the HI and LI areas by 33% (Table 3). Figures 3 and 4 show PM10 and PM2.5 daily average concentrations in HI and LI areas. These values are often above the law limit of 50 µg/m^3 for PM10 and above 25 for PM2.5 µg/m^3 compared to the LI area. Compared to the LI, the HI area had higher exceedances over the law limit for both PM10 and PM2.5 observed over the years.

Table 3. PM10 and PM2.5 air quality indexes for HI and LI areas during 2016.

		Station			
	2016	**HI Area**			**LI Area**
		AS	**AC**	**AI**	**OL**
PM10	Annual average, µg/m^3	39.4	42.5	33.6	25.7
	Standard deviation, µg/m^3	26.6	24.1	20.6	13.6
	Minimum, µg/m^3	1.8	5	2.8	3.2
	Maximum, µg/m^3	159.2	150.3	139.6	83.5
PM2.5	Annual average, µg/m^3	18.0	—	17.8	18.8
	Standard deviation, µg/m^3	25.2	—	18.7	10.5
	Minimum, µg/m^3	1.8	—	0.4	0.9
	Maximum, µg/m^3	104.5	—	84.5	45.5

As shown in Figure 3, the PM10 concentrations measured in the HI area exceed the law limit 66 times, but only 17 times in the LI area. In both areas the PM10 concentrations are high during the winter period due to the usage of domestic heating. Moreover, as previously said, the HI impact area is characterized by a high industrial development; therefore, the difference in PM10 levels between the two areas could be attributed to it. Figures 3 and 4 show low PM10 and PM2.5 concentration values during the summer due to reduced use of domestic heating systems. The pollution levels of the two areas are different, with the LI area levels being lower than those of the HI area. The minimum pollution level is reached in August for the HI area and October for the LI area. Given the seasonality of pollution, it is generally lower during the summer period, but considering that the LI area is located 10 km from the coastal area and is crossed by a busy highway, the pollution levels in summer are higher.

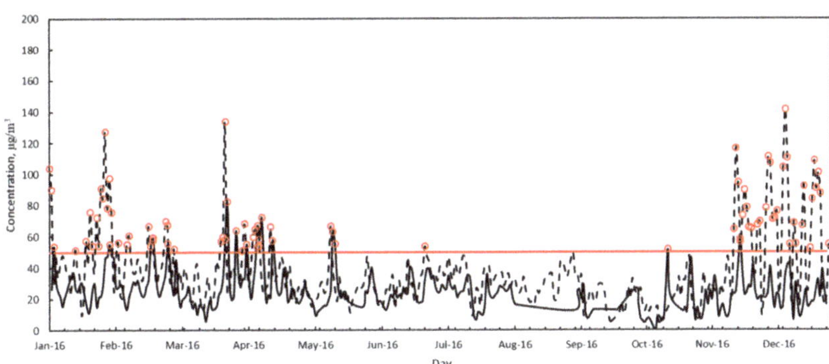

Figure 3. PM10 daily average concentrations measured in the HI (dashed black line) and LI (solid black line) area during 2016 from the ARPAC network. Exceedance from the law limit according to the D.Lgs. 55/2010 (solid red line) is represented by red dots.

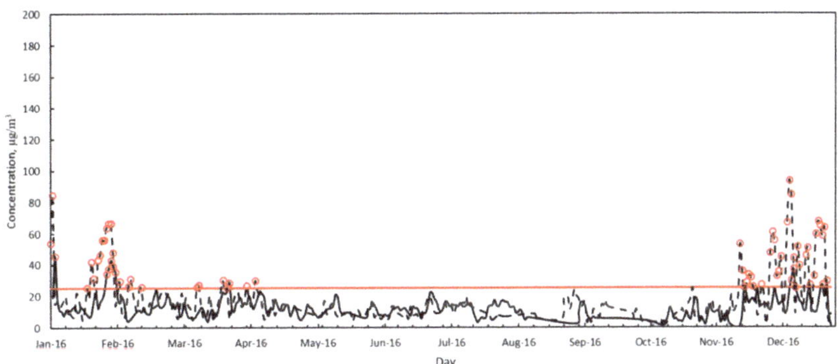

Figure 4. PM2.5 daily average concentrations measured in the HI (dashed black line) and LI (solid black line) area during 2016 from the ARPAC network. Exceedance from the law limit according to the D.Lgs. 55/2010 (solid red line) is represented by red dots.

The metals concentration values obtained using the percentages shown in Table 4 are shown in Figures 5–8. Considering the metals percentages found in PM10, a first difference is noted between the two areas. The metals values in the HI zone are higher than in the LI zone. Consequently, Figures 5–8 show higher metals concentrations for the HI zone than the L1 zone.

Table 4. PM10 metal concentration for HI and LI area during 2016.

		HI Area	LI Area
As	Annual average/± StandDev, µg/m^3	0.6 ± 0.4	0.4 ± 0.2
	% in PM10	1.7	1.5
Cd	Annual average/± StandDev, µg/m^3	0.5 ± 0.3	0.3 ± 0.2
	% in PM10	14	13
Ni	Annual average/± StandDev, µg/m^3	5.1 ± 3.2	3.0 ± 2.1
	% in PM10	14	12
Pb	Annual average/± StandDev, µg/m^3	0.08 ± 0.00	0.1 ± 0.003
	% in PM10	0.25	0.25

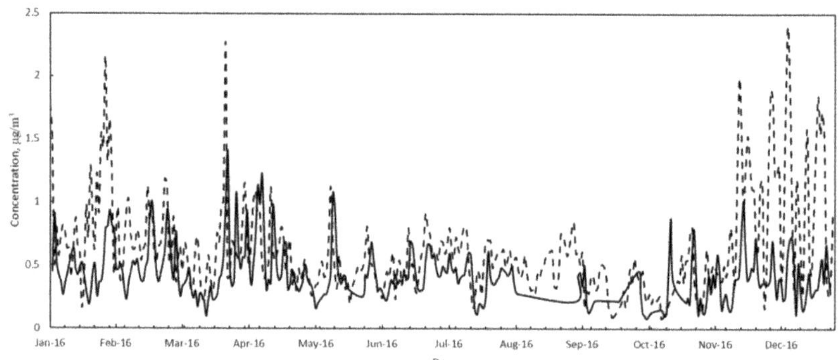

Figure 5. Arsenic daily average concentrations measured in the HI (dashed black line) and LI (solid black line) area during 2016 from the ARPAC network.

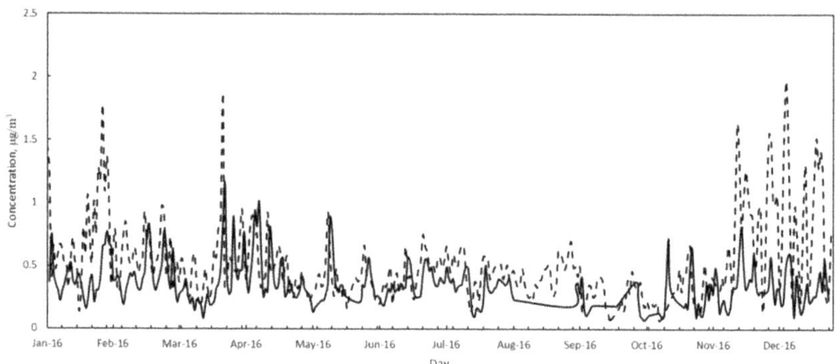

Figure 6. Cadmium daily average concentrations measured in the HI (dashed black line) and LI (solid black line) area during 2016 from the ARPAC network.

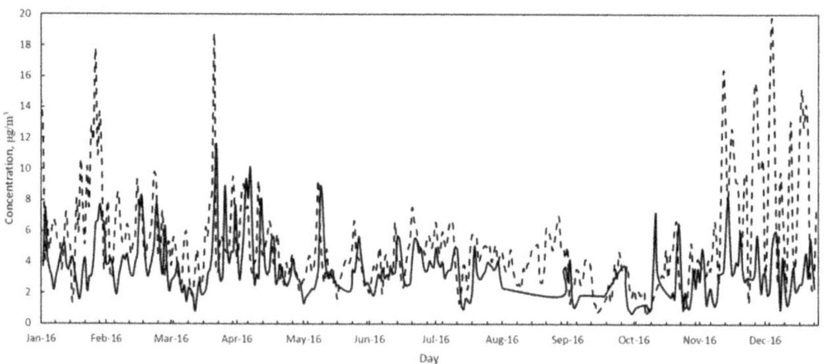

Figure 7. Nickel daily average concentrations measured in the HI (dashed black line) and LI (solid black line) area during 2016 from the ARPAC network.

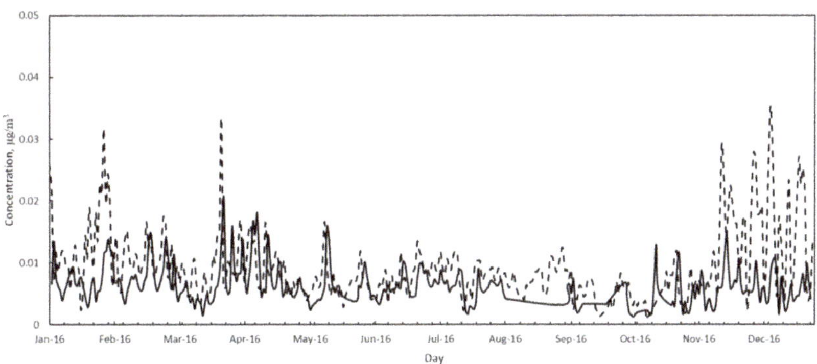

Figure 8. Lead daily average concentrations measured in the HI (dashed black line) and LI (solid black line) area during 2016 from the ARPAC network.

The difference between metals in the winter months (Sept-Dec) corresponds to the increase in domestic heating use. However, the HI area has values in metals higher than those measured in LI, which remain low and constant for most of the year. The difference between the values measured in the two zones is, therefore, attributable to a source of additive pollution with respect to the basic pollution from heating and the problem also encountered in LI.

3.3. Pollutants Adsorption

The CDI was calculated monthly (values are reported in Table 5) for 2016, using Equation (1). Figure 9 shows that the absorption levels in the HI zone are, on average, 74% higher than in the LI zone. Furthermore, absorption is greater in the winter when the concentrations of pollutants are notoriously higher. The minimum absorption occurs in the HI zone in October (linked with the minimum PM10 levels described in Figure 3). This stems from the high amount of rainfall recorded in October, which significantly reduced the measured pollution levels (as reported in the ARPAC reports).

Table 5. Chronicle Daily Intake Expressed in µg/(kg*day).

	Arsenic (As)		Cadmium (Cd)		Nickel (Ni)		Lead (Pb)	
	LI Area	HI Area	LI Area	HI Area	LI Area	HI Area	LI Area	HI Area
January	1.58	3.24	1.30	2.67	13.02	26.70	0.02	0.05
February	1.77	2.43	1.46	2.01	14.61	20.05	0.03	0.04
March	1.48	2.15	1.22	1.77	12.18	17.67	0.02	0.03
April	1.89	2.19	1.56	1.81	15.56	18.06	0.03	0.03
May	1.19	1.66	0.98	1.37	9.80	13.65	0.02	0.02
June	1.45	1.70	1.19	1.40	11.91	14.00	0.02	0.03
July	1.41	2.04	1.16	1.68	11.64	16.83	0.02	0.03
August	0.26	1.80	0.21	1.49	2.15	14.85	0.00	0.03
September	0.36	1.20	0.29	0.99	2.94	9.86	0.01	0.02
October	0.74	1.00	0.61	0.82	6.07	8.23	0.01	0.01
November	1.20	2.59	0.99	2.13	9.89	21.32	0.02	0.04
December	1.33	3.97	1.10	3.27	10.98	32.69	0.02	0.06

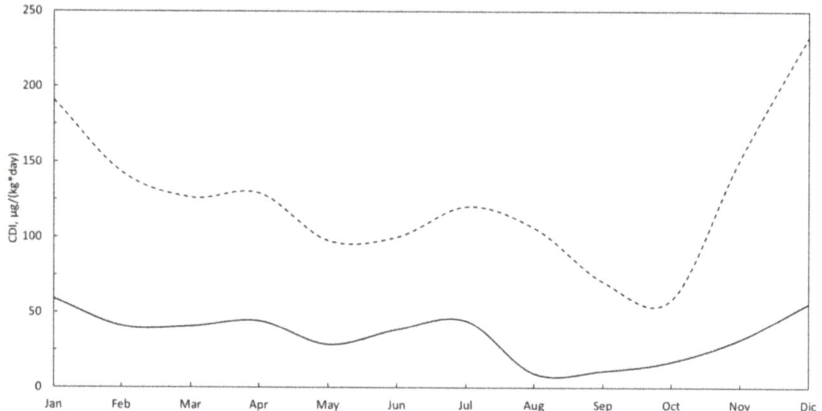

Figure 9. PM10 CDI for the HI (solid black line) and LI (dashed black line) area for every month in 2016.

The CDI calculated for the metals considered (As, Cd, Ni, Pb) are reported in Figures 10–13, respectively. The curves representing the HI and LI areas show a higher percentage average difference for all the metals. In detail, for all the metals considered, the difference between the HI and LI areas is around 40–50%.

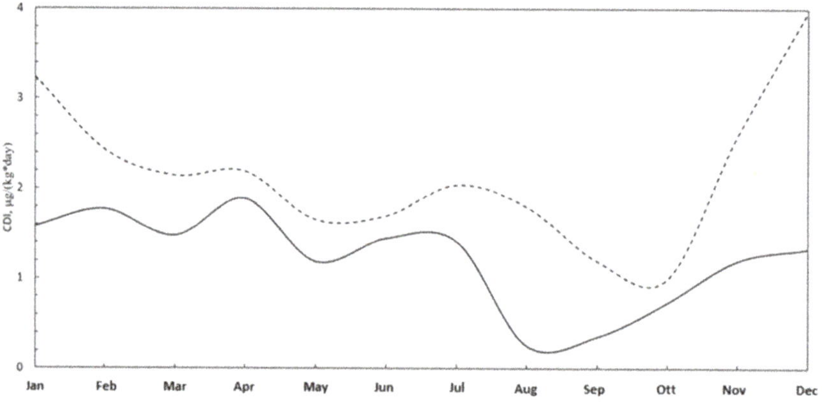

Figure 10. Arsenic (As) CDI for the HI (solid black line) and LI (dashed black line) area for every month in 2016.

The lead CDI values were below the oral reference dose (RfD) for lead (0.00014 mg/kg·day) for all the months in both areas. CDI values of cadmium were above the oral reference dose (RfD) (0.001 mg/kg·day) during 10 months in the HI area, achieving a maximum in December of 3.27 µg/(kg·day). In contrast, the values for the LI area are above the reference dose for 7 months. The HI area is highly industrialized, so the environmental contamination from industrial cadmium and human exposure has increased [58]. Considering the CDI values for Nickel, they are below the reference dose (RfD) (0.02 mg/kg·day) for all months and above the reference dose during the winter period (January, February, November, and December). The CDI values for arsenic exceed the oral reference dose (Rfd) (0.003 mg/kg·day) for all the months considered in the LI and HI area. The oral reference dose (RfD) is assumed as the threshold value (obtained from US EPA) [55]. The metal values were in line with those found in men exposed to road dust and high pollutants levels [59,60]. The test of the CDI average values calculated for each metal and the *p*-value obtained by the Poisson correlation were significantly different from zero. This highlights that the two distributions come from different populations. The CDI values for each metal indicate a proportionality between them and the measured pollutant concentrations. Above all, this proportionality is expected, considering the pollution seasonality and its variability during the year. When pollution levels are low, exposure is reduced, and consequently, the amount of pollutants inhaled by the human body decreases [61]. The high heavy metals concentration values measured in the HI area are probably linked to the heavy metals aero-dispersed. The causes of the high pollution levels are found in the characteristics of the areas considered. Industrialization, urbanization, traffic levels, and domestic heating technology predominantly affect the pollutants emissions. The population considered is also essential for the definition of pollutants absorption: age, attitudes to smoking, and the amount of time spent in the open air are further factors that massively influence the CDI value. A drastic decrease in pollution levels would be desirable to reduce the pollutants absorption by inhalation, which would significantly reduce the CDI values. Knowledge of the pollutants' impact on human health is a fundamental tool for the conscious management of the environment. The high values of heavy metals found in the bloodstream of the patients examined must alarm the institutions to apply policies that significantly reduce the concentrations of these metals. It is certain that installing air quality measurement systems closer to the population with a greater space–time resolution would lead to a more detailed and real-time analysis of pollution levels.

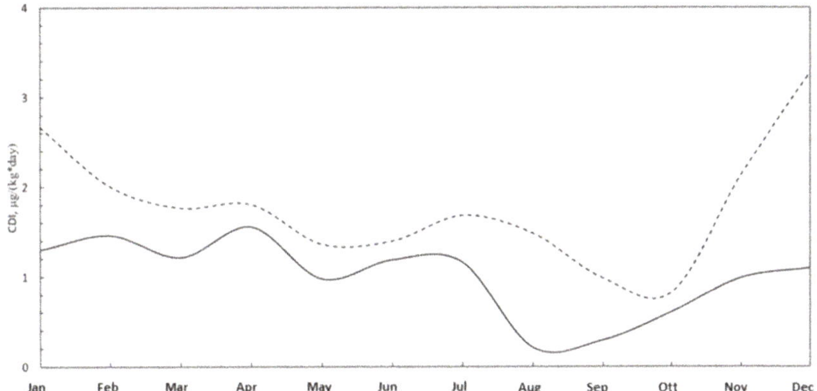

Figure 11. Cadmium (Cd) CDI for the HI (solid grey line) and LI (dashed grey line) area for every month in 2016.

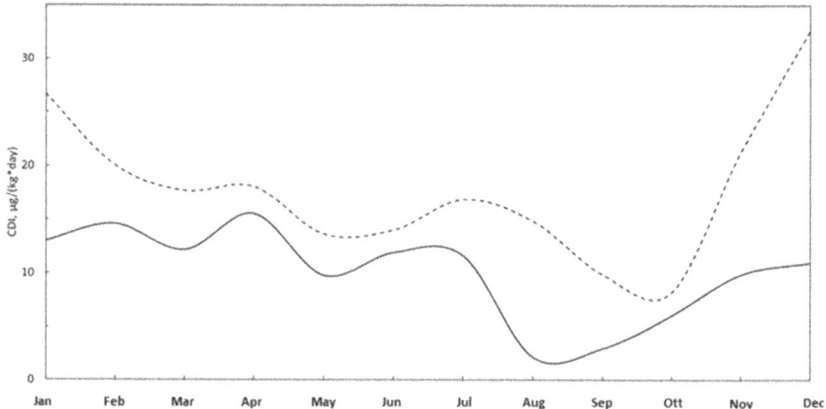

Figure 12. Nickel (Ni) CDI for the HI (solid black line) and LI (dashed black line) area for every month in 2016.

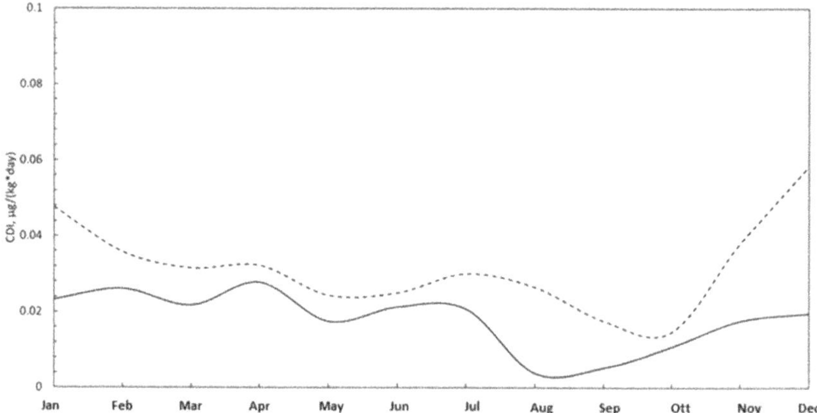

Figure 13. Lead (Pb) CDI for the HI (solid black line) and LI (dashed black line) area for every month in 2016.

4. Conclusions

A survey of 75 male patients was conducted in high and low environmental impact areas in 2016. The heavy metals blood concentrations measured in this study were higher in the high impact area than in the low impact area. Starting from the air quality data was calculated the impact of the heavy metals aero-dispersed on human health. The pollution expressed as PM10 concentration is higher in the HI area than in the LI area. In the area with high environmental impact, there is a variation in the concentration of PM10 during the year, which on average reaches a minimum in the July–November period. The potential pollutant adsorption in the HI area is 72% higher than in the LI area. The metals considered (As, Cd, Ni, Pb) are part of the PM10 measured and are higher in the HI area, meaning that people living in the HI area are more exposed to the pollutant adsorption. Further development of this work includes implementing an air quality monitoring network in the LI area to obtain more detailed information on the air characteristics. The results of this study can be supported by the measurement of metal concentrations in the human body. The next step of the study will be the urine analysis of the subjects belonging to the two investigated areas to verify the effective human absorption of pollutants. Furthermore, it will be interesting to investigate whether the absorption is also linked to PM2.5, which largely constitutes the airborne particulate.

Author Contributions: Conceptualization, N.L., P.T. and D.S.; methodology, N.L., D.S.; software, N.L., D.S.; validation, N.L., P.T. and D.S.; formal analysis, N.L., P.T. and D.S.; data curation, N.L., I.M.B. and T.G., writing—original draft preparation, N.L., D.S.; writing—review and editing, N.L., P.T., D.S., T.G., I.M.B. and L.M.; supervision, I.M.B. and T.G., project administration, D.S. and L.M. All authors have read and agreed to the published version of the manuscript.

Funding: This research received no external funding.

Institutional Review Board Statement: Not applicable.

Informed Consent Statement: Informed consent was obtained from all subjects involved in the study.

Data Availability Statement: Data are available at https://www.arpacampania.it/aria (accessed on 11 January 2021).

Acknowledgments: This paper was performed in the frame of the EcoFoodFertility project.

Conflicts of Interest: The authors declare no conflict of interest.

References

1. European Environment Agency (EEA). *Healthy Environment, Healthy Lives: How the Environment Influences Health and Well-Being in Europe*; Publications Office: Luxembourg, 2020. [CrossRef]
2. Lim, S.S.; Vos, T.; Flaxman, A.D.; Danaei, G.; Shibuya, K.; Adair-Rohani, H.; Amann, M.; Anderson, H.R.; Andrews, K.G.; Aryee, M.; et al. A comparative risk assessment of burden of disease and injury attributable to 67 risk factors and risk factor clusters in 21 regions, 1990–2010: A systematic analysis for the Global Burden of Disease Study 2010. *Lancet* **2012**, *380*, 2224–2260. [CrossRef]
3. Cohen, A.J.; Anderson, H.R.; Ostro, B.; Pandey, K.V.; Krzyzanowski, M.; Künzli, N.; Gutschmidt, K.; Pope, A.; Romieu, I.; Samet, J.M.; et al. The Global Burden of Disease due to Outdoor Air Pollution. *J. Toxicol. Environ. Health A* **2005**, *68*, 1301–1307. [CrossRef] [PubMed]
4. Ezzati, M.; Lopez, A.D.; Rodgers, A.; Vander Hoorn, S.; Murray, C.J. Selected major risk factors and global and regional burden of disease. *Lancet* **2002**, *360*, 1347–1360. [CrossRef]
5. Krewski, D.; Jerrett, M.; Burnett, R.T.; Ma, R.; Hughes, E.; Shi, Y.; Turner, M.C.; Pope, C.A., 3rd; Thurston, G.; Calle, E.E.; et al. Extended Follow-Up and Spatial Analysis of the American Cancer Society Study Linking Particulate Air Pollution and Mortality. *Res. Rep. Health Eff. Inst.* **2009**, *140*, 5–114; discussion 115.
6. Pope, C.A.; Ezzati, M.; Dockery, D.W. Fine-Particulate Air Pollution and Life Expectancy in the United States. *N. Engl. J. Med.* **2009**, *360*, 376–386. [CrossRef]
7. Laden, F.; Schwartz, J.; Speizer, F.; Dockery, D. Reduction in Fine Particulate Air Pollution and Mortality: Extended Follow-up of the Harvard Six Cities Study. *Am. J. Respir. Crit. Care Med.* **2006**, *173*, 667–672. [CrossRef]
8. Kim, D.; Chen, Z.; Zhou, L.-F.; Huang, S.-X. Air pollutants and early origins of respiratory diseases. *Spec. Issue Air Pollut. Chronic Respir. Dis.* **2018**, *4*, 75–94. [CrossRef]
9. Mitova, M.I.; Cluse, C.; Goujon-Ginglinger, C.G.; Kleinhans, S.; Rotach, M.; Tharin, M. Human chemical signature: Investigation on the influence of human presence and selected activities on concentrations of airborne constituents. *Environ. Pollut.* **2020**, *257*, 113518. [CrossRef]
10. Cambra-López, M.; Aarnink, A.J.A.; Zhao, Y.; Calvet, S.; Torres, A.G. Airborne particulate matter from livestock production systems: A review of an air pollution problem. *Environ. Pollut.* **2010**, *158*, 1–17. [CrossRef]
11. Hsu, A.; Reuben, A.; Shindell, D.; de Sherbinin, A.; Levy, M. Toward the next generation of air quality monitoring indicators. *Atmos. Environ.* **2013**, *80*, 561–570. [CrossRef]
12. Whelpdale, D.M. Particulate residence times. *Water. Air. Soil Pollut.* **1974**, *3*, 293–300. [CrossRef]
13. Anderson, J.O.; Thundiyil, J.G.; Stolbach, A. Clearing the Air: A Review of the Effects of Particulate Matter Air Pollution on Human Health. *J. Med. Toxicol.* **2012**, *8*, 166–175. [CrossRef]
14. Kim, K.-H.; Kabir, E.; Kabir, S. A review on the human health impact of airborne particulate matter. *Environ. Int.* **2015**, *74*, 136–143. [CrossRef]
15. Hussain, S.; Rengel, Z.; Qaswar, M.; Amir, M.; Zafar-ul-Hye, M. Arsenic and Heavy Metal (Cadmium, Lead, Mercury and Nickel) Contamination in Plant-Based Foods. In *Plant and Human Health, Volume 2: Phytochemistry and Molecular Aspects*; Ozturk, M., Hakeem, K.R., Eds.; Springer International Publishing: Cham, Switzerland, 2019; pp. 447–490. [CrossRef]
16. Sun, Z.; Zhu, D. Exposure to outdoor air pollution and its human health outcomes: A scoping review. *PLoS ONE* **2019**, *14*, e0216550. [CrossRef]
17. Satarug, S.; Gobe, G.C.; Vesey, D.A.; Phelps, K.R. Cadmium and lead exposure, nephrotoxicity, and mortality. *Toxics* **2020**, *8*, 86. [CrossRef]

18. Alengebawy, A.; Abdelkhalek, S.T.; Qureshi, S.R.; Wang, M.-Q. Heavy metals and pesticides toxicity in agricultural soil and plants: Ecological risks and human health implications. *Toxics* **2021**, *9*, 42. [CrossRef]
19. Massányi, P.; Massányi, M.; Madeddu, R.; Stawarz, R.; Lukáč, N. Effects of cadmium, lead, and mercury on the structure and function of reproductive organs. *Toxics* **2020**, *8*, 94. [CrossRef]
20. Satarug, S. Dietary cadmium intake and its effects on kidneys. *Toxics* **2018**, *6*, 15. [CrossRef]
21. Sijko, M.; Kozłowska, L. Influence of dietary compounds on arsenic metabolism and toxicity. Part i—Animal model studies. *Toxics* **2021**, *9*, 258. [CrossRef]
22. Silvera, S.A.N.; Rohan, T.E. Trace elements and cancer risk: A review of the epidemiologic evidence. *Cancer Causes Control.* **2007**, *18*, 7–27. [CrossRef]
23. Sofia, D.; Giuliano, A.; Gioiella, F. Air quality monitoring network for tracking pollutants: The case study of salerno city center. *Chem. Eng. Trans.* **2018**, *68*, 67–72. [CrossRef]
24. Sofia, D.; Lotrecchiano, N.; Trucillo, P.; Giuliano, A.; Terrone, L. Novel air pollution measurement system based on ethereum blockchain. *J. Sens. Actuator Netw.* **2020**, *9*, 49. [CrossRef]
25. Sofia, D.; Lotrecchiano, N.; Giuliano, A.; Barletta, D.; Poletto, M. Optimization of number and location of sampling points of an air quality monitoring network in an urban contest. *Chem. Eng. Trans.* **2019**, *74*, 277–282. [CrossRef]
26. Lotrecchiano, N.; Sofia, D.; Giuliano, A.; Barletta, D.; Poletto, M. Spatial Interpolation Techniques For innovative Air Quality Monitoring Systems. *Chem. Eng. Trans.* **2021**, *86*, 391–396. [CrossRef]
27. Lotrecchiano, N.; Gioiella, F.; Giuliano, A.; Sofia, D. Forecasting Model Validation of Particulate Air Pollution by Low Cost Sensors Data. *J. Model. Optim.* **2019**, *11*, 63–68. [CrossRef]
28. Lotrecchiano, N.; Sofia, D.; Giuliano, A.; Barletta, D.; Poletto, M. Pollution dispersion from a fire using a Gaussian plume model. *Int. J. Saf. Secur. Eng.* **2020**, *10*, 431–439. [CrossRef]
29. Sofia, D.; Gioiella, F.; Lotrecchiano, N.; Giuliano, A. Mitigation strategies for reducing air pollution. *Environ. Sci. Pollut. Res.* **2020**, *27*, 19226–19235. [CrossRef] [PubMed]
30. Sofia, D.; Gioiella, F.; Lotrecchiano, N.; Giuliano, A. Cost-benefit analysis to support decarbonization scenario for 2030: A case study in Italy. *Energy Policy* **2020**, *137*, 111137. [CrossRef]
31. Kunugi, Y.; Arimura, T.H.; Iwata, K.; Komatsu, E.; Hirayama, Y. Cost-efficient strategy for reducing PM 2.5 levels in the Tokyo metropolitan area: An integrated approach with air quality and economic model. *PLoS ONE* **2018**, *13*, e0207623. [CrossRef] [PubMed]
32. Singh, V.; Meena, K.K.; Agarwal, A. Travellers' exposure to air pollution: A systematic review and future directions. *Urban Clim.* **2021**, *38*, 100901. [CrossRef]
33. Hormigos-Jimenez, S.; Padilla-Marcos, M.Á.; Meiss, A.; Gonzalez-Lezcano, R.A.; Feijó-Muñoz, J. Assessment of the ventilation efficiency in the breathing zone during sleep through computational fluid dynamics techniques. *J. Build. Phys.* **2019**, *42*, 458–483. [CrossRef]
34. Becerra, J.A.; Lizana, J.; Gil, M.; Barrios-Padura, A.; Blondeau, P.; Chacartegui, R. Identification of potential indoor air pollutants in schools. *J. Clean. Prod.* **2020**, *242*, 118420. [CrossRef]
35. Mendes, A.; Bonassi, S.; Aguiar, L.; Pereira, C.; Neves, P.; Silva, S.; Teixeira, J.P. Indoor air quality and thermal comfort in elderly care centers. *Urban Clim.* **2015**, *14*, 486–501. [CrossRef]
36. Environmental Protection Agency (EPA). *Proposed Guidelines for Exposure-Related Measurements*; Federal Register: Washington, DC, USA, 1988.
37. Ott, W.R. Total human exposure. *Environ. Sci. Technol.* **1985**, *19*, 880–885. [CrossRef]
38. Sexton, K.; Barry Ryan, P. Assessment of Humen Exposure to Air Pollution: Methods, Measurements, and Models. In *Air Pollution, the Automobile, and Public Health*; Health Effects Institute, National Academy Press: Washington, DC, USA, 1988.
39. Asante-Duah, K. Exposure Assessment: Analysis of Human Intake of Chemicals. In *Public Health Risk Assessment for Human Exposure to Chemicals*; Asante-Duah, K., Ed.; Springer: Dordrecht, The Netherlands, 2017; pp. 189–229. [CrossRef]
40. Triassi, M.; Alfano, R.; Illario, M.; Nardone, A.; Caporale, O.; Montuori, P. Environmental Pollution from Illegal Waste Disposal and Health Effects: A Review on the "Triangle of Death". *Int. J. Environ. Res. Public. Health* **2015**, *12*, 1216–1236. [CrossRef] [PubMed]
41. Maisto, G.; Baldantoni, D.; De Marco, A.; Alfani, A.; Virzo De Santo, A. Biomonitoring of trace element air contamination at sites in Campania (Southern Italy). *J. Trace Elem. Med. Biol.* **2003**, *7* (Suppl. 1), 51–55.
42. Posters. *Andrology* **2014**, *2*, 45–117. [CrossRef]
43. Pacini, N.; Abate, V.; Brambilla, G.; De Felip, E.; De Filippis, S.P.; De Luca, S.; di Domenico, A.; D'Orsi, A.; Forte, T.; Fulgenzi, A.R.; et al. Polychlorinated dibenzodioxins, dibenzofurans, and biphenyls in fresh water fish from Campania Region, southern Italy. *Chemosphere* **2013**, *90*, 80–88. [CrossRef]
44. EcoFoodFertility project. Available online: http://www.ecofoodfertility.it/the-project.html. (accessed on 18 January 2021).
45. Montano, L.; Iannuzzi, L.; Rubes, J.; Avolio, C.; Pistos, C.; Gatti, A.; Raimondo, S.; Notari, T. EcoFoodFertility—Environmental and food impact assessment on male reproductive function. *Andrology* **2014**, *2* (Suppl. 2), 69.
46. World Medical Association. World Medical Association Declaration of Helsinki: Ethical Principles for Medical Research Involving Human Subjects. *JAMA* **2013**, *310*, 2191–2194. [CrossRef]

47. Vecoli, C.; Montano, L.; Borghini, A.; Notari, T.; Guglielmino, A.; Mercuri, A.; Turchi, S.; Andreassi, M.G. Effects of Highly Polluted Environment on Sperm Telomere Length: A Pilot Study. *Int. J. Mol. Sci.* **2017**, *18*, 1703. [CrossRef]

48. Montano, L.; Ceretti, E.; Donato, F.; Bergamo, P.; Zani, C.; Viola, G.; Notari, T.; Pappalardo, S.; Zani, D.; Ubaldi, S.; et al. Effects of a Lifestyle Change Intervention on Semen Quality in Healthy Young Men Living in Highly Polluted Areas in Italy: The FASt Randomized Controlled Trial. *Eur. Urol. Focus* **2022**, *8*, 351–359. [CrossRef]

49. Lettieri, G.; D'Agostino, G.; Mele, E.; Cardito, C.; Esposito, R.; Cimmino, A.; Giarra, A.; Trifuoggi, M.; Raimondo, S.; Notari, T.; et al. Discovery of the Involvement in DNA Oxidative Damage of Human Sperm Nuclear Basic Proteins of Healthy Young Men Living in Polluted Areas. *Int. J. Mol. Sci.* **2020**, *21*, 4198. [CrossRef]

50. Bosco, L.; Notari, T.; Ruvolo, G.; Roccheri, M.C.; Martino, C.; Chiappetta, R.; Carone, D.; Lo Bosco, G.; Carrillo, L.; Raimondo, S.; et al. Sperm DNA fragmentation: An early and reliable marker of air pollution. *Environ. Toxicol. Pharmacol.* **2018**, *58*, 243–249. [CrossRef]

51. Longo, V.; Forleo, A.; Ferramosca, A.; Notari, T.; Pappalardo, S.; Siciliano, P.; Capone, S.; Montano, L. Blood, urine and semen Volatile Organic Compound (VOC) pattern analysis for assessing health environmental impact in highly polluted areas in Italy. *Environ. Pollut.* **2021**, *286*, 117410. [CrossRef]

52. Senior, K.; Mazza, A. Italian "Triangle of death" linked to waste crisis. *Lancet Oncol.* **2004**, *5*, 525–527. [CrossRef]

53. US Environmental Protecgtion Agency (EPA). Users' Guide and Background Technical Document for USEPA Region 9—Preliminary Remediation Goals (PRG) Table. 2013. Available online: https://semspub.epa.gov/work/02/103453.pdf (accessed on 30 April 2021).

54. I International Commission on Radiological Protection (ICRP). Human Respiratory Tract Model for Radiological Protection. A report of a Task Group of the International Commission on Radiological Protection. *Ann. ICRP* **2002**, *32*, 307–309.

55. US Environmental Protecgtion Agency (EPA). *Risk Assessment Guidance for Superfund, Volume I: Human Health Evaluation Manual (Part A)*; Environmental Protection Agency: Washington, DC, USA, 1989.

56. US Environmental Protecgtion Agency (EPA). *Risk Assessment Guidance for Superfund Volume I: Human Health Evaluation Manual (Part F, Supplemental Guidance for Inhalation Risk Assessment)*; Environmental Protection Agency: Washington, DC, USA, 2003.

57. Alimonti, A.; Bocca, B.; Mattei, D.; Pino, A. Biomonitoraggio della popolazione italiana per l'esposizione ai metalli: Valori di riferimento 1990–2009. *Rapp. Istisan-Ist. Super. Di Sanità* **2010**, *10*, 58.

58. Elinder, C.G.; Järup, L. Cadmium Exposure and Health Risks: Recent Findings. *Ambio* **1996**, *25*, 370–373.

59. Baralić, K.; Javorac, D.; Marić, D.; Đukić-Ćosić, D.; Bulat, Z.; Antonijević Miljaković, E.; Anđelković, M.; Antonijević, B.; Aschner, M.; Buha Djordjevic, A. Benchmark dose approach in investigating the relationship between blood metal levels and reproductive hormones: Data set from human study. *Environ. Int.* **2022**, *165*, 107313. [CrossRef]

60. Magnano, G.C.; Marussi, G.; Pavoni, E.; Adami, G.; Larese Filon, F.; Crosera, M. Percutaneous metals absorption following exposure to road dust powder. *Environ. Pollut.* **2022**, *292*, 118353. [CrossRef] [PubMed]

61. Chen, Y.; Luo, X.-S.; Zhao, Z.; Chen, Q.; Wu, D.; Sun, X.; Wu, L.; Jin, L. Summer–winter differences of PM2.5 toxicity to human alveolar epithelial cells (A549) and the roles of transition metals. *Ecotoxicol. Environ. Saf.* **2018**, *165*, 505–509. [CrossRef] [PubMed]

Article

Prediction of PM2.5 Concentration on the Basis of Multi-Time Scale Fusion

Jianfei Zhang * and Wangui Xia

School of Computer & Information Engineering, Heilongjiang University of Science & Technology, Harbin 150027, China; xiaofangdui@163.com
* Correspondence: zjfnefu2008@163.com

Abstract: Long-term prediction of hour-concentration of PM2.5 (particles in atmospheric suspension with effective dimensions equal or lower than 2.5 microns) is of great significance for environmental protection and people's health. At present, the prediction of hour-concentration of PM2.5 is mostly single-step prediction, which is to predict PM2.5 concentration at a future time point based on a period of historical data. In this paper, a model based on multi-time scale fusion is proposed to study single-step prediction and multi-step prediction, respectively. Experimental results show that the proposed model is better than stacked LSTM and CNN-LSTM in predicting PM2.5 hour-concentration.

Keywords: PM2.5 concentration prediction; multi-time scale fusion; time series

1. Introduction

At present, the traditional methods in the field of PM2.5 concentration prediction research mainly combine four conventional methods formed by meteorology, environmental science, mathematics, and computational science. That is, empirical model prediction based on historical data and statistical methods, probability model prediction based on statistical and mathematical methods or models, prediction based on synthetic methods, and prediction based on conventional machine learning models. With the rapid development of deep learning, Fan et al. used the recurrent neural network model for predicting PM2.5 concentration in the future 1 hour based on air quality and meteorological data in the past 48 h [1]. Qi et al. proposed the model GCN-LSTM (A model based on Graph Convolutional Network and Long Short-Term Memory), which proved that the model was superior to CNN (Convolutional Neural Networks) and LSTM in predicting the air quality in the future one hour [2]. He et al. combined the wavelet transform with the LSTM (Long Short-Term Memory) model and took the daily average concentration as the input to predict the pollutant concentration of the next day, and proved that the proposed model was superior to MLR (Mixed Logistic Regression), LSTM and WT-MLR (Mixed Logistic Regression based on Wavelet Transform) [3]. Huang and Kuo constructed the model APNet (Attention-based Parallel Networks) and proved through experiments that the model was superior to CNN and LSTM in predicting PM2.5 concentration in the future one hour [4].

However, the above studies only focus on PM2.5 concentration of single-step prediction, did not predict the PM2.5 concentration for a period of time in the future, that is, did not make a multi-step prediction. To solve this problem, this article builds a CNN-LSTM network on the basis of the combination of attention mechanism, the multi-time scale fusion model of multi-time scale features is integrated. The aim is to accurately predict the value of PM2.5 corresponding to each hour in a continuous period of time in the future. Through experiments, the validity and superiority of the method proposed in this paper are verified.

Citation: Zhang, J.; Xia, W. Prediction of PM2.5 Concentration on the Basis of Multi-Time Scale Fusion. *Processes* **2022**, *10*, 171. https://doi.org/10.3390/pr10010171

Academic Editors: Daniele Sofia and Paolo Trucillo

Received: 20 December 2021
Accepted: 11 January 2022
Published: 17 January 2022

Publisher's Note: MDPI stays neutral with regard to jurisdictional claims in published maps and institutional affiliations.

2. PM2.5 Prediction Model Based on Multi-Time Scale Fusion

2.1. LSTM (Long Short-Term Memory)

LSTM is a variant produced to solve the long-term dependence problem that RNN (Recurrent Neural Network) cannot solve, which effectively alleviates the gradient explosion problem that RNN cannot avoid and can better predict the time series [5]. The architecture of an LSTM memory cell is shown in Figure 1, where each cell has three "gate" structures, include, the input gate, the forget gate, and the output gate. A chain of repeating cells forms the LSTM layer. The calculation process of the spatiotemporal feature matrix $X = [x_1, x_2, \dots , x_t]$ in the LSTM layer is given in Equations (1)–(6). Equation (1) represents the forget gate and it decides what information should be thrown away from the cell state. The directions are: Input h_{t-1} and x_t into the forget gate, and calculate the output value ft of the forget gate through the sigmoid activation function. Equations (2) and (3) represent the input gate, which decides what new information should be stored in the state of cell. The directions are: Input h_{t-1} and x_t into the input gate, and get i_t and $\tilde{c}_t$ through the sigmoid activation function and tanh activation function respectively. Equation (4) uses the output of the forget gate and the input gate to update the current cell state. Equations (5) and (6) together constitute the output of the current cell. The directions are: First, input h_{t-1} and x_t into the output gate, and calculate the output o_t of the output gate through the sigmoid activation function. Then get the current cell output h_t by calculating the output of the output gate and the state of the current cell.

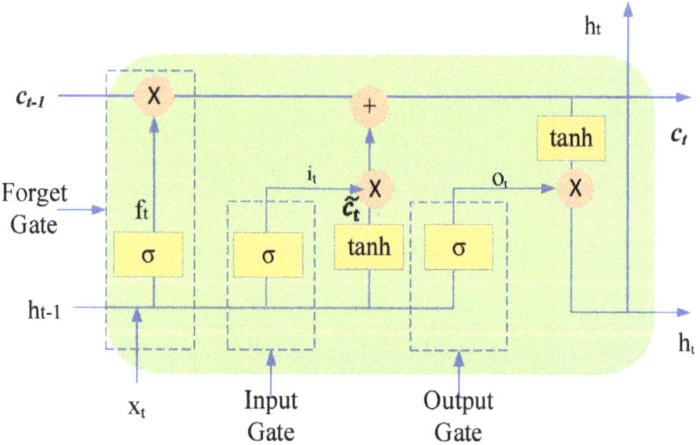

Figure 1. The structure of LSTM neurons.

The following Equations (1)–(6) describe the internal calculation process of an LSTM neural unit:

$$f_t = \sigma(w_f \cdot [h_{t-1}, x_t] + b_f) \tag{1}$$

$$i_t = \sigma(w_i \cdot [h_{t-1}, x_t] + b_i) \tag{2}$$

$$\tilde{c}_t = \tan h(w_c \cdot [h_{t-1}, x_t] + b_c) \tag{3}$$

$$c_t = f_t \odot c_{t-1} + i_t \odot \tilde{c}_t \tag{4}$$

$$o_t = \sigma(w_o \cdot [h_{t-1}, x_t] + b_o) \tag{5}$$

$$h_t = o_t \odot \tan h(c_t) \tag{6}$$

where f_t is the output of forget gate, the value range of f_t is (0,1); i_t is the output of input gate, the value range of i_t is (0,1); c_t is the state of the current cell; o_t is the output of output gate, the value range of o_t is (0,1); h_{t-1} is the output of the previous cell; h_t is the output of the current cell; w_f, w_i, w_c, and w_o are the weight matrices for input vector x_t at time step t; b_f, b_i, b_c, and b_o are the bias vectors; σ is sigmoid activation function; tanh is hyperbolic sine function; $\odot$ stands for element-wise multiplication of the matrix; $\otimes$ stands for multiplication; $\oplus$ stands for the sum operation;

2.2. Ensemble Empirical Mode Decomposition (EEMD)

As a noise-assisted signal decomposition method, EEMD adds white noise to the original signal and performs EMD decomposition on it, and finally calculates lumped average using the results of multiple decomposition [6].

The specific operation steps are as follows:

(1) Set the overall average times M;

(2) Add a white noise $n_i(t)$ with standard normal distribution to the original signal $x(t)$ to generate a new signal:

$$x_i(t) = x(t) + n_i(t) \tag{7}$$

where $n_i(t)$ is i-th additive white noise sequence; $x_i(t)$ is the additional noise signal of the i-th test, $i = 1, 2, 3, \ldots M$.

(3) EMD decomposition is performed on the obtained signal $x_i(t)$ containing noise to obtain the form of their respective IMF (Intrinsic Mode Function) sum:

$$x_i(t) = \sum_{j=1}^{J} c_{i,j}(t) + r_{i,j}(t) \tag{8}$$

where $c_{i,j}(t)$ is the J-th IMF obtained by decomposing after adding white noise for the i-th time. $r_{i,j}(t)$ is the residual term represents the average trend of the signal, and j is the number of IMF;

(4) Repeat steps (2) and (3) for M times, decompose and add white noise signals with different amplitudes each time, and the set of IMF is: $c_{1,j}(t), c_{2,j}(t), \ldots c_{M,j}(t)$, where $j = 1, 2, 3, \ldots J$;

(5) Based on the principle that the statistical average value of unrelated sequences is zero, the above IMF is calculated by aggregate average to obtain the final IMF, namely:

$$c_j(t) = \frac{1}{M} \sum_{i=1}^{M} c_{i,j}(t) \tag{9}$$

where $c_j(t)$ is the j-th IMF, $i = 1, 2, \ldots M$, $j = 1, 2, \ldots J$;

2.3. Attention Mechanism

The attention mechanism mimics the internal process of biological observation behavior [7]. His principle is through a set of weights $\alpha_{T_{s-e}}^{T_t} = \left[\alpha_{T_e}^{T_t}, \alpha_{T_{e+1}}^{T_t}, \ldots \alpha_{T_s}^{T_t} \right]$ to express the value of a certain time slice in the target sequence x_{T_t} and the dependent sequence $x_{T_{s-e}} = [x_{T_e}, x_{T_{e+1}}, \ldots, x_{T_s}]$ relevance. Eeach element in x_{T_t} and $x_{T_{s-e}}$ has the same dimension. Map x_{T_t} and $x_{T_{s-e}}$ to the parameter space:

$$Query = x_{T_t} W_Q \tag{10}$$

$$Key = x_{T_{s-e}} W_k \tag{11}$$

$$Value = x_{T_{s-e}} W_v \tag{12}$$

where W_Q is dx*dq dimensional Query parameter matrix; W_k is dx*dk dimensional Key parameter matrix; W_v is dx*dv dimensional Value parameter matrix;

The attention mechanism is divided into three stages: in the first stage, the target sequence is mapped from x_{T_t} map of dx dimension to Query of dq dimension, and similarly transformed $x_{T_{s-e}}$ into matrix mapping to Key matrix with dk element dimension and Value matrix with dv element dimension, calculating the similarity between Query and Key; In the second stage, the original score of the first stage is normalized, and the $\alpha_{T_{s-e}}^{T_t}$ weight of Value is calculated by Softmax. In the third stage, the Value is weighted and summed according to the weight coefficient to obtain the attention Value.

2.4. Multi Time Scale Fusion Model

In this paper, the multi-time scale fusion model is applied to the prediction of PM2.5 hour-concentration for the first time, and the model process is shown in Figure 2. EEMD (Ensemble Empirical Mode Decomposition) decomposition can decompose the original PM2.5 sequence into new sequences with different time scales. CNN-LSTM was employed to extract characteristic information of time series. Attention_layer pays attention to important features and ignores non-important features through attention mechanism to improve prediction accuracy.

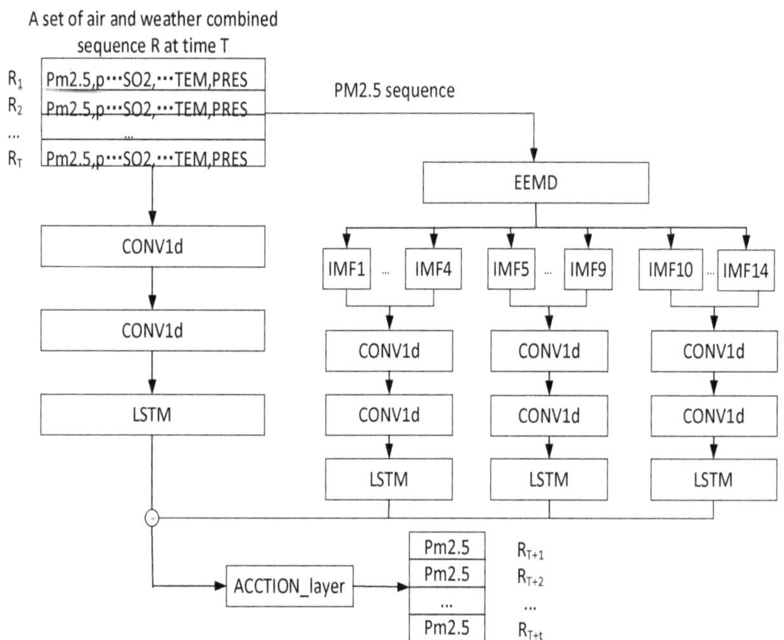

Figure 2. System flow of multi-time scale network model.

The specific steps are as follows:

(1) Input the original PM2.5 sequence into the EEMD model, and perform EEMD decomposition on the original PM2.5 concentration data. This is the first improvement made by the model in this paper on the basis of CNN-LSTM model. Compared with the original sequence, the decomposed sequence can more precisely express the period of the original sequence and better obtain information of different time scales.

(2) The original PM2.5 data sequence and the decomposed PM2.5 sequence were input into CNN-LSTM network composed of two layers of Conv1d and one layer of LSTM respectively for feature extraction. As convolutional neural network has excellent feature extraction and feature expression capabilities, LSTM has natural advantages in processing time sequence. Therefore, CNN and LSTM are used in combination in feature extraction in this paper. In this paper, the decomposed sequences are recombined into new sequences according to different time scales and used as the input of different network layers respectively with the original sequence.

(3) The outputs of different LSTM layers output the prediction results through the attention mechanism layer. Attention mechanism is another improvement based on CNN-LSTM. Through attention mechanism, more important feature information can be paid attention to in features of different time scales to improve the accuracy of prediction.

3. Results and Discussion

3.1. Experimental Configuration and Data Set Description

The experimental environment of this paper uses TensorFlow + Keras framework, Python 3.7 development language, the system uses Windows, with multiple Python library functions for code implementation and result analysis.

The data in this paper are the monitoring data from ground stations in Harbin, mainly including AQI, PM2.5, PM10, O3, and other data. The update frequency is one hour, and the time span is from May 2014 to April 2021. PM2.5 is shown in Figure 3.

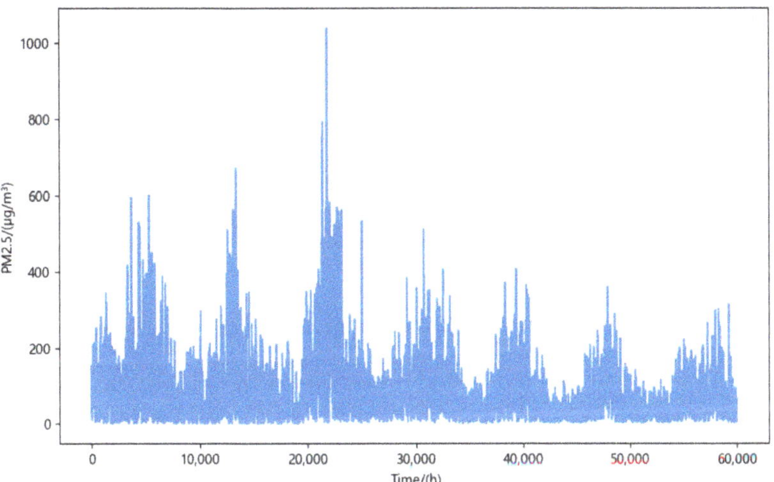

Figure 3. Changes in PM2.5 concentration over time.

3.2. Data Pre-Processing

In this paper, data pre-processing includes data cleaning and data normalization. During data cleaning, clear redundant data. When the pollutant data is missing, this paper uses 8 h moving average data to replace it. After processing, the short-term missing values that still exist are supplemented by simple linear interpolation of adjacent values, and the missing data that are too long are deleted.

The normalization of maximum and minimum values is used in this paper, as follows:

$$f* = \frac{f - f_{min}}{f_{max} - f_{min}} \tag{13}$$

where f_{max} is the maximum value of sample data; f_{min} is the minimum value of sample data.

3.3. EEMD Decomposition of PM2.5 Concentration

In this paper, the pre-treated TIME series of PM2.5 value is decomposed into 14 IMF series and one trend item, as shown in Figure 4.

For the period calculation of IMF components, this paper uses the average period as the period of IMF components. The calculation results are shown in Table 1 below. According to the cycle calculation results, imF1-IMF4 is hour scale, IMF5-IMF9 is day scale, IMF10-IMF12 is month scale, and IMF13-IMF14 is year scale.

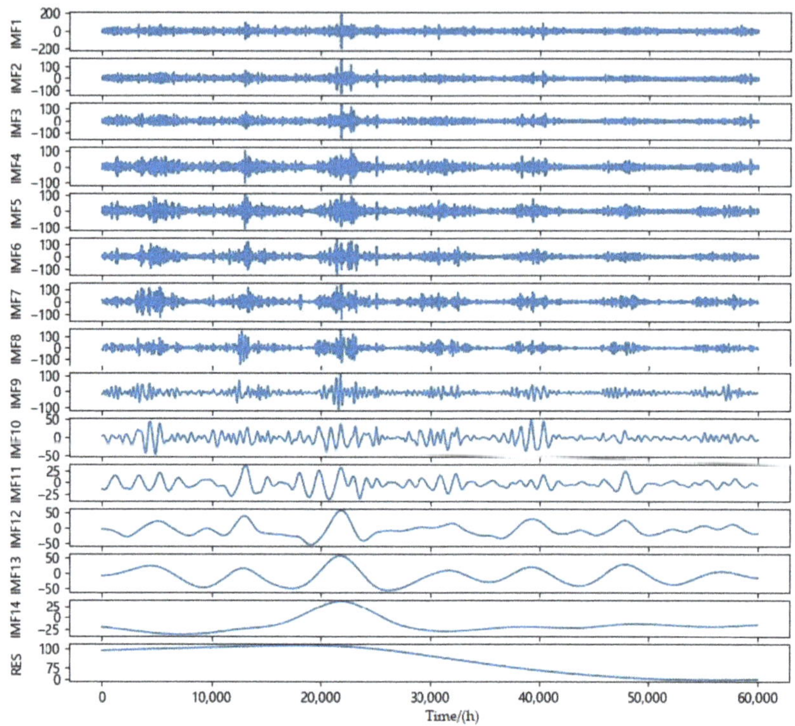

Figure 4. EEMD decomposition results of PM2.5 concentration.

Table 1. The period of each IMF component of PM2.5 concentration.

IMF Component	Period/h
IMF1	3
IMF2	5
IMF3	8
IMF4	15
IMF5	25
IMF6	46
IMF7	89
IMF8	168
IMF9	321
IMF10	659
IMF11	1395
IMF12	4000
IMF13	8572
IMF14	20,000
RES	–

3.4. Evaluation Index

The following indicators are selected as the evaluation criteria in this paper:

(1) RMSE (Root Mean Square Error)

$$RMSE = \sqrt{\frac{1}{M} \sum_{m=1}^{M} (y_m - y'_m)^2} \tag{14}$$

where y_m is the true value in the test set; y_m' is the predicted value.

(2) MAE (Mean Absolute Error)

$$MAE = \frac{1}{m} \sum_{m=1}^{M} |Y' - Y| \tag{15}$$

where Y' is predicted results; Y is true value.

(3) $R2_{adj}$ (Adjusted R-Square)

$$R2 = 1 - \frac{\sum_{m=1}^{M} \left(y_m - \bar{y}_m\right)^2}{\sum_{m=1}^{M} \left(y_m - \bar{y}\right)^2} \tag{16}$$

$$R2_{adj} = 1 - \frac{(1 - R2)(n - 1)}{n - p - 1} \tag{17}$$

where y_m is the true value in the test set; $\bar{y}_m$ is the predicted value; $\bar{y}$ is the average of the true values in the test set; R2 is R-Square; n is the number of samples; p is the number of features; $R2_{adj}$ offsets the impact of the number of samples on R2, so that the value of $R2_{adj}$ is between zero and one, and the larger the value of $R2_{adj}$, the better the performance of the model.

3.5. Comparison of Experimental Results

3.5.1. Impact of Historical Time Windows on Model Performance

PM2.5 data is affected by a variety of related time series, but the change of each time series value does not immediately affect PM2.5 concentration value, which means that the variable value at the previous moment has a lag effect on the PM2.5 concentration value at the next moment, which may be strong in the short term and weak in the long term [8]. A smaller window size cannot guarantee sufficient long-term memory input for LSTM model, while a larger window size will increase the input of irrelevant information and increase the unnecessary computational complexity of the model [9]. In order to determine the appropriate historical time window, the historical time window in this study starts from 12 h, and every 12 h is a time interval. The prediction scale is the concentration of 1 h PM2.5 in the future. The results are shown in Table 2 below. When the historical time window is 36 h, the RMSE, MAE and R2 of the model in this paper are 9.66, 6.95, and 0.95, respectively, which are the best. For LSTM model, when the history time window is 24 h, RMSE 14.0 is the best. When the historical time window is 36 h, MAE is 7.63 and R2 is 0.89. For CNN-LSTM model, when the historical time window is 24 h, RMSE is 13.66, MAE is 9.88, and R2 is 0.91. The model in this paper is superior to the comparison model in terms of indicators. The RMSE of the model is 31% lower than that of LSTM and 25% lower than that of CNN-LSTM. For the index MAE, it is 24% lower than LSTM and 22% lower than CNN-LSTM. For index R2, it is 5% higher than LSTM and 3% higher than CNN-LSTM.

3.5.2. Performance Comparison of Multi-Step Prediction

In order to test the multi-step prediction performance of the model in this paper for PM2.5 hour-concentration, experiments were carried out on the three models for 1 h, 4 h, 8 h, 12 h, and 24 h in the future, respectively, and the results are shown in Table 3. It can be seen from Table 3 that: (1) each model achieves the best effect when the prediction step size is one hour, and the evaluation indexes of the model proposed in this paper are better. (2) With the increase of prediction step size, the accuracy of prediction decreases, but the prediction evaluation index of the model proposed in this paper is superior to LSTM and CNN-LSTM in each prediction time scale. Therefore, it indicates that the model proposed in this paper is effective in improving the long-term prediction accuracy.

Table 2. Performance comparison of models in different historical time windows.

Historical Window Time	LSTM			CNN-LSTM			Model of This Paper		
	RMSE	MAE	Adjusted R2	RMSE	MAE	Adjusted R2	RMSE	MAE	Adjusted R2
12 h	14.65	9.39	0.90	13.85	9.56	0.91	10.62	7.27	0.94
24 h	14.0	9.23	0.91	12.90	8.80	0.92	9.79	7.02	0.95
36 h	14.24	9.15	0.90	12.96	8.91	0.92	9.66	6.95	0.95
48 h	16.87	10.57	0.86	13.30	9.20	0.91	10.37	7.25	0.94
60 h	17.16	11.07	0.86	14.90	10.34	0.89	10.99	7.56	0.94
72 h	17.41	11.45	0.85	14.92	10.40	0.89	11.06	7.42	0.94

Table 3. Comparison of the performance of the three methods for different time step predictions.

Time Step (Predicted)	LSTM			CNN-LSTM			Model of This Paper		
	RMSE	MAE	Adjusted R2	RMSE	MAE	Adjusted R2	RMSE	MAE	Adjusted R2
1 h	14.25	9.15	0.90	12.96	8.91	0.92	9.96	6.95	0.95
4 h	14.95	9.72	0.89	14.51	9.70	0.90	11.68	8.10	0.93
8 h	16.88	11.16	0.86	15.20	10.26	0.89	14.25	9.69	0.90
12 h	17.65	11.27	0.84	17.60	11.32	0.85	15.00	9.85	0.89
24 h	21.21	13.48	0.78	20.80	13.63	0.79	18.48	11.78	0.83

In order to display the forecast results intuitively, the forecast data from 26 February 2021 to 18 March 2021 are selected for display, as shown in Figures 5–9 below. The blue represents the real data value, the yellow is the predicted value of the LSTM model, the green is the predicted value of the CNN-LSTM model, and the red is the predicted value of the model in this article. It can be seen from Figures 5 and 6 that when the prediction step length is short, although the prediction results of the other two models and the predicted future trend can be well consistent with the real data, the model proposed in this article has achieved better results. At the same time, the model proposed in this article is also superior to the other two models in peak prediction. It can be seen from Figures 7–9 that as the prediction duration increases, the accuracy of the peak prediction and the prediction of the future trend of each model decreases. When the prediction time step is 24 h, the prediction trend of LSTM and CNN-LSTM starts to be opposite to that of the real data, as shown in the predicted value between 400 h and 450 h in Figure 9. The prediction results and future trends of the model in this article can be better agreement with the real data. Therefore, the model in this article can better simulate the long-term forecast of PM2.5.

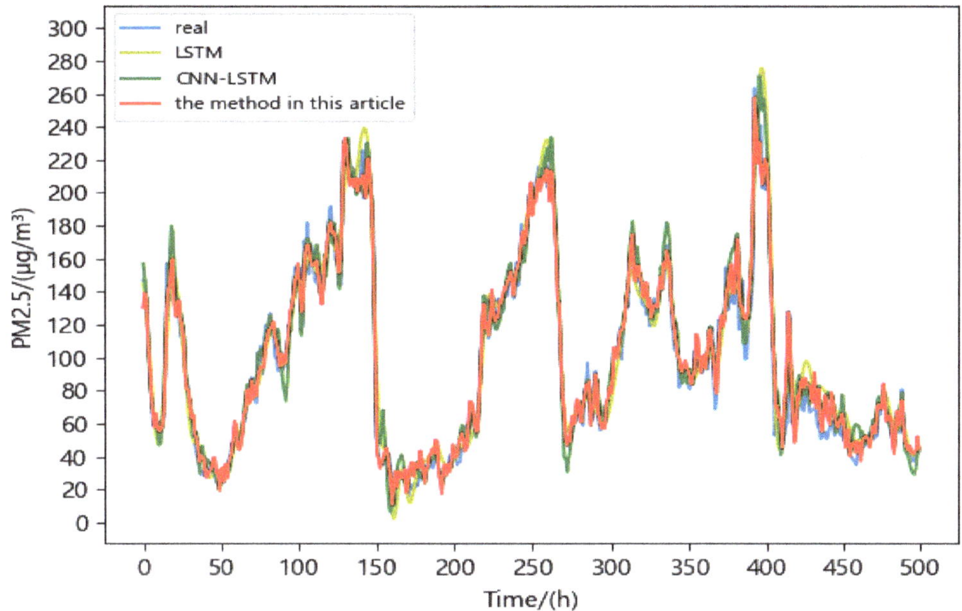

Figure 5. The prediction results of the three methods with a prediction step of 1 h.

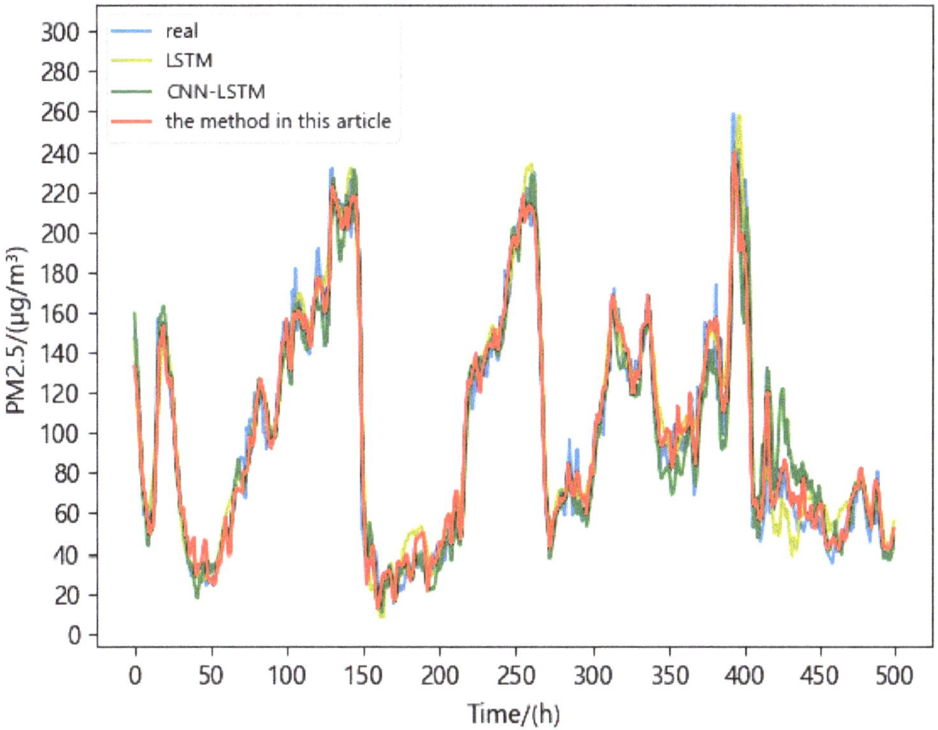

Figure 6. The prediction results of the three methods with a prediction step of 4 h.

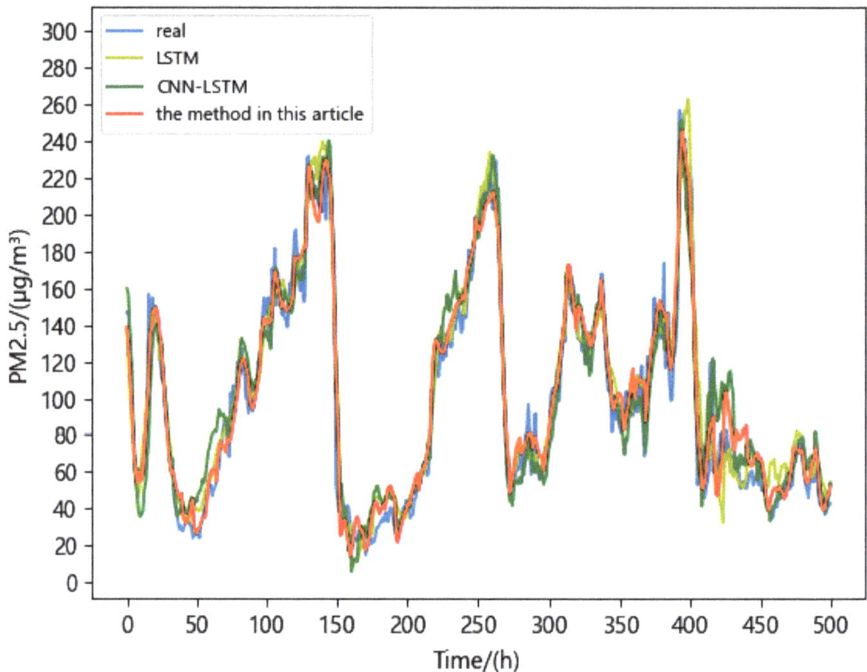

Figure 7. The prediction results of the three methods with a prediction step of 8 h.

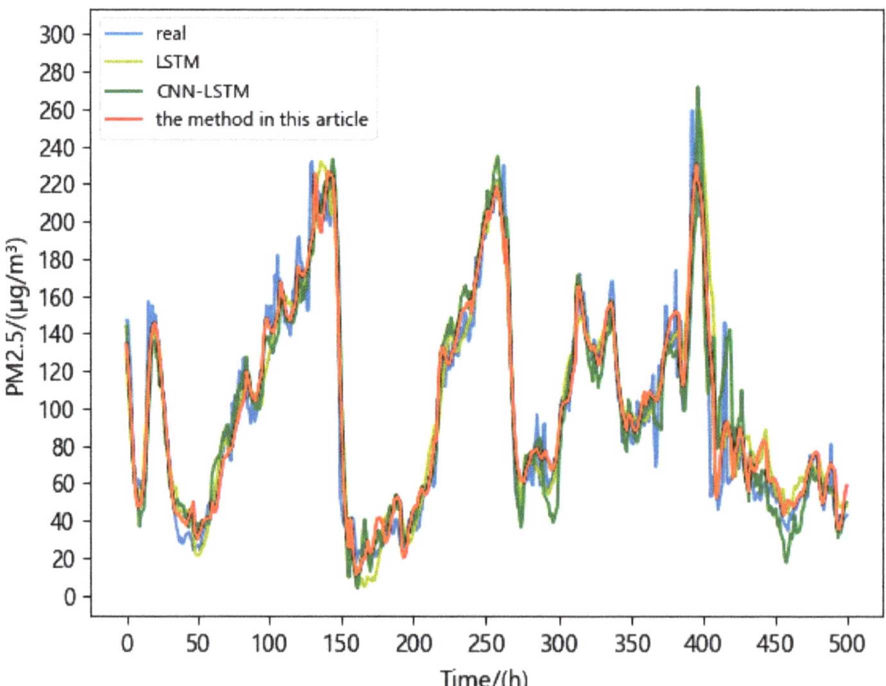

Figure 8. The prediction results of the three methods with a prediction step of 12 h.

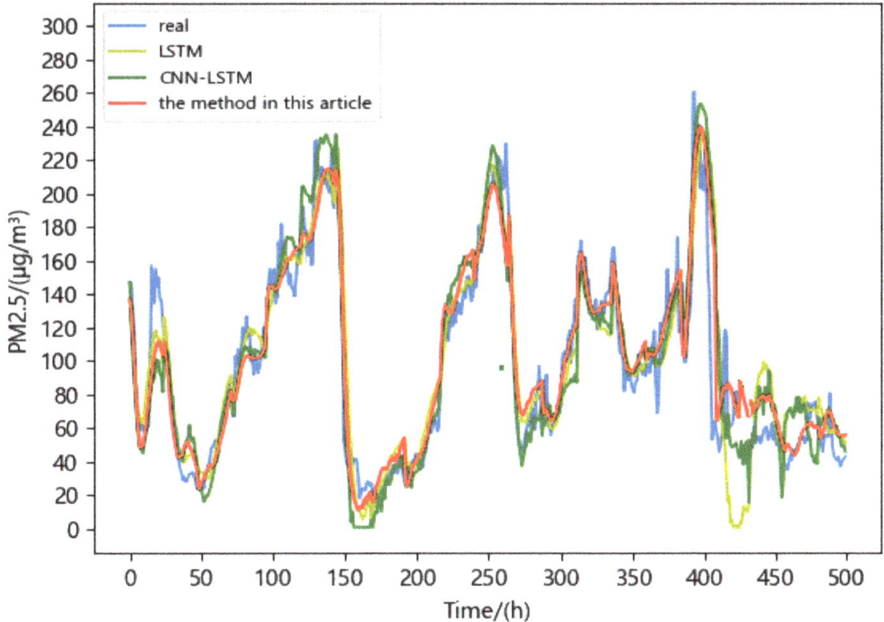

Figure 9. The prediction results of the three methods with a prediction step of 24 h.

4. Conclusions

The prediction of PM2.5 concentration is of great significance for People's Daily life and environmental governance. Because the characteristic information of different time scales has different influence on the prediction results, a multi-time scale fusion model is proposed in this paper. The experimental results show that the proposed multi-time scale fusion model is superior to the comparison model in single and multi-step prediction, indicating that the multi-time scale fusion is effective for long-term prediction. In addition, in this paper, only the data of one site is used for the experiment, the amount of data is too small, and the influence between sites is not taken into account. In the future, PM2.5 between adjacent stations will be studied and analyzed, and the accuracy of prediction will be improved by studying the spatial correlation between stations.

Author Contributions: Writing—original draft preparation, W.X.; writing—review and editing, J.Z.; funding acquisition, J.Z. All authors have read and agreed to the published version of the manuscript.

Funding: This work is supported by National Natural Science Foundation of China under Grant NSFC-61803148.

Institutional Review Board Statement: Not applicable.

Informed Consent Statement: Not applicable.

Data Availability Statement: The datasets used and/or analyzed during the current study are available from the corresponding author on reasonable request.

Conflicts of Interest: The authors declare no conflict of interest.

References

1. Fan, J.X.; Li, Q.; Zhu, Y.J.; Hou, J.X.; Feng, X. Research on time and space prediction model of air pollution based on RNN. *Sci. Surv. Mapp.* **2017**, *42*, 76–83.
2. Qi, B.L.; Guo, K.P.; Yang, B.; Du, Y.M.; Liu, M.; Wang, J.N. Air quality prediction based on GCN-LSTM. *Appl. Comput. Syst.* **2021**, *30*, 208–213.

3.	He, Z.X.; Li, L. A prediction model of air pollutant concentration based on wavelet transform and LSTM. *Environ. Eng.* **2021**, *39*, 111–119.
4.	Huang, C.J.; Kuo, P.H. A deep CNN-LSTM model for particulate matter (PM2.5) forecasting in smart cities. *Sensors* **2018**, *18*, 2220. [CrossRef]
5.	Liu, J.; Shahroudy, A.; Xu, D.; Kot, A.C.; Wang, G. Skeleton-based action recognition using spatio-temporal LSTM network with trust gates. *IEEE Trans. Pattern Anal. Mach. Intell.* **2017**, *40*, 3007–3021. [CrossRef]
6.	Wu, Z.; Huang, N.E. Ensemble empirical mode decompo-sition: A noise-assisted data analysis method. *Adv. Adapt. Data Anal.* **2009**, *1*, 1–41. [CrossRef]
7.	Lu, Y.; Yang, J.; Shao, Z.J.; Zhu, C.C. Robust prediction of PM (2.5) based on staged temporal attention network. *Environ. Eng.* **2021**, *39*, 1–9.
8.	Huang, W.J.; Li, D.Y.; Huang, Y. Long-term prediction of PM2.5 concentration based on deep learning. *Appl. Res. Comput.* **2021**, *38*, 1809–1814.
9.	Li, X.; Peng, L.; Yao, X.J.; Cui, S.L.; Hu, Y.; You, C.Z.; Chi, T. Long short-term memory neural network for air pollutant concentration predictions: Method development and evaluation. *Environ. Pollut.* **2017**, *231*, 997–1004. [CrossRef] [PubMed]

Article

Impact of Air Pollution on Global Burden of Disease in 2019

Meghnath Dhimal [1,2,*,†], Francesco Chirico [3,4,*,†], Bihungum Bista [1], Sitasma Sharma [1], Binaya Chalise [5], Mandira Lamichhane Dhimal [2,6], Olayinka Stephen Ilesanmi [7,8], Paolo Trucillo [9] and Daniele Sofia [10]

[1]   Nepal Health Research Council, Ramshah Path, Kathmandu 44600, Nepal; bistabihungum@gmail.com (B.B.); sitasmasharma4019@gmail.com (S.S.)
[2]   Global Institute for Interdisciplinary Studies, Kathmandu 44600, Nepal; ldmandira@gmail.com
[3]   Health Service Department, State Police, Ministry of Interior, 20162 Milan, Italy
[4]   Post-Graduate Specialization School of Occupational Health, Università Cattolica del Sacro Cuore, 00100 Rome, Italy
[5]   Graduate School for International Development and Cooperation, Hiroshima University, Hiroshima 739-8528, Japan; binayachalise@gmail.com
[6]   Policy Research Institute, Sano Gaucharan, Kathmandu 44600, Nepal
[7]   Department of Community Medicine, College of Medicine University of Ibadan, Ibadan 200284, Oyo State, Nigeria; ileolasteve@yahoo.co.uk
[8]   Department of Community Medicine, University College Hospital, Ibadan 200284, Oyo State, Nigeria
[9]   Department of Chemical, Material and Industrial Production Engineering, University of Naples Federico II, Piazzale Tecchio, 80, 80125 Napoli, Italy; paolo.trucillo@unina.it
[10]   Department of Industrial Engineering, University of Salerno, Via Giovanni Paolo II, 132, 84084 Fisciano, Italy; dsofia@unisa.it
*   Correspondence: meghdhimal@gmail.com (M.D.); medlavchirico@gmail.com (F.C.)
†   First contributorship.

Citation: Dhimal, M.; Chirico, F.; Bista, B.; Sharma, S.; Chalise, B.; Dhimal, M.L.; Ilesanmi, O.S.; Trucillo, P.; Sofia, D. Impact of Air Pollution on Global Burden of Disease in 2019. *Processes* **2021**, *9*, 1719. https://doi.org/10.3390/pr9101719

Academic Editors: Zhiming Liu and Ki-Youn Kim

Received: 23 July 2021
Accepted: 22 September 2021
Published: 25 September 2021

Publisher's Note: MDPI stays neutral with regard to jurisdictional claims in published maps and institutional affiliations.

Abstract: Air pollution consisting of ambient air pollution and household air pollution (HAP) threatens health globally. Air pollution aggravates the health of vulnerable people such as infants, children, women, and the elderly as well as people with chronic diseases such as cardiorespiratory illnesses, little social support, and poor access to medical services. This study is aimed to estimate the impact of air pollution on global burden of disease (GBD). We extracted data about mortality and disability adjusted life years (DALYs) attributable to air pollution from 1990 to 2019. The extracted data were then organized and edited into a usable format using STATA version 15. Furthermore, we also estimated the impacts for three categories based on their socio-demographic index (SDI) as calculated by GBD study. The impacts of air pollution on overall burden of disease by SDI, gender, type of pollution, and type of disease is estimated and their trends over the period of 1990 to 2019 are presented. The attributable burden of ambient air pollution is increasing over the years while attributable burden of HAP is declining over the years, globally. The findings of this study will be useful for evidence-based planning for prevention and control of air pollution and reduction of burden of disease from air pollution at global, regional, and national levels.

Keywords: air pollution; gender; burden of disease; non-communicable diseases; deaths; DALYs; policy; socio-demographic index

1. Introduction

Globally, air pollution is reported as a major environmental risk factor that poses a substantial threat to human health [1]. Approximately 90% of the global population is at risk of air pollution and the disease burden attributable to air pollution continues to grow, posing a serious threat to global health [2–5]. As per global estimates, air pollution, both ambient air pollution (AAP) and household air pollution (HAP), accounts for 7 million premature deaths worldwide [2]. Moreover, the burden of disease attributed to air pollution is considerably higher in low-and-middle-income countries (LMICs), where more than 90% of deaths occur, compared to high-income countries [1,6,7].

Particularly, the morbidity and mortality due to AAP have escalated steadily over time, with an estimated 4.2 million annual deaths [3,8,9]. For this reason, it is important to monitor pollution data clearly and actuate optimized solutions in order to establish a new equilibrium among sustainability and human productivity and wellness [10]. Particulate matter (PM), nitrogen oxides (NO and NO_2), sulfur dioxide (SO_2), and ozone (O_3) are the major established ambient air pollutants responsible for adverse health effects [2]. Albeit in a declining trend, household air pollution (HAP) accounts for around 3.8 million deaths per year [11]. About 3 billion people, particularly in developing countries, continue to rely on inefficient and polluting fuels for various purposes such as cooking, heating, and lighting [2,5,12]. Therefore, HAP remains a great environmental health threat in low-and middle-income countries [13].

Air pollution, both AAP and HAP, aggravates health problems and escalates the likelihood of severe health impacts. Respiratory diseases and conditions are the most known and established health effects of air pollution; nevertheless, cardiovascular, cerebrovascular, reproductive, neurologic, and carcinogenic effects have also been associated with air pollution [14–18]. Non-communicable diseases (NCDs) contribute most of the global disease burden (GBD) [5] and air pollution is one of the top-ranking risk factors of NCDs. Air pollution accounted for 16% of all NCD deaths among those aged 30 to 69 in 2016 [13,19].

Every year, around 7 million people die prematurely because of an increased number of deaths from stroke, chronic obstructive pulmonary disease (COPD), heart disease, lung cancer, and acute respiratory infections (ARIs) caused by air pollution [2]. As per the estimation of the World Health Organization (WHO), ischemic heart disease (IHD) and strokes accounted for 58% of AAP-related premature deaths in 2016, while COPD and ARIs accounted for 18% and lung cancer for 6%, respectively [2]. On the other hand, pneumonia and ischemic heart disease accounted for 27% of HAP-related deaths, while COPD, stroke, and lung cancer accounted for 20%, 18%, and 8% of deaths, respectively [11]. This descriptive study, which is based on secondary data drawn from the GBD study, aims to estimate the impact of air pollution on the GBD in 2019.

2. Materials and Methods

This descriptive study accessed the data from GBD Study 2019 led by the Institute for Health Metrics and Evaluation (IHME) at University of Washington and publicly available via the GBD compare website [20]. The GBD study is a systematic way to quantify the comparative amount of health loss due to diseases, injuries, and risk factors by age, sex, and geographies for specific points in time. IHME makes this information freely available so that appropriate decisions can be made based on available evidence.

The GBD study uses various sources of data such as censuses, household surveys, health service use, civil registration and vital statistics disease registries, air pollution monitors, satellite imaging, disease notifications, and other sources from 1990 to 2019 in making estimates of the burden of disease (BoD) [21]. Several studies have shown the attribution of air pollution on BOD especially affecting several NCDs such as cardiorespiratory diseases [1,14,19,22].

For this article, we extracted data about mortality and disability adjusted life years (DALYs) which are attributable to air pollution from 1990 to 2019. The extracted data were then organized and presented in a usable format using STATA version 15. Furthermore, we also display the results for three categories based on their Socio-demographic Index (SDI) as calculated by GBD. This SDI is a composite indicator of development status, which ranges from 0 to 1 which is calculated as a geometric mean of the values of the indices of lag-distributed per capita income, mean education in people aged 15 years or older, and total fertility rate in people younger than 25 years. Hence, the SDI is a summary measure of a certain geographic or administrative boundary' socio-economic development was developed by GBD researcher.

3. Results

The impacts of air pollution on overall BOD by SDI, gender, type of pollution, and type of disease is estimated and their trends over the period of 1990 to 2019 are presented. Figure 1 illustrates the DALYs rate over a population of 100,000 attributable to air pollution in various SDI regions including a global DALYs trend over 29 years from 1990 to 2019. The global trend in DALYS rate attributable to air pollution declined from 1990 to 2019.

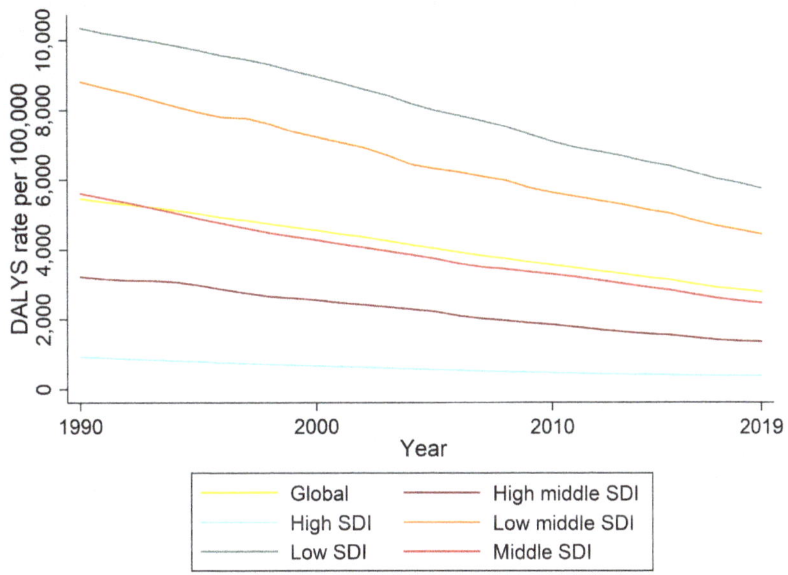

Figure 1. DALYs attributable to air pollution from 1990 to 2019.

Similarly, the DALYs rate in low SDI and low middle SDI regions fell gradually over time, albeit these regions continued to have the highest DALYs rates due to air pollution, respectively. From 1990 to 2019, the low SDI region accounted for most of the DALYs attributed to air pollution. In 1990, the middle SDI region had a greater rate of DALYs than the global average rate, but this rate began to decline slowly in the mid-1990s. Between 1990 and 2008, the high middle SDI region showed somewhat undulating trends, but the DALYs rate gradually declined between 2010 and 2019. From 1990 through 2019, the high SDI region experienced the lowest DALYs rate and remained considerably stable between 2008 and 2019.

Figure 2 shows the mortality rate per 100,000 population attributable to air pollution in various SDI regions of the word from 1990 to 2019. During a 29-year timeframe, death rate due to air pollution decreased in all five regions. However, three regions—middle SDI, low middle SDI, and low SDI—had higher death rates in comparison to the global average deaths rate, with the low SDI region having the highest death rates since 1990. In the low middle SDI region, the falling mortality trend rate showed a slight increase in the late 1990s, followed by an undulating trend from 1997 to 2015, and then a modest decline since then. On the other hand, the high SDI region had the lowest death rate per 100,000 attributable to air pollution from 1990 to 2019; nonetheless, the declining trend remained steady from 2009 to 2019.

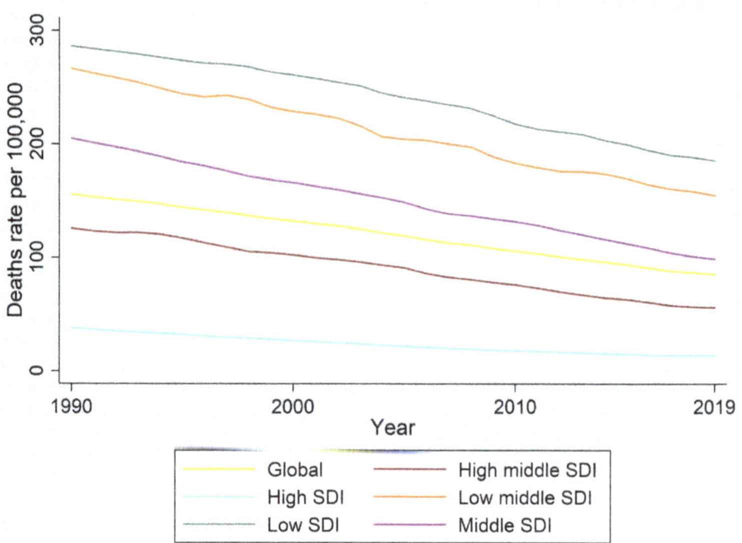

Figure 2. Deaths attributable to air pollution from 1990 to 2019.

The Figure 3 shows the gender-wise death rate attributable to air pollution from 1990 to 2019. Air pollution attributable death rate per 100,000 decreased considerably in both gender groups in the 29-year timeframe. However, in comparison to the female population, males were characterized by a greater mortality rate from 1990 to 2019. For the female population, the death rate attributed to air pollution was 135 per 100,000 in 1990, and the rate fell to 70 in 2019. For males, the mortality rate dropped from 185 in 1990 to 110 in 2019.

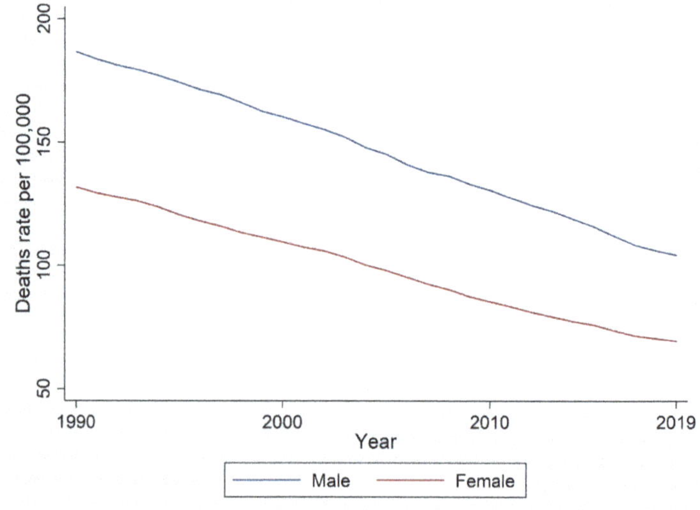

Figure 3. Gender-wise death rate attributable to air pollution from 1990 to 2019.

Figure 4 shows the death rate per 100,000 population attributable to ambient air pollution and HAP from 1990 to 2019. The graph illustrates that the death rate attributable to HAP plummeted from 81 per 100,000 population in 1990 to 30 in 2019; meanwhile, death rate attributable to AAP was just 39 per 100,000 population in 1990 but increased over the last 20 years. The death rate attributable to AAP somewhat fluctuated in between and reached a peak of 52 per 100,000 in 2015 and 54 per 100,000 in 2019.

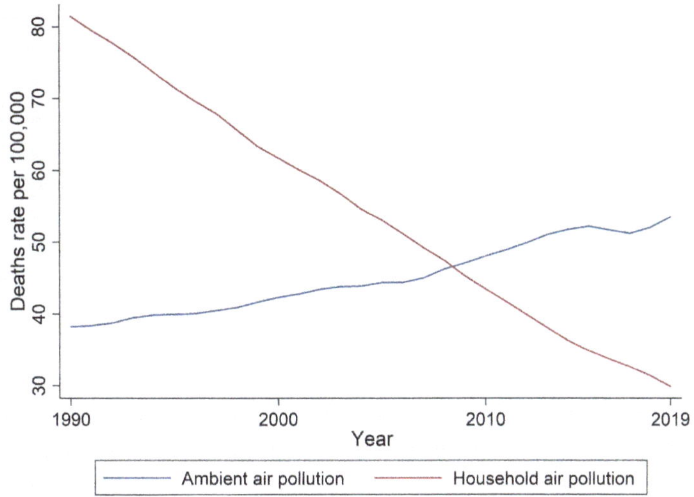

Figure 4. Different types of air pollution and attributable deaths.

Figure 5 shows the communicable maternal neonatal and nutritional diseases (CMN-NDs) and NCD-related deaths rate attributable to HAP from 1990 to 2019. Both CMNNDs and NCD-related deaths per 100,000 due to HAP plummeted in the 29-year period. In this period, the death rates from CMNNDs and NCDs attributable to HAP lowered by 28%. Household air pollution accounted for a greater number of deaths from NCDs in comparison to CMNNDs diseases.

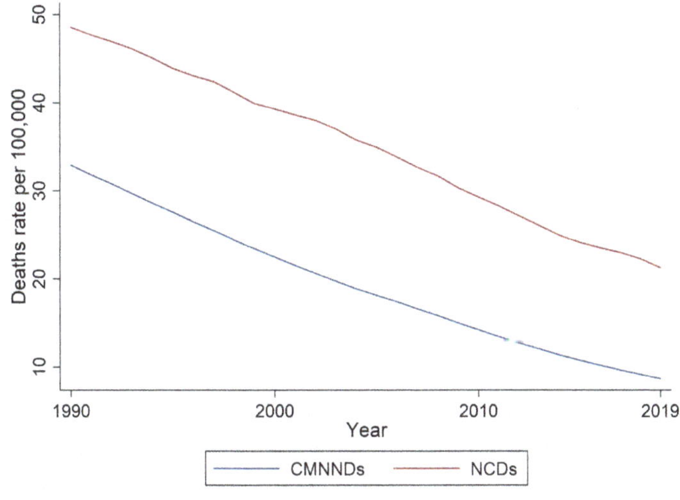

Figure 5. Deaths attributable to household air pollution from 1990 to 2019.

The Figure 6 illustrates CMNNDs and NCD-related deaths per 100,000 population attributable to AAP from 1990 to 2019. In comparison to CMNNDs, AAP significantly accounted for a higher death rate from NCDs. In 1990, AAP attributed to the death rate from NCDs was around 30 per 100,000 population, but by 2019 it surged up to 48. Meanwhile, the death rate from CMNNDs remained considerably low and stable in the same period.

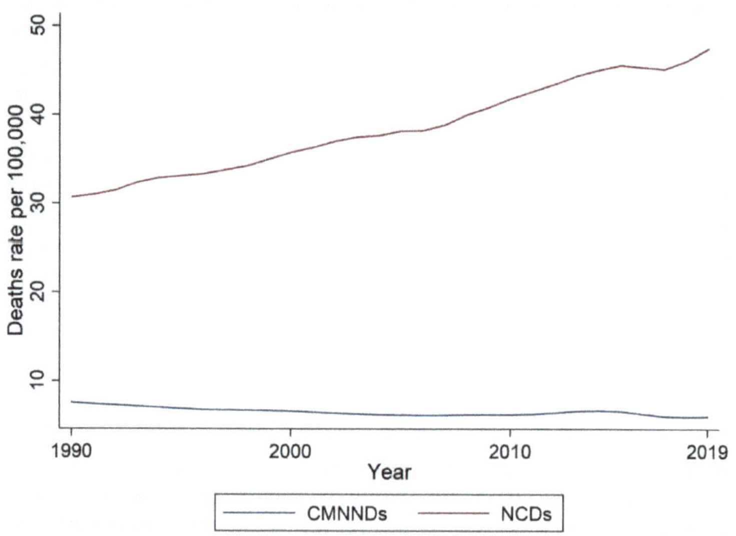

Figure 6. Deaths attributable to ambient air pollution from 1990 to 2019.

4. Discussion

Findings from this study revealed that air pollution declined between 1990 and 2019. The decline in air pollution could be due to two reasons. First, the millennium signalled increased awareness of unintended health outcomes due to air pollution. Rapid urbanization and globalization were ongoing that could increase air pollution. However, efforts towards reducing the risk of humans to air pollution secondary to energy generation were strengthened [23]. Although increased waste generation due to overcrowding in industrialized and commercial centres was a risk factor for the increased rate of waste generation, several studies investigated the measures through which air pollution could be minimized [24–26]. These measures, therefore, helped to identify that the combustion of fossil fuels and the generation of air pollutants due to industrial activities resulted in a trade-off of global health with financial sustainability. As a result, a large portion of the government's budgetary plan was channeled towards curative and preventive health costs since many productive hours were lost while seeking treatment for disability [27]. Thus, overall, air pollution was indeed contradictory to the government's aim of promoting the wellbeing of its citizens. As a result, increased involvement of experts in the medical and allied professions helped to understand the strategies for reducing ambient and household air production. The replacement of kerosene lamps and stoves with chargeable fluorescent lamps and gas cylinders for cooking alludes to this fact in many settings. Although a drastic change may be unattainable at the same point globally, all regions recorded a decline in DALYs due to air pollution.

We identified a decline in death rates due to air pollution across the five regions, with the low SDI region recording the highest death rates. Given the ongoing rapid commercialization, it is likely that environmental policy stringency measures such as the Clean Air Act contributed significantly to the reported decline in death rates due to air pollution between 1990 and 2019 [28]. The Clean Air Act is a comprehensive federal law under the auspices of the Environmental Protection Agency to establish National Air quality standards to promote environmental and public health [29]. As a result, the release of common air pollutants such as nitrogen oxides, particulate matter (PM2.5 and PM10) and ozone are placed under regular checks [30]. Agriculture, forestry, and land-use laws have helped to regulate ambient air pollution in many countries in the low SDI region whose economy primarily thrives on agriculture. Since ambient air pollution caused many NCD-related deaths, this elucidates that ambient air pollution is one of the drivers of the development of NCDs. If left unchecked, there is a high likelihood to record a surge in DALYs due to NCDs. To ensure that a steady decline in death rates and development of NCDs due to air pollution, policymakers are required to have high stringency regarding indiscriminate air pollution.

This study found that males had a higher mortality rate due to air pollution between 1990 and 2019. This finding shows that males are more likely to be users of technologies that emit air pollutants than females. Literature has reported that females are more likely to die due to CVDs than primarily from air pollution compared to males. This was confirmed from a study conducted in Poland where it was found that the effect of SO_2 inhalation was more pronounced in males [31,32]. As a result, the inhalation of this pollutant could cause pulmonary inflammation, resulting in thrombosis, endothelial dysfunction, imbalance of the autonomic nervous system, increased arterial pressure, and ultimately death [32–36]. The exact pathophysiologic mechanisms that underlay gender differences in the death rate due to air pollution are unknown. Pollution control measures should be considered and implemented promptly to avert preventable deaths.

5. Conclusions

Globally, air pollution is still a major risk factor which continues to pose a significant threat to environmental safety and public health. If the drivers and sources of air pollution are not immediately controlled, more air pollution-related deaths are likely to result. In addition, higher DALYs and NCDs are bound to occur frequently. Despite the decline of DALYS attributable to HAP, the attributable burden of AAP is increasing over the years globally. Hence, air pollution prevention and control policies are urgently required to reduce burden of diseases attributable to air pollution at the global, regional, and national level.

Author Contributions: Conceptualization, M.D., B.B., B.C., M.L.D. and F.C.; methodology, M.D., B.C., B.B.; software, B.B. and B.C.; validation, M.D., F.C. and B.C.; formal analysis, B.B.; investigation, M.D., F.C.; resources, M.D., F.C.; data curation, B.C. and B.B.; writing—original draft preparation, S.S., M.D., B.B., M.L.D. and F.C.; writing—review and editing, M.D., M.L.D., F.C., P.T., D.S., O.S.I.; B.B. and B.C. All authors have read and agreed to the published version of the manuscript.

Funding: This research received no external funding.

Data Availability Statement: The data for this study are publicly available at http://www.healthdata.org/gbd/2019 accessed on 15 June 2021.

Conflicts of Interest: The authors declare no conflict of interest.

References

1. Cohen, A.J.; Brauer, M.; Burnett, R.; Anderson, H.R.; Frostad, J.; Estep, K.; Balakrishnan, K.; Brunekreef, B.; Dandona, L.; Dandona, R.; et al. Estimates and 25-year trends of the global burden of disease attributable to ambient air pollution: An analysis of data from the Global Burden of Diseases Study 2015. *Lancet* **2017**, *389*, 1907–1918. [CrossRef]
2. World Health Organization. Ambient (Outdoor) Air Pollution. Available online: https://www.who.int/news-room/fact-sheets/detail/ambient-(outdoor)-air-quality-and-health (accessed on 15 June 2021).
3. Shaddick, G.; Thomas, M.L.; Mudu, P.; Ruggeri, G.; Gumy, S. Half the world's population are exposed to increasing air pollution. *NPJ Clim. Atmos. Sci.* **2020**, *3*, 1–5. [CrossRef]

4. World Health Organization. Health Benefits Far Outweigh the Costs of Meeting Climate Change Goals. Available online: https://www.who.int/news/item/05-12-2018-health-benefits-far-outweigh-the-costs-of-meeting-climate-change-goals (accessed on 15 June 2021).
5. GBD 2019 Risk Factors Collaborators. Global burden of 87 risk factors in 204 countries and territories, 1990–2019: A system-atic analysis for the Global Burden of Disease Study 2019. *Lancet* **2020**, *396*, 1223–1249. [CrossRef]
6. Landrigan, P.J.; Fuller, R.; Acosta, N.J.R.; Adeyi, O.; Arnold, R.; Basu, N.N.; Baldé, A.B.; Bertollini, R.; Bose-O'Reilly, S.; Boufford, J.I.; et al. The Lancet Commission on pollution and health. *Lancet* **2018**, *391*, 462–512. [CrossRef]
7. World Health Organization. Burden of Disease from the Joint Effects of Household and Ambient Air Pollution for 2016. Available online: https://www.who.int/airpollution/data/AP_joint_effect_BoD_results_May2018.pdf (accessed on 15 June 2021).
8. Forouzanfar, M.H.; Afshin, A.; Alexander, L.T.; Anderson, H.R.; Bhutta, Z.A.; Biryukov, S.; Brauer, M.; Brunett, R.; Cercy, K.; Charlson, F.; et al. Global, regional, and national comparative risk assessment of 79 behavioural, environmental and oc-cupational, and metabolic risks or clusters of risks, 1990–2015: A systematic analysis for the Global Burden of Disease Study 2015. *Lancet* **2016**, *388*, 1659–1724. [CrossRef]
9. Lancet, T. *Air Pollution—Crossing Borders*; Elsevier: Amsterdam, The Netherlands, 2016.
10. Sofia, D.; Lotrecchiano, N.; Trucillo, P.; Giuliano, A.; Terrone, L. Novel Air Pollution Measurement System Based on Ethereum Blockchain. *J. Sens. Actuator Netw.* **2020**, *9*, 49. [CrossRef]
11. World Health Organization. Household Air Pollution and Health. Available online: https://www.who.int/news-room/fact-sheets/detail/household-air-pollution-and-health (accessed on 15 June 2021).
12. Stanaway, J.D.; Afshin, A.; Gakidou, E.; Lim, S.S.; Abate, D.; Abate, K.H.; Abbafati, C.; Abbasi, N.; Abbastabar, H.; Abd-Allah, F.; et al. Global, regional, and national comparative risk assessment of 84 behavioural, environmental and occu-pational, and metabolic risks or clusters of risks for 195 countries and territories, 1990–2017: A systematic analysis for the Global Burden of Disease Study 2017. *Lancet* **2018**, *392*, 1923–1994.
13. World Health Organization. WHO's First Global Conference on Air Pollution and Health. Available online: https://www.who.int/news-room/events/detail/2018/10/30/default-calendar/who-s-first-global-conference-on-air-pollution-and-health (accessed on 16 June 2021).
14. Hoek, G.; Krishnan, R.M.; Beelen, R.; Peters, A.; Ostro, B.; Brunekreef, B.; Kaufman, J.D. Long-term air pollution exposure and cardio-respiratory mortality: A review. *Environ. Health* **2013**, *12*, 43. [CrossRef]
15. Thurston, G.D.; Kipen, H.; Annesi-Maesano, I.; Balmes, J.; Brook, R.D.; Cromar, K.; De Matteis, S.; Forastiere, F.; Forsberg, B.; Frampton, M.W.; et al. A joint ERS/ATS policy statement: What constitutes an adverse health effect of air pollution? An analytical framework. *Eur. Respir. J.* **2017**, *49*, 1600419. [CrossRef]
16. Loomis, D.; Grosse, Y.; Lauby-Secretan, B.; El Ghissassi, F.; Bouvard, V.; Benbrahim-Tallaa, L.; Guha, N.; Baan, R.; Mattock, H.; Straif, K. The carcinogenicity of outdoor air pollution. *Lancet Oncol.* **2013**, *14*, 1262–1263. [CrossRef]
17. Chirico, F.; Magnavita, N. Letter to the editor (January 1, 2019) concerning the paper "Impact of air pollution on depression and suicide. *Int. J. Occup. Med. Environ. Health* **2019**, *32*, 413–414. [CrossRef]
18. Chirico, F. Comments on "Climate Change and Public Health: A Small Frame Obscures the Picture". *NEW SOLUTIONS A J. Environ. Occup. Health Policy* **2018**, *28*, 5–7. [CrossRef] [PubMed]
19. Prüss-Ustün, A.; Van Deventer, E.; Mudu, P.; Campbell-Lendrum, D.; Vickers, C.; Ivanov, I.; Forastiere, F.; Gumy, S.; Dora, C.; Adair-Rohani, H.; et al. Environmental risks and non-communicable diseases. *BMJ* **2019**, *364*, l265. [CrossRef]
20. Institute for Health Metrics and Evaluation. Global Burden of Disease Study 2019. Available online: http://vizhub.healthdata.org/gbd-compare/ (accessed on 7 April 2021).
21. Vos, T.; Lim, S.S.; Abbafati, C.; Abbas, K.M.; Abbasi, M.; Abbasifard, M.; Abbasi-Kangevari, M.; Abbastabar, H.; Abd-Allah, F.; Abdelalim, A.; et al. Global burden of 369 diseases and injuries in 204 countries and territories, 1990–2019: A systematic analysis for the Global Burden of Disease Study 2019. *Lancet* **2020**, *396*, 1204–1222. [CrossRef]
22. Dhimal, M.; Neupane, T.; Dhimal, M.L. Understanding linkages between environmental risk factors and noncommunicable diseases—A review. *FASEB BioAdvances* **2021**, *3*, 287–294. [CrossRef] [PubMed]
23. Huang, J.; Pan, X.; Guo, X.; Li, G. Impacts of air pollution wave on years of life lost: A crucial way to communicate the health risks of air pollution to the public. *Environ. Int.* **2018**, *113*, 42–49. [CrossRef]
24. Van Vliet, E.D.S.; Kinney, P.L. Impacts of roadway emissions on urban particulate matter concentrations in sub-Saharan Africa: New evidence from Nairobi, Kenya. *Environ. Res. Lett.* **2007**, *2*. [CrossRef]
25. Pope, C.A., III; Ezzati, M.; Dockery, D.W. Fine-Particulate Air Pollution and Life Expectancy in the United States. *N. Engl. J. Med.* **2009**, *360*, 376–386. [CrossRef] [PubMed]
26. Young, M.T.; Sandler, D.P.; DeRoo, L.A.; Vedal, S.; Kaufman, J.; London, S.J. Ambient Air Pollution Exposure and Incident Adult Asthma in a Nationwide Cohort of U.S. Women. *Am. J. Respir. Crit. Care Med.* **2014**, *190*, 914–921. [CrossRef]
27. Oliva, P.; Alexianun, M.; Nasir, R. Suffocating Prosperity: Air Pollution and Economic Growth in Developing Countries. International Growth Centre, 2019. Available online: https://www.theigc.org/wp-content/uploads/2019/12/IGCJ7753-IGC-Pollution-WEB_.pdf (accessed on 16 June 2021).
28. Sofia, D.; Gioiella, F.; Lotrecchiano, N.; Giuliano, A. Mitigation strategies for reducing air pollution. *Environ. Sci. Pollut. Res.* **2020**, *27*, 19226–19235. [CrossRef] [PubMed]

29. Sofia, D.; Lotrecchiano, N.; Giuliano, A.; Barletta, D.; Poletto, M. Optimization of number and location of sampling points of an air quality monitoring network in an urban contest. *Chem. Eng. Trans.* **2019**, *74*, 277–282.
30. U.S. Department of the Interior, Bureau of Ocean Energy Management. Air Quality Act (1967) or the Clean Air Act (CAA). Available online: https://www.boem.gov/air-quality-act-1967-or-clean-air-act-caa (accessed on 17 June 2021).
31. Kuźma, Ł.; Struniawski, K.; Pogorzelski, S.; Bachórzewska-Gajewska, H.; Dobrzycki, S. Gender Differences in Association between Air Pollution and Daily Mortality in the Capital of the Green Lungs of Poland–Population-Based Study with 2,953,000 Person-Years of Follow-Up. *J. Clin. Med.* **2020**, *9*, 2351. [CrossRef] [PubMed]
32. Arias-Pérez, R.D.; Taborda, N.A.; Gómez, D.M.; Narvaez, J.F.; Porras, J.; Hernandez, J.C. Inflammatory effects of particulate matter air pollution. *Environ. Sci. Pollut. Res.* **2020**, *27*, 42390–42404. [CrossRef]
33. Tainio, M.; Andersen, Z.J.; Nieuwenhuijsen, M.J.; Hu, L.; de Nazelle, A.; An, R.; Garcia, L.M.; Goenka, S.; Zapata-Diomedi, B.; Bull, F.; et al. Air pollution, physical activity and health: A mapping review of the evidence. *Environ. Int.* **2021**, *147*, 105954. [CrossRef] [PubMed]
34. Lee, K.K.; Bing, R.; Kiang, J.; Bashir, S.; Spath, N.; Stelzle, D.; Mortimer, K.; Bularga, A.; Doudesis, D.; Joshi, S.S.; et al. Adverse health effects associated with household air pollution: A systematic review, meta-analysis, and burden estimation study. *Lancet Glob. Health* **2020**, *8*, e1427–e1434. [CrossRef]
35. Norbäck, D.; Wang, J. Household air pollution and adult respiratory health. *Eur. Respir. J.* **2021**, *57*, 2003520. [CrossRef]
36. Eguiluz-Gracia, I.; Mathioudakis, A.G.; Bartel, S.; Vijverberg, S.J.H.; Fuertes, E.; Comberiati, P.; Cai, Y.S.; Tomazic, P.V.; Diamant, Z.; Vestbo, J.; et al. The need for clean air: The way air pollution and climate change affect allergic rhinitis and asthma. *Allergy* **2020**, *75*, 2170–2184. [CrossRef]

Article

Air Pollution Analysis during the Lockdown on the City of Milan

Nicoletta Lotrecchiano [1,2], Paolo Trucillo [3], Diego Barletta [1], Massimo Poletto [1] and Daniele Sofia [1,2,*]

[1] DIIN—Department of Industrial Engineering, University of Salerno, Via Giovanni Paolo II 132, 84084 Fisciano, Italy; nlotrecchiano@unisa.it (N.L.); dbarletta@unisa.it (D.B.); mpoletto@unisa.it (M.P.)

[2] Research Department, Sense Square Srl, 84084 Salerno, Italy

[3] Department of Chemical, Material and Industrial Production Engineering, University of Naples Federico II, Piazzale V. Tecchio 80, 80125 Napoli, Italy; paolo.trucillo@unina.it

* Correspondence: dsofia@unisa.it or danielesofia@sensesquare.eu

Abstract: From February 2020, the progressive adoption of measures to contain coronavirus's contagion has resulted in a sudden change in anthropogenic activities in Italy, especially in Lombardy. From a scientific point of view, this situation represents a unique laboratory for understanding and predicting the consequences of specific measures aimed at improving air quality. In this work, the lockdown effect on Milan's (Italy) air quality was analyzed. The PM10 and PM2.5 values were measured by the ARPA Lombardia, and the real-time on-road (ROM) air quality monitoring network indicates the seasonality of these pollutants, which typically record the highest values in the coldest months of the year. The 10-year particulate matter concentrations analysis shows a PM10 reduction of 35% from 2010 to 2020. March 2020 data analysis shows an alternation of days with higher and lower particulate matter concentrations; values decrease in pollutants concentrations of 16%, respective to 2018. The complexity of the phenomena related to the atmospheric particulates formation, transport, and accumulation is highlighted by some circumstances, such as the Sahara dust events. The study showed that the trend of a general pollutant concentration reduction should be attributed to the decrease in emissions (specifically, from the transport sector) from the variation of meteorological and environmental conditions.

Keywords: pollution; dynamic monitoring; micro monitoring; atmospheric pollution; coronavirus; COVID-19; lockdown; anthropic impact; saharan dust

Citation: Lotrecchiano, N.; Trucillo, P.; Barletta, D.; Poletto, M.; Sofia, D. Air Pollution Analysis during the Lockdown on the City of Milan. *Processes* **2021**, *9*, 1692. https://doi.org/10.3390/pr9101692

Academic Editors: Avelino Núñez-Delgado and Cherng-Yuan Lin

Received: 20 July 2021
Accepted: 17 September 2021
Published: 22 September 2021

Publisher's Note: MDPI stays neutral with regard to jurisdictional claims in published maps and institutional affiliations.

1. Introduction

As similar to many other countries in the rest of the world, starting with the identification of the first case of SARS-CoV2 on 31 January 2020, Italy was hit by the pandemic wave with the consequent lockdown of all activities [1]. During this period, which took place between February and April 2020, the world has become open-air and it was possible to verify the effects of the most polluting anthropogenic activities reduction on air quality [2].

The air quality improvements caused by the 2020 lockdown were unprecedented in many parts of the world. The restrictions imposed by the pandemic on transportation and many production sectors have not prevented the values of particulate matter (PM) from exceeding the limits set by the World Health Organization. Only 24 countries have managed to stay below the safety threshold set by the WHO, out of the 106 countries surveyed by a new report prepared by IQAir and the United Nation agency [3]. According to the NU report, the worst countries for air quality are Bangladesh, Pakistan, India, Mongolia, and Afghanistan, where the pollution surveys record average values of PM2.5 ranging from 47 to 77 $\mu g/m^3$. The UN report indicates that in 2020, all Indian cities improved in air quality compared to 2018, while 63% recorded improvements compared to 2019. However, India continues to occupy a prominent place among the most polluted cities, and New Delhi is confirmed as the urban area with the worst air pollution in the world. The report highlights that in 2020, half of the European cities exceeded the WHO parameters for PM2.5, recording the highest levels in Eastern and Southern Europe, with Bosnia and Herzegovina, Macedonia, and Bulgaria. Italy suffered the lockdown effect in a diversified way in the various regions [4], remaining in the average of European cities.

Italy is characterized by a complex orography by different climatic zones, where, depending on the seasonal period, very favorable conditions for the pollutants accumulation and formation in the atmosphere can occur [5]. A striking example is the river Po basin and some areas of the central-southern region in the winter period, when the generated pollutants accumulate and coastal areas that, despite of the population, present conditions that generally favor dispersion and reduce the possibility of secondary pollutants forming.

To assess the lockdown effects, it should be considered that the period of March is usually less favorable to the accumulation of pollutants than the months of January and February, when conditions of thermal inversion at low altitude and atmospheric stability often occur with high values of the main pollutants [6]. In this perspective, the lockdown effect should be found by comparing the pollutant levels observed during the lockdown with the observations in the same late winter–spring period of the previous year. The report from the National Service for Environmental Protection (SNPA) [6] 2020 revealed that, throughout the Italian peninsula, there has been a reduction in the concentrations of nitrogen monoxide (emitted directly) and dioxide (partly emitted directly and partly formed in the atmosphere), carbon monoxide, and benzene. The nitrogen dioxide reduction was around 40% [7], ranging from a few percentage points to values over 70% in some sites. If it was easy to define the changes for nitrogen oxides, the effects on the levels of particulate matter (PM10 and PM2.5) are less easy to read. Particulate matter is a complex mixture of solid and liquid particles dispersed in the atmosphere that may have different sources, depending on the season and the geographical area. Some are of natural origin, such as the marine aerosol and the particles that originate from the long-distance transport of desert sands [8]. Particulates are generated locally and may have natural origin, such as wind-lifted terrigenous particles, or anthropic origin, such as the particles emitted from vehicles due to combustion, re-suspended emissions, and friction phenomena. They could be produced by industrial and energetic combustion processes or other primary (agriculture and mining) or secondary (civil and industrial construction) activities. Particulate matter concentrations vary, not only for changes in meteorology or changes in anthropogenic emissions but also due to the influence of natural emissions. These emissions are hard to predict, due to the high variability over the years and the complex relationship to its gas precursors emitted from different sources. Hence, the correlation between PM observed levels and emission sources during lockdown is complex. In the regions where people have been forced to stay indoors, there may have been an increase in primary PM emissions from the domestic burning of coal or wood, while traffic PM emissions have been significantly reduced. Particulate matter from agricultural emissions was probably not affected by the lockdown, while some industrial emissions (e.g., PM, NO_X, and SO_X) were reduced in various sites due to the temporary stop of many factories [9].

Certainly, during the lockdown period starting from the middle of March, very significant reductions in traffic flows were observed on a national basis [10,11], down to 70% for light vehicles and 38% for heavy vehicles. These levels gradually returned to previous levels in the first half of June. This traffic reduction consequently affected the particulate levels, but sometimes was compensated by emission due to domestic heating, especially for a month of March that was, on average, colder than usual. This may explain the slight increase in the particulate level observed in same areas, with respect to the season average. The contributions related to industrial and livestock activities should also be taken into account. The particulate transported by the Saharan dust should be added in some cases to the one already produced in the urban context. All these considerations may constitute a first motivation for the insignificant PM10 and PM2.5 reduction levels recorded during the lockdown period in Italy.

The lockdown resulted in a significant reduction in air pollution that has always been a peculiarity of northern Italy, due to the large concentration of industrial activities. In this perspective, the metropolitan city of Milan (Italy) appears to be particularly representative, due to the characteristics of the urban context and variety of the industrial activities. Data analysis in Milan is possible due to the public and private air quality monitoring networks installed in the area, allowing for the monitoring of the pollution levels with high space and time resolution. The need to install air quality monitoring networks arise from the need to adequately know the state of the air breathed by the population. Information on air quality is useful to public authorities, such as the government, which deals with environmental monitoring with the help of regional environmental protection agencies (ARPA). Municipalities, individuals, or companies support the government and the ARPA installing their networks for air quality monitoring [12] in the optimal points of the city context [13]. To carry out detailed analysis it is necessary to have a good quantity of measured data distributed over the territory with a high spatial resolution, otherwise the analysis is limited [14]. The monitoring networks installed in Milan have low cost but were developed according to the internet of things (IoT) standards and can provide real-time data with a high temporal resolution that is easily accessible by users. The possibility of relying on data from air quality monitoring networks is fundamental for the definition of pollution levels but also for implementing models that allow, for example, the pollution evolution forecast [15,16] and the analysis of accidental events, such as explosions or fires [17]. The data reliability in the latest generation of smart measuring devices can be ensured with the use of blockchain technology [18]. Monitoring network data is also useful to the institutions for the development of air quality improvement [19,20].

In the present study, air quality, high time–space resolution network data, and ARPA network air quality data in Milan are analyzed to assess on the lockdown on the air quality. To this end, the concentrations of the different pollutants in the pre- and post-lockdown scenario are compared and discussed.

The Case Study

The metropolitan city of Milan is one of the largest and most populous cities in northern Italy, with approximately 1,400,000 inhabitants in the municipality alone and more than 3 million inhabitants in the metropolitan area. The air quality is strongly affected by the town activities of different nature, by which it is daily interested. The many industrial poles and massive urbanization (together with an orography that is not favorable to the pollutants dispersion) determine the frequent exceedances of the PM10 levels, with respect to the law limits. In Italy, a condition is shared with other industrial conurbations such as Turin, Venice, Naples, and Cagliari. The mountains surround the valley to the north and west (Alps), as well as south (Apennines), while the east side is open to the Adriatic Sea. This topography favors air stagnation and is accompanied by frequent winter thermal inversions and diffuse fog events that lead to the accumulation of particulate (PM) pollution [21,22].

2. Materials and Methods

2.1. Data Sources

2.1.1. ARPA Lombardia

ARPA Lombardia is the regional authority for environmental protection. Its air quality monitoring network consists of 85 fixed stations, which, by means of automatic analyzers, provide continuous data at regular time intervals, generally on an hourly basis. The pollutant species monitored continuously are NO_X, SO_2, CO, O_3, PM10, PM2.5, and benzene. Depending on the environmental context (urban, industrial, traffic, and rural), in which the monitoring is active, the list of pollutants considered relevant is different. Therefore, not all stations are equipped with the same analytical instrumentation. ARPA stations are distributed throughout the region according to the population density and the territory type, respecting the criteria defined by the national Legislative Decree 155/2010.

In detail, 5 monitoring stations are active throughout the Milan urban area, as shown in Figure 1, and their characteristics are listed in Table 1.

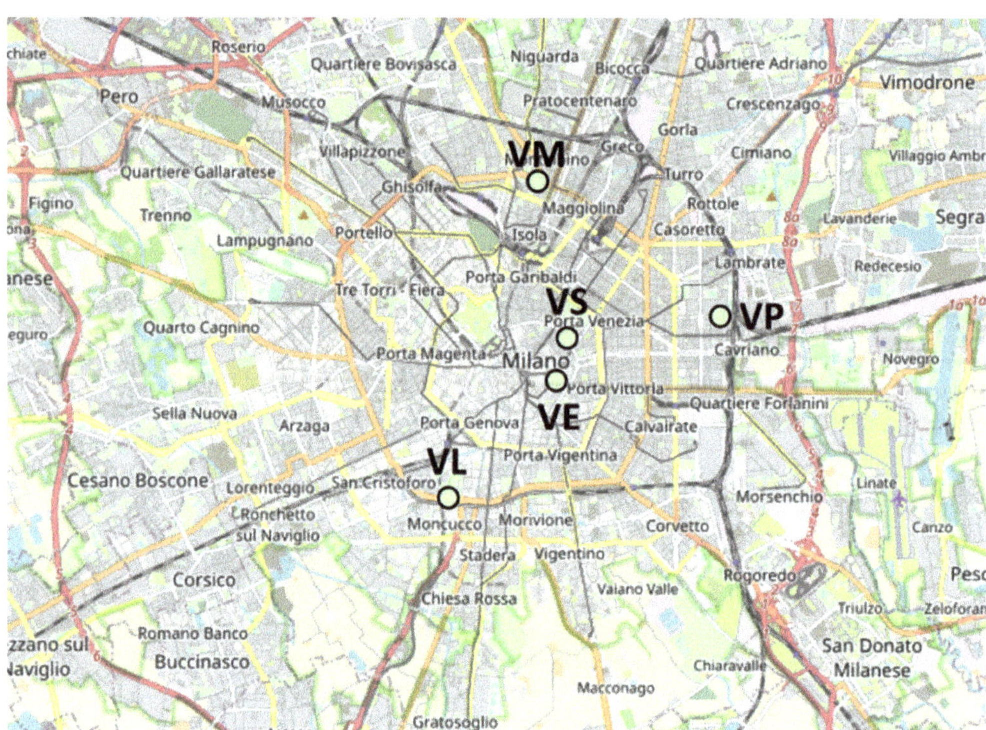

Figure 1. Location of ARPA Lombardia air quality monitoring stations in Milan urban area (Italy).

Table 1. ARPA Lombardia air quality monitoring stations characteristics.

Station	Acronym	Type	Pollutants Measured
Viale Liguria	VL	Urban Traffic	NO_2, CO
Viale Marche	VM	Urban Traffic	PM10, NO_2, CO
Via Pascal	VP	Urban Background	SO_2, PM10, PM2.5, CO, O_3, Benzene
Via Senato	VS	Urban Traffic	PM10, PM2.5, CO, Benzene
Verziere	VE	Urban Traffic	PM10, PM2.5, CO, O_3

The data used for this study includes all PM10 and PM2.5 values measured by the regional air quality network from 2010 to 2020.

2.1.2. Real-Time On-Road Mobile Monitoring Network

The real-time on-road network for air quality monitoring (ROMS) has been active in Milan since 2018. This new technology makes it possible to map pollution in real-time through measurement devices located on moving vehicles. This latest generation of monitoring system, described by Lotrecchiano et al. [23], provides data on the main airborne pollutants with a high temporal resolution, with deeper knowledge on the area covered. The considered measuring system can return the measurements of the average particulate levels of three parameters PM10, PM2.5, and PM1 (particulates with diameters less than 10 μm, 5 μm, and 1 μm, respectively). In addition, the measurement system provides data on the concentration of gaseous pollutants of air, such as NO_2, SO_2, O_3, CO, and VOC, as well as weather parameters, such as the temperature, pressure, and relative humidity of the air. Information on wind direction and intensity is provided to the system via a network of API keys (application programming interface key, i.e., a simple encrypted string that identifies an application) provided by external sites. In this study, only the particulate matter PM10 and PM2.5 have been considered. However, since the measurement systems are linked to the vehicle on which they are located, this technology has limits, in terms of spatial coverage. Some areas are excluded from monitoring, as they are outside the predetermined routes of vehicles or because they are forbidden from vehicular traffic, such as green and pedestrian areas.

Data used in this study includes all measurements made by the ROM network between January and June 2020. To overcome the aforementioned limits of spatial coverage, a model of spatial expansion was applied to the measured data, in order to estimate the missing ones, as proposed by Lotrecchiano et al. [24].

The Milan area was divided into square cells of 1 km^2 and the proposed expansion model was applied to each cell containing valid measures.

2.2. Data Aggregation

Data aggregation was carried out using an increasing degree of detail. The first data aggregation level is a macro-visualization, as the whole area to be investigated has been divided into zones corresponding to the 9 Milan municipalities (Figure 2a). By increasing the detail level, the area to be monitored was divided according to the 88 local identity nuclei (NIL) (Figure 2b). NILs are areas that can be defined as districts of Milan, in which it is possible to recognize historical and project districts, with different characteristics from each other. They are introduced by the PGT (territory governance plan) as a set of areas that are connected by infrastructures and services for mobility and greenery. They are systems of urban vitality (local commercial activities, gardens, meeting places, and services), but they are also 88 NILs to be strengthened and designed, through which small and large services can be organized (service plan). The maximum degree of detail is obtained by dividing the entire Milan area into cells with a surface area of 1 km^2, whose centroids form a regular grid (Figure 2c). This degree of detail is guaranteed by the mobile sensor network. As previously described, each sensor housed on a vehicle allows for the measurement and tracking of pollutants in many areas of the city at the same time. Furthermore, since they are housed in commercial vans, they have not been subject to the lockdown, thus allowing for the measurement of pollutants.

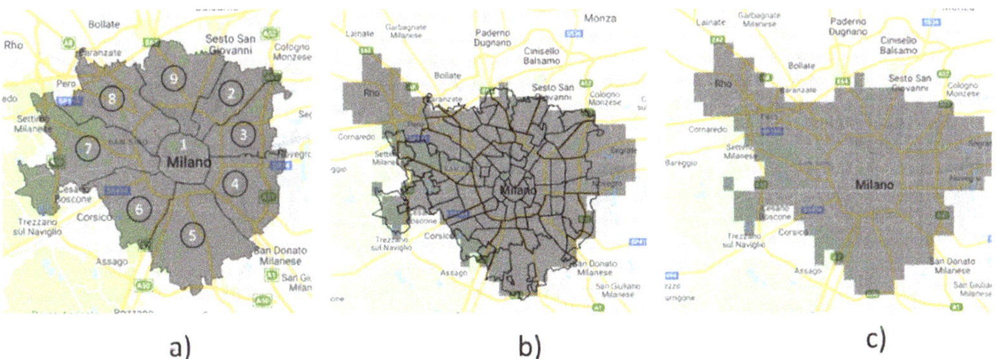

Figure 2. (**a**) Territory division for the data aggregation into 9 municipalities. (**b**) Territory division for the data aggregation into 88 NIL. (**c**) Territory division for the data aggregation into 306 squares of 1 km^2.

Traffic Information

A fundamental part of air quality analysis is the analysis of the traffic influence. Information on the traffic flows recorded in the city of Milan was obtained by implementing an algorithm for data acquisition. The algorithm takes data from the server via an API (application programming interface) key, in order to collect traffic data, process them, and insert them into a database. In this work, the REST API key was used; it returns the traffic flow through the jam factor. The traffic jam factor shows the traffic condition in numerous ways. It is a value in the range [0.0, 10.0]. A large jam factor value means more traffic jams, in general. Specifically, 0.0 means free traffic and 10.0 means stationary traffic.

3. Results and Discussion

3.1. Annual Trend 2010–2020

To understand the pollutant values measured during the March 2020 lockdown, it is necessary to extend the analysis to previous years to compare the pollution levels. Therefore, the first analysis was carried out using the historical data of PM10 measured in Milan by the air quality monitoring network of ARPA Lombardia. The annual PM10 concentration values were taken from the ARPA Lombardia reports for 2010–2020.

The analysis starts from the trends relating to PM10 concentrations on an annual basis. The annual average limit of 40 $\mu g/m^3$ (according to Legislative Decree 155/2010) was not exceeded in the years 2010, 2013, 2014, and 2016, with a decreasing trend between 2017 and 2019. (Figure 3).

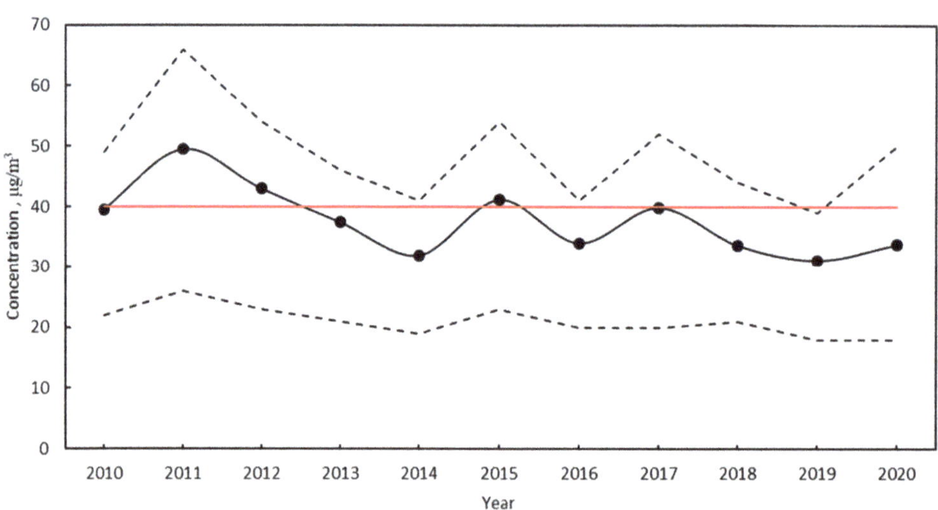

Figure 3. The 2010–2020 PM10 annual averages in the Milan stations measured by ARPA Lombardia (black solid line), 25th and 75th percentile of the annual averages (black dashed line), and PM10 annual law limit of 40 $\mu g/m^3$ according to Lgs. 155/2010 (red solid line).

In September 2010, Legislative Decree 155/2010 was adopted in Italy, which implemented the European Directive 2008/50/CE for cleaner air in Europe. Since then, in Italy, policies to improve air quality have been implemented, leading to a decrease in the PM10 annual average values measured. This decrease has settled in recent years, reaching slightly different values between 2018 and 2020 (Figure 3).

To better define the multi-year trend, eliminating the variability between contiguous years due to different meteorological conditions, the five-year moving average was calculated (Figure 4). In this way, considering that the meteorological difference over five-year periods is less evident, the trend linked to the emissions is highlighted more clearly. The Figure 4 analysis clearly shows the decreasing pollution trend over the years. If all the variables that influence the phenomenon have been excluded, it is possible to verify its actual evolution.

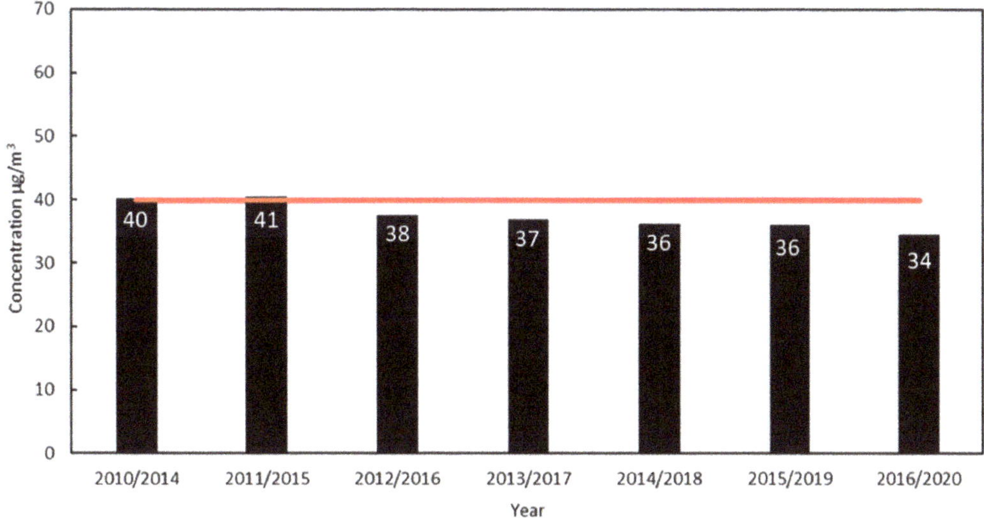

Figure 4. PM10 five-year moving annual average concentrations measured in Milan by the ARPA Lombardia network and PM10 annual law limit of 40 µg/m³ according to Lgs. 155/2010 (red solid line).

3.2. Trend Analysis of March 2020

To analyze the lockdown pollution trend in Milan, it is useful to divide the time series into two periods. The first was between 23 February and 8 March, in which the restrictive measures were gradually introduced. The second was from 8 March to 4 May, in which the measures of containment were stronger, since a strict lockdown was imposed. It can be seen from Figure 5 that, up to 9 March, the particulate pollution levels were low and slightly decreasing. An increase in pollution appeared in the period between 17 and 20 March. The increased values in this period were due to an accumulation of pollutants, connected to the high atmospheric pressure and the consequent low wind intensity, both in the period records. In fact, these are the typical atmospheric phenomena which favor air stagnation and pollutants accumulation Moving forward in the month, a very high peak is found in the period between 28 and 29 March. In order to understand the origin of this peak, we can refer to the ratio between the concentrations of PM2.5 and PM10 (black solid line in Figure 5), the low value of the ratio for this period suggests the external origin of these particulates. In fact, according to the global aerosol "Copernicus Atmosphere Monitoring Service" forecast, those days when a mass of dust from Central Asia hit Europe determined a raise of the measured values of the larger particulates (map showing the mineral aerosol covering Italy is available in the Supplementary Materials). The transported particulate was of desert and sandy origin, mainly composed of siliceous materials. This represents a further difference from PM10 of urban origin, mostly characterized by the presence of carbon- and metal-based particles.

From a climatic point of view, the Mediterranean atmosphere (MED) is characterized by rainy winter seasons and hot and dry summer seasons that affect the continental areas. It is characterized by the particulate of anthropogenic origin (coming from the large industrialized European regions), the natural particulate of crustal origin (coming from the extensive arid and semi-arid areas of Africa and the Middle East), and of the aerosols of marine origin (generated by the MED), that of volcanic dust, emitted by the main volcanoes of the basin, such as Etna and Stromboli [25].

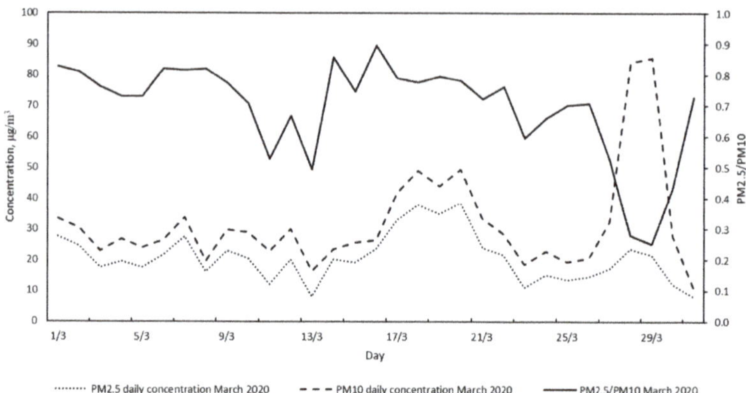

Figure 5. PM10 daily average (dashed black line), PM2.5 daily average (dotted black line), ratio between PM2.5 and PM10 (solid black line) measured in Milan by ARPA Lombardia, March 2020.

Some methodologies to estimate the contribution of Saharan dust on urban pollution were reported in the literature, such as the analysis by Bonasoni et al. [26], which confirms that the mineral aerosol contributions to the urban PM10 values can be very critical, favoring overcoming critical thresholds. The PM2.5/PM10 ratio curve, reported in Figure 5, shows a minimum during 28 and 29 March that clearly defines the Saharan dust event. During these days, the PM10 and PM2.5 are not proportional so, an external PM10 contribution should be added to the urban particulate. To evaluate the urban PM10 value, the one measured during the Saharan event should be cut by the mineral aerosol contribution. In the proposed case study, to estimate the contribution of Saharan dust to urban PM10, the average ratio between PM2.5 and PM10 was calculated over March 2020, excluding the Saharan dust events days. The average PM2.5/PM10 value obtained was used to estimate the urban value of PM10 on 28 and 29, with respect to the PM2.5 values measured. The estimated PM10 values for 28 and 29 are reported in Figure 6. In Figure 6, it is possible to see how the peak between 28 and 29 March smooths down, with a consequent increase in the value of the PM25/PM10. This estimated PM10 value will be used for all PM10 analysis in March 2020. The contribution of the Saharan dust to urban pollution is a phenomenon to be taken into account, in order to define the actual pollution levels. Saharan dust being of natural origin is beyond human control and independent from anthropogenic activity [27].

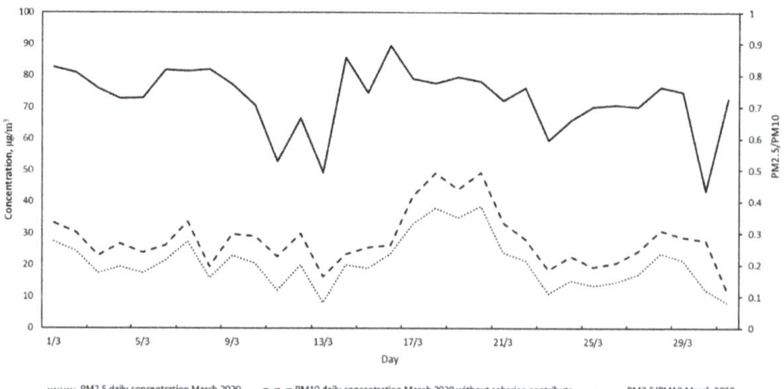

Figure 6. PM10 daily average without saharan contribute estimated (dashed black line), PM2.5 daily average (dotted black line), ratio between PM2.5 and PM10 (solid black line) measured in Milan by ARPA Lombardia, March 2020.

3.3. Trend Analysis of the First Six Months of the Years between 2010–2020

Provided that the data from March, without exceptional meteorological contributions, is useful to analyze, in detail, the differences in the 2010–2020 period, especially between the months of January, February, March, and April, which characterized the lockdown. From the values shown in Figure 7, it can be seen that, considering the month of March, years 2016 and 2020 showed the lowest concentration value of PM10. In particular, it appears that March 2020 was the month with the lowest pollution levels in the last ten years (net of Saharan dust contribute). The Figure also clearly shows the well-known seasonal nature of pollution, higher during the winter season, for the high emission from domestic heating. Coherently, during the spring/summer season, pollution levels decrease, and mostly, the contribution of vehicular traffic and industrial activities plays a role. The particulate matter concentration measured in January is generally greater than that measured in February. Only in 2011, 2012, and 2017, the concentrations measured in January were lower than in February. These inversions could be related to the well-known permanence phenomenon of particulates that can remain in the atmosphere for several weeks, accumulating and, therefore, showing a maximum in concentration shift forward, concerning the maximum the emission rates. April 2020 appears to have quite low values, similar to those recorded in 2012 and 2019. The month of June 2020 had the lowest concentration ever measured since 2010. Overall, considering that data in the period from March–June 2020, the generally very low values are evidence of the lockdown's positive effect on the air quality.

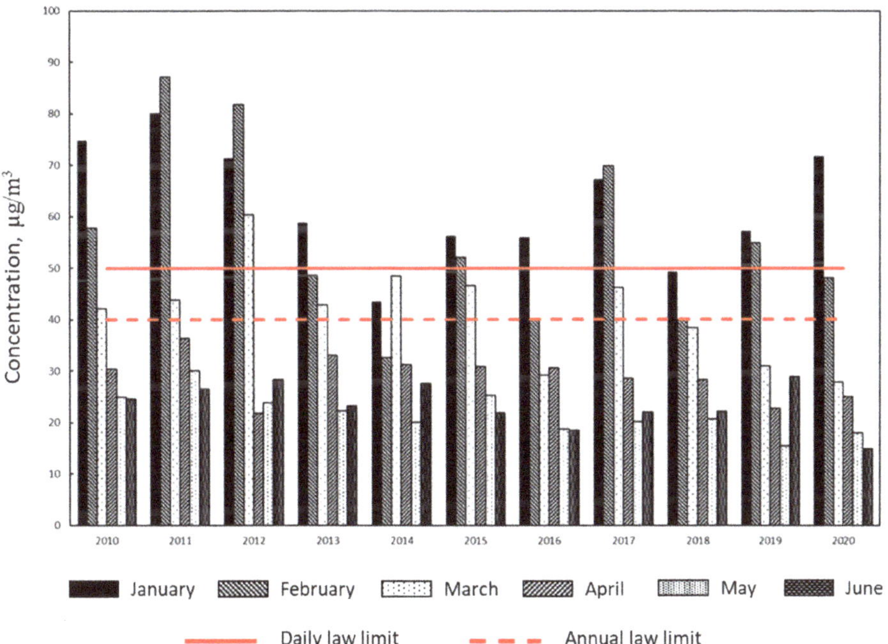

Figure 7. PM10 average monthly concentration in the first six months of the year in the period 2010–2020, measured in Milan by ARPALombardia, without considering the contribution of Saharan dust in March 2020; daily law limit for PM10 according to Lgs. 155/2010 (red solid line) and annual law limit for PM10 according to Lgs. 155/2010 (red dashed line).

3.4. Trend Analysis of March–April 2018/2020

The evidence of the lockdown's effects on the air quality and the PM10 data of the March and April period was aggregated for 2018–2020 and reported in Figure 8. Inspection of the Figure indicates that PM10 progressively decreased during the observed period, while the value of PM2.5 had a minimum in the year 2019.

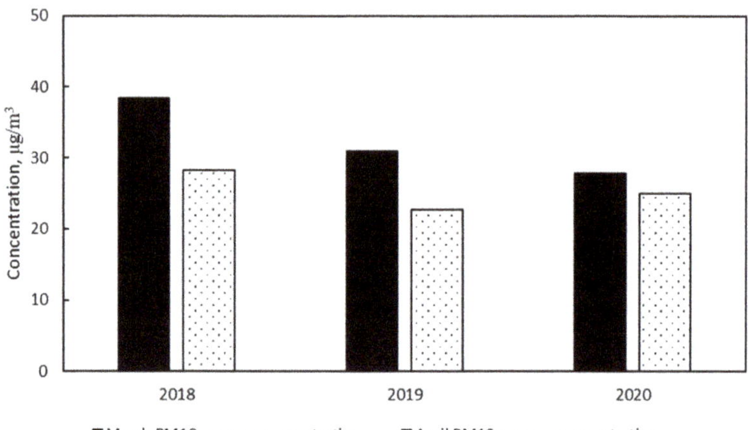

Figure 8. Monthly average concentrations trends of March and April in the years 2018–2020, measured in Milan by ARPA Lombardia.

3.5. Traffic Analysis

One of the variables that affects air particulates levels is vehicular traffic. In particular, it contributes both with the emissions (due to the fuels used) and the wear of the cars mechanical parts (coming from brakes and tires). The weekly traffic behaviour, with a minimum corresponding to Sundays, is evident from Figure 9, which shows the traffic flow recorded in the period January–June 2020. The figure shows that, since the start of the lockdown in March, there was a progressive reduction in traffic levels, reaching a minimum in the month of April. Even during the lockdown, finite levels of traffic flow were observed due to residual private traffic, public transport, couriers, and all other vehicles in service for public utility. The evident reduction of traffic during the lockdown was likely to have positively influenced the reduction of pollution. With the resumption of anthropic activities, traffic flow rose again (May and June); however, it did not reach January and February levels.

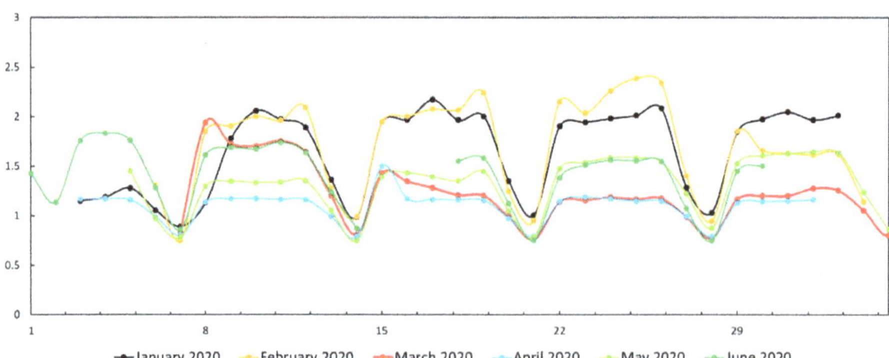

Figure 9. Daily traffic flows recorded in the city of Milan in the months of January–June 2020.

3.6. Meteorological Aspects

All of the 2020 measured data should be interpreted while also taking into account the meteorological variables that can affect pollution. Precipitations as the rainfall cumulative daily value (Figure 10), as well as the number of rainy days, together with the average monthly temperature (Figure 11) strongly influenced the pollution levels. The seasonality of pollution indicates the months of December to March as those when pollution levels are higher. Despite the high probability of rain, the high atmospheric pressure that characterizes the winter months favors the accumulation and stagnation of pollutants in the air. January 2020 was characterized by minimum temperatures close to the period average and maximum temperatures significantly above the average. This difference was determined by favourable conditions: despite a rather mild air mass, the minimum temperatures at the ground were affected by continuous thermal inversions, which kept the values rather low. The lowest temperature in January was recorded on day 13, with −3.6 °C, while the highest high was recorded on day 29, with 14.8 °C. There were 20 days with temperatures below zero degrees, compared to an average of 17 in the last decade. From the pluviometric point of view, the month of January 2020 recorded lower than average rainfall values, with almost all rainfall falling in a single event on day 18.

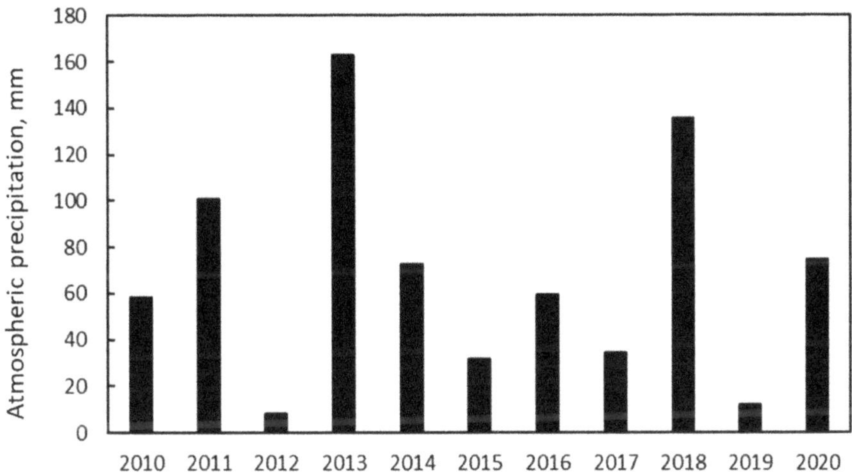

Figure 10. Cumulative daily value of rainfall measured in Milan by the ARPA Lombardia meteorological network in March.

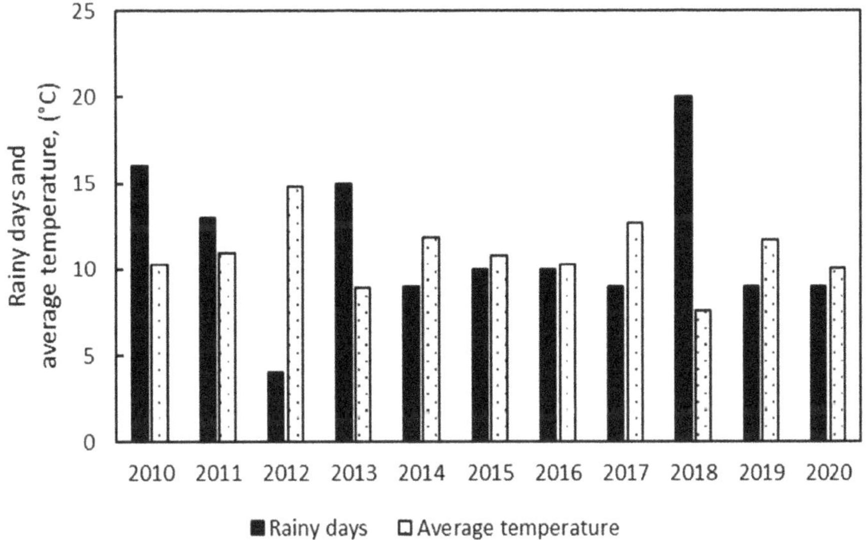

Figure 11. Rainy days and average temperature measured in Milan by the ARPA Lombardia meteorological network in March.

February 2020 was characterized by significantly higher than average temperatures and is considered the mildest February of the last 20 years. It was the ninth consecutive month showing temperatures above the average of the last decade. The lowest value in February was recorded on day 6, with $-2.5\ ^\circ$C, while the highest temperature was recorded on day 11, with 19.3 $^\circ$C. There were only five days with temperatures below zero degrees, compared to an average of 9 in the last decades and 13 in 1981–2010. From the pluviometric point of view, the month of February recorded much lower than average rainfall values, bringing the annual deficit to about 100 mm. In particular, the day 26 was characterized by a strong Föhn wind (about 31 km/h maximum), which massively affected the lowering of particulate concentrations.

March 2020 was characterized by temperatures very close to the averages: the recorded temperatures were slightly lower than what was typical of the last decades but still slightly higher than averages, in reference to the last century. The lowest temperature in March was recorded on 24, with $-1.3\ ^\circ$C, while the highest on 19, with 20.7 $^\circ$C. There was a single day with negative minimum temperatures, in line with the last decade and below 3 $^\circ$C in the 1981–2010 period. From the pluviometric point of view, March recorded rainfall values in line with the averages.

The mild weather in the months of February and March certainly reduced the heating power of domestic heaters, but emission might have been compensated by a longer period of operation, due to the prolonged presence of people at home.

Figures 10 and 11 show that March 2013 and 2018 were characterized by numerous rainy days and high atmospheric precipitations. This affects the quantity of particulate matter measured with low PM10 levels in those years.

High atmospheric precipitation favors pollution abatement. As seen in Figure 7, the years with high rainy days correspond to low PM10 concentrations. The temperature measured during March (Figure 11) indicates the characteristics of the season. It is logical to assume that the lower the temperatures, the higher the pollution, due to the intense utilization of domestic heating. Combining the effect of temperature and rain, new scenarios can be considered. A year with low temperature but with a high number of rainy days will reasonably show a low pollution level. The high pollution levels, reached by the intense usage of domestic heating, are compensated by the rain that decreases the pollution level. These conditions characterized March 2013 and 2018; Figure 7 shows low PM10 concentrations, despite the low temperature measured.

3.7. Comparison of Mobile Network Data with ARPA Lombardia Network

Data from ROMS are compared with those measured by the ARPA Lombardia air quality network in Figure 12. ARPA air quality data are taken from a single station. ROM network data are taken by averaging all the measured data coming from cells within the same NIL where the ARPA station is located. Data measured by both kinds of monitoring networks are in agreement if referred to the trends, even if reporting different absolute values. The differences between the values most certainly lie in the different measurement methods, which, therefore, lead to intrinsic variations of the measurements, but above all, of how data are aggregated. Moreover, ARPA values refer to point measurement, while ROM values are averages of data taken throughout the territory. From Figure 12, it is clear that from March to June the agreement between the two measurements was greater. The other months (Figure 12a,b) showed a reduction of PM10 pollution at the urban traffic station (VE), due to the traffic abatement. In Figure 12b,d, such reductions are not seen, mainly because of a minor traffic impact. The VP station is located in an area that was not busy (it was classified as an urban background station, as reported in Table 1), and the increase of PM2.5 concentrations probably results from the domestic heating increase.

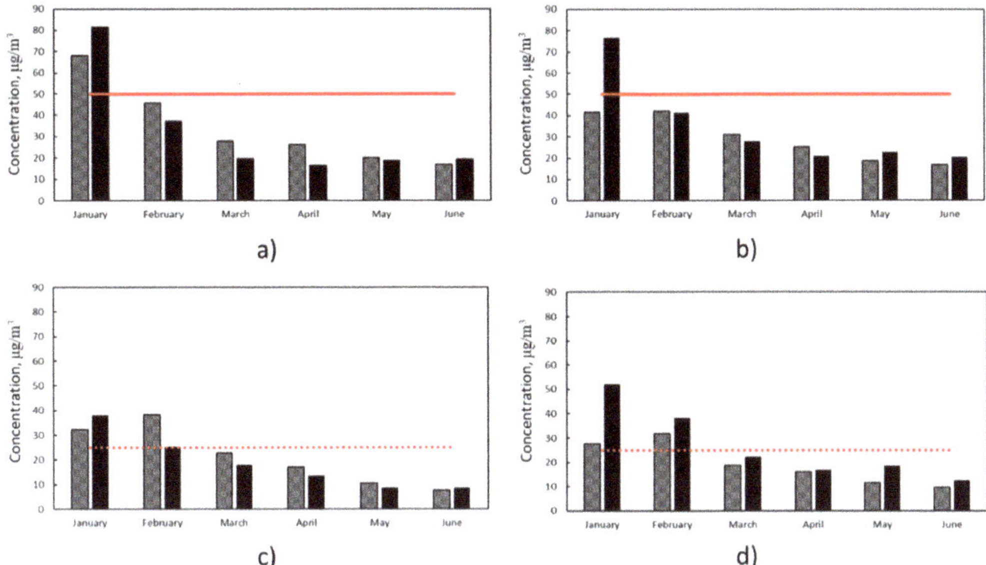

Figure 12. PM10 average monthly concentrations, measured by the ROM network (black) and the ARPALombardia (grey) in the months of January–April 2020 in Milan. Data for ARPA were taken from a single station, while data for the ROM network were taken by averaging all the data coming from all the cells within the same NIL, in which the ARPA station is located. The red solid line and the red dotted lines are the PM10 and PM2.5 daily law limits, according Lgs 155/2010, respectively. (**a**) PM10 monthly averages, relevant to the VE station; (**b**) PM10 monthly averages, relevant to the VP station; (**c**) PM2.5 monthly averages, relevant to the VE station; (**d**) PM2.5 monthly averages, relevant to the VP station.

3.8. ROM Data Analysis between January and June 2020

Data measured by the ROM network in Milan, in the months of January–June 2020, confirmed the previously analyzed data provided by ARPA Lombardia (Figure 13). It appears that a strong reduction in concentrations between January and March can be quantified in 68.4% for PM10 and 70% for PM2.5. Referring to February instead, the reduction, with respect to March, was 42.2% for PM10 and 42% for PM2.5. Milan pollution levels, due to particulate matter PM10 and PM2.5 between January and February, were already decreasing, but with the contribution of the lockdown, they dropped further. PM10 and PM2.5 monthly averages values measured during March and April were even below the law limit, according to D.Lgs, with 155/2010 of 50 $\mu g/m^3$ and 25 $\mu g/m^3$, respectively. May and June confirm the typical seasonal values; however, as previously mentioned, the month of June recorded the lowest pollution levels since 2010.

3.9. Pollution Analysis on a NIL Basis

The PM10 and PM2.5 pollution levels were calculated for NIL. For sake of brevity, the four most polluted NILs in the months January–April 2020 have been reported (Table 2). By comparing the average, monthly PM10 and PM2.5 values for each NIL, it can be observed that they have drastically decreased over the months. It is interesting to note that there are NILs that always maintain high values. For example, the NIL 23-Lambrate is always present among the worst four NILs of the month. Despite the high PM10 and PM2.5 values measured in January, this NIL is characterized by a reduction in pollution of 70% for PM10 and PM2.5 in March. The reduction between February and March was 25% for PM10 and 23% for PM2.5. The decrease in pollution, largely attributable to the reduction in activity during the lockdown in Milan was supported by favorable weather conditions.

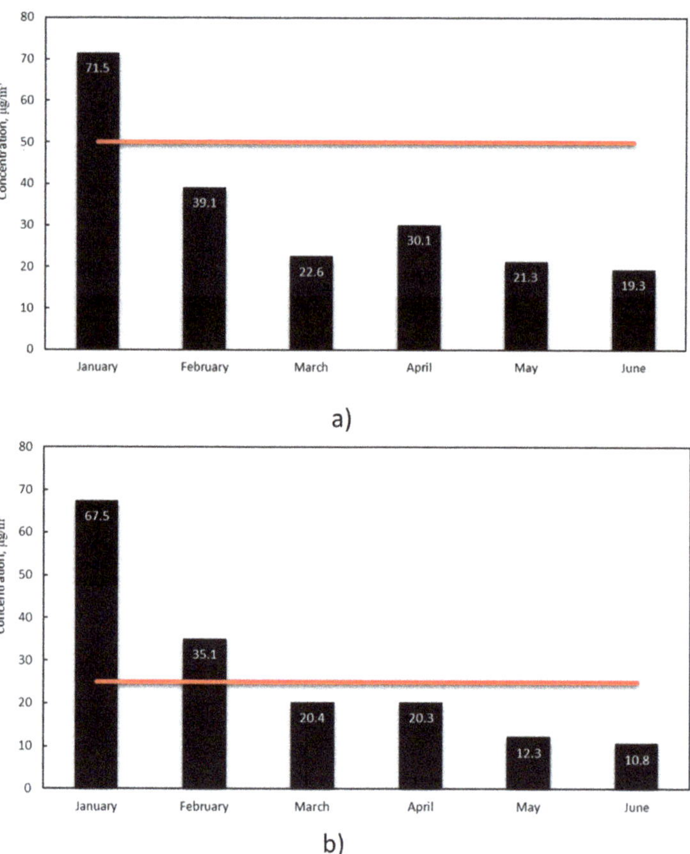

a)

b)

Figure 13. (**a**) PM10 average monthly concentrations measured by the ROM network in Milan in the months of January–June 2020. PM10 daily law limit according to D.Lgs 155/2010 (red solid line). (**b**) PM2.5 average monthly concentrations measured by the ROM network in Milan in the months of January–June 2020. PM2.5 daily law limit according to D.Lgs, 155/2010 (red solid line).

Table 2. Ranking of the worst 4 NILs for the months January–April 2020.

Month	NIL Name	PM10 Monthly Average, $\mu g/m^3$	PM2.5 Monthly Average, $\mu g/m^3$
January	23	110	101
	73	107	95
	17	101	94
	25	101	86
February	44	65	60
	73	65	58
	41	59	57
	63	58	54
March	46	32	28
	23	31	29
	17	30	28
	73	29	25
April	38	39	35
	42	37	32
	46	35	31
	23	32	30

An in-depth analysis of the Lambrate NIL, considering its urbanization characteristics, traffic, and public parks, is possible. Figure 14 reports daily values measured by the ROM.

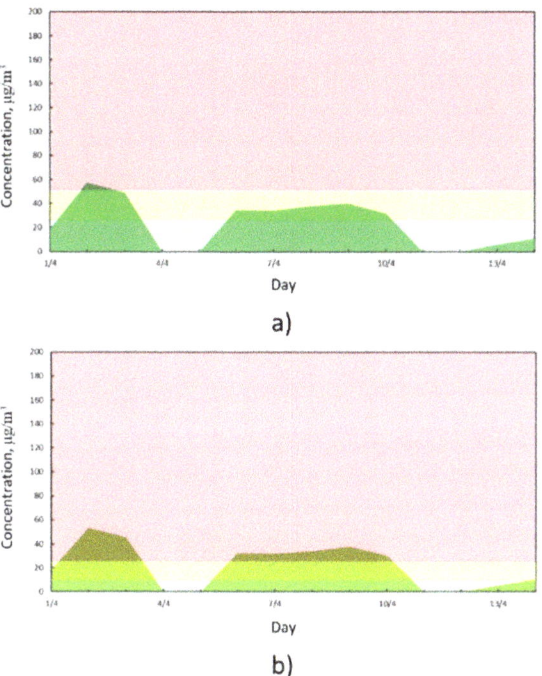

Figure 14. (a) PM10 daily average concentration for NIL 23—Lambrate measured in April 2020 from the ROM network in Milan. (b) PM2.5 daily average concentration for NIL 23—Lambrate measured in April 2020 from the ROM network in Milan.

Figure 14 allows the definition of local pollution and underlines the days with higher pollution levels. In particular, for the PM10, only had two days that were recorded concentration values over the law limit (Figure 14a), and for the PM2.5, seven days exceeded the law limit (Figure 14b).

The high spatial resolution also allows for providing the pollution levels measurement to the detail of the 1 km^2 cells. The pollution maps on a kilometric scale are shown in Figure 15 for the PM10 concentrations. The adopted chromatic scale uses a traffic light code; for PM10, green indicates the level of good air quality between 0 and 25 µg/m^3, yellow indicates the alarm threshold with concentrations in the range 25–50 µg/m^3, and red indicates that the limit value has exceeded that of the 50 µg/m^3 defined by Legislative Decree 155/2010. For PM2.5, green indicates the level of good air quality between 0 and 10 µg/m^3, yellow indicates the alarm threshold with concentrations in the range 10–25 µg/m^3, and red indicates limit value of 25 µg/m^3, as defined by Legislative Decree 155/2010.

As can be seen from Figures 15c and 16c, March shows an improvement in air quality, with respect to January and February, especially in the city center. The city lockdown led to the gradual suspension of almost all of the city's daily activities. In particular, the reduction in daily vehicular traffic was significant (reducing by 23% in just one week). This reduction contributed to the decrease in the PM10 and PM2.5 concentrations produced by cars. The concentration of particulate in urban areas was due to the 58% [6] emissions from domestic heating that, in any case, remained on during the lockdown period. In fact, constantly staying at home induces a greater use of heating. Despite this, the mitigating temperatures for a few weeks in March and the favorable weather conditions have greatly reduced their use and, consequently, the particulate matter associated with their operation.

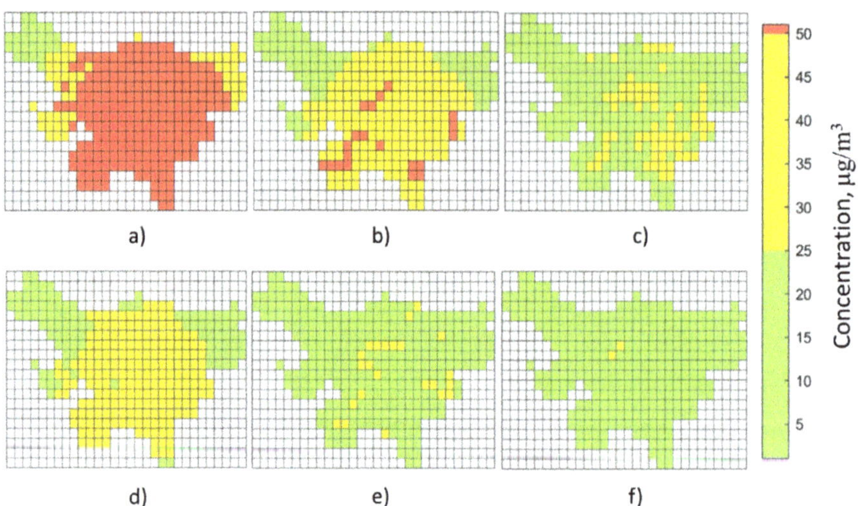

Figure 15. PM10 monthly average measured by the ROM network in Milan: (**a**) January 2020; (**b**) February 2020; (**c**) March 2020; (**d**) April 2020; (**e**) May 2020; (**f**) June 2020.

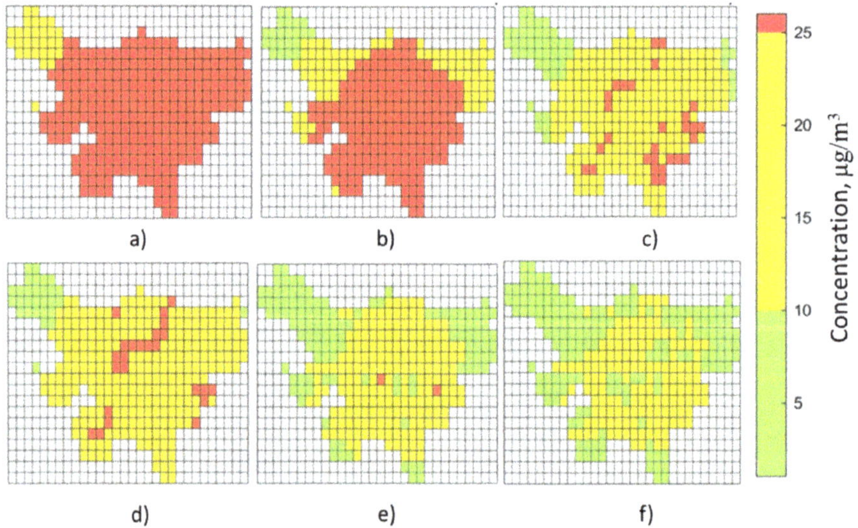

Figure 16. PM2.5 monthly average measured by the ROM network in Milan: (**a**) January 2020; (**b**) February 2020; (**c**) March 2020; (**d**) April 2020; (**e**) May 2020; (**f**) June 2020.

Tables 3–6 report PM10 and PM2.5 data for the NILs that registered the highest reduction measured between January and March, as well as between February and March. Some NILs, detailed in Table 3, reached a maximum reduction of 77% between January and March. In particular, if we consider the Milan city center, with reference to the NIL, it can be seen that in the NIL referred to the Duomo between January and March, and PM10 pollution decreased by 75%. Due to the lockdown, the major cities' thoroughfares have slowed down and the associated NILs, such as Tibaldi, had a 75% reduction in pollution due to traffic, when compared to January. Furthermore, the north-western area of the city, which converges towards the Rho/Pero areas where many industrial activities have been progressively reduced, due to the city lockdown, appeared to be less affected by pollution in March, when compared to January and February. In detail, the NIL Sacco and Cascina Triulza-EXPO saw a decrease of PM10 concentrations of 76 and 74%, respectively, compared to January (Table 3).

Table 3. Top 10 NILs with the highest PM10 reduction rate between March and January.

Ranking	Nil Name	Best PM10 Daily Average Concentration, $\mu g/m^3$		Worst PM10 Daily Average Concentration, $\mu g/m^3$		PM10 Monthly Average Concentration, $\mu g/m^3$		Reduction %
		January	March	January	March	January	March	
1	25	18	7	174	50	101	23	77
2	57	6	<5	200	34	59	13	77
3	74	6	<5	192	84	78	18	76
4	1	7	6	168	41	81	20	75
5	59	7	5	155	46	81	20	75
6	43	6	<5	183	28	55	14	75
7	54	5	<5	82	21	43	11	74
8	68	5	5	134	44	71	19	74
9	70	5	<5	84	23	44	12	74
10	22	11	6	190	62	92	24	74

Table 4. Top 10 NILs with the highest PM10 reduction rate between March and February.

Ranking	Nil Name	Best PM10 Daily Average Concentration, $\mu g/m^3$		Worst PM10 Daily Average Concentration, $\mu g/m^3$		PM10 Monthly Average Concentration, $\mu g/m^3$		Reduction %
		February	March	February	March	February	March	
1	57	<5	<5	136	34	48	13	72
2	41	6	5	162	39	59	19	67
3	65	<5	5	149	33	54	19	65
4	44	5	6	169	52	65	26	60
5	63	<5	<5	195	101	58	23	60
6	73	<5	9	166	60	65	29	56
7	62	<5	<5	140	44	39	18	54
8	55	<5	<5	123	36	36	17	52
9	77	5	6	139	47	49	24	51
10	43	<5	<5	71	28	28	14	50

Table 5. Top 10 NILs with the highest PM2.5 reduction rate between March and January.

Ranking	Nil Name	Best PM2.5 Daily Average Concentration, $\mu g/m^3$		Worst PM2.5 Daily Average Concentration, $\mu g/m^3$		PM2.5 Monthly Average Concentration, $\mu g/m^3$		Reduction %
		January	March	January	March	January	March	
1	57	6	<5	166	29	56	12	79
2	54	5	<5	82	19	43	10	77
3	70	5	<5	84	21	44	10	77
4	43	6	<5	161	26	54	13	76
5	74	6	<5	153	73	68	16	76
6	25	22	6	156	46	86	21	76
7	1	5	5	149	38	72	18	75
8	53	7	<5	166	41	73	18	75
9	59	5	<5	137	42	73	18	75
10	56	7	<5	153	38	75	19	75

For completeness, the list of 10 NILs, in which PM10 pollution decreased the most in March, compared to January and February, is shown (Tables 3 and 4). Tables 3 and 4 explain the improvement in air quality that followed the progressive blocking of activities in Milan.

PM2.5 concentrations show a drastic reduction, which reaches a maximum of 79% between January and March, as well as 75% (Table 5) between February and March (Table 6). The city center is confirmed to be among those that have undergone an improvement of air quality, with, for example, the NIL Duomo, where PM2.5 recorded a decrease of 75%, when compared to January.

Table 6. Top 10 NILs with the highest PM2.5 reduction rate between March and February.

Ranking	Nil Name	Best PM2.5 Daily Average Concentration, µg/m³		Worst PM2.5 Daily Average Concentration, µg/m³		PM2.5 Monthly Average Concentration, µg/m³		Reduction %
		February	March	February	March	February	March	
1	57	5	<5	137	29	48	12	75
2	41	6	<5	155	36	57	17	69
3	65	4	4	141	30	53	17	68
4	63	<5	<5	164	90	54	20	63
5	44	5	5	155	47	60	23	62
6	77	5	5	157	43	55	22	61
7	62	<5	<5	119	38	35	15	56
8	43	<5	<5	76	26	29	13	56
9	73	<5	8	146	50	58	25	56
10	55	<5	<5	109	32	34	17	51

The decreasing PM10 and PM2.5 monthly average concentrations measured in the months of January, February, and March (Figures 15a–c and 16a–c) means a progressive improvement in air quality. Compared to the PM10 concentrations measured in the same period, the decrease in PM2.5 would seem slightly lower (Figure 16); considering that the PM2.5 has a greater variability and a lower law limit than PM10, and that only the 8% of PM2.5 is produced by vehicular traffic, it can be easily understood why the chromatic scale that represents it would contain some red cells. In particular, between January and March, there is a progressive movement of the pollution towards the city center-north/south, with an improvement in air quality, especially in the central area, as can be seen in Figure 15a,c and Figure 16a,c. A deep analysis of March PM10 and PM2.5 levels (Figures 15d and 16d) revealed that the northeast area is characterized by low pollution levels. The city center has medium pollution levels and only in few cells, the pollution reaches high levels for PM2.5 (Figure 16d). This means that the red cells (Figure 16d) present some characteristics as intense traffic flow, high urbanization degree, and low green areas that favour the PM2.5 formation and accumulation.

3.10. Comparison of Significant Day Maps before and during the Lockdown by COVID-19

To complete the analysis made so far, the two most significant days of the February-March period were considered to appreciate the positive influence of the lockdown on air quality. Corresponding data are reported in Figure 17. It can be seen how, under the same meteorological conditions, the PM10 concentration measured by the ROM had drastically decreased, passing from a medium-high pollution level (Figure 17a) to a low level (Figure 17b). The days used as an example both presented the meteorological characteristics typical of a day favorable to the pollutants accumulation, i.e., high atmospheric pressure and low wind intensity.

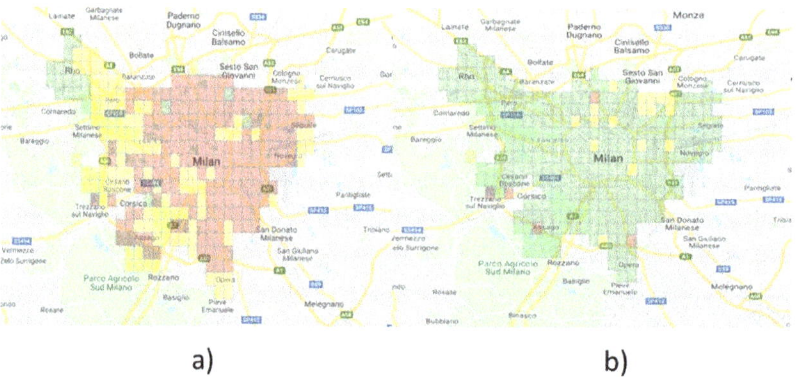

a) b)

Figure 17. Pollution levels (**a**) before and (**b**) during lockdown.

4. Conclusions

The lockdown influence on air quality, due to the multiple influencing factors, is not simple to analyze. To correctly understand how the current situation has affected pollution, it is necessary to investigate these multiple factors. The analysis of pollution levels evolution in 2010–2020 highlighted the general trend of air quality improvement. In more detail, it was possible to define the PM10 specific trends for the years 2010–2020 from January to June, the period including the first lockdown of the city. Considering that February was partially involved by the 2020 lockdown, the last three years were analyzed in detail for the variations in particulate concentrations for March and April. From the analysis, is it clear that there was a progressive decrease in the PM10 average concentration of 8.15% in March, from 2019 to 2020, and 16%, compared to 2018. April, compared with 2018 and 2019, showed a reduction of 11%, with respect to 2018, but an increase of about 10%, compared to 2019. The analysis of vehicular traffic flows showed a decrease of 60%, in line with the measured particulate concentrations values, underlining the reduction of the contribution of vehicular traffic to pollution. ROM network data were used to define pollutant trends on a NIL basis, reporting the worst and identifying the km^2 and NIL, where the greatest pollution reduction occurred. The maximum reduction measured in March, compared to January and February, is higher than 70% for both PM10 and PM2.5. Ultimately, the data from the ROM network were compared with the ARPA Lombardia air quality monitoring network data, showing a congruence between the datasets.

Supplementary Materials: The following is available online at https://www.mdpi.com/article/10.3390/pr9101692/s1, Figure S1: Dust aerosol optical depth at 550 nm (provided by CAMS, the Copernicus Atmosphere Monitoring Service, for 28 March 2020), accessed on 10 August 2020.

Author Contributions: Conceptualization, N.L., P.T. and D.S.; methodology, N.L. and D.S.; software, N.L. and D.S.; validation, N.L., P.T. and D.S.; formal analysis, N.L., P.T. and D.S.; writing—original draft preparation, N.L., P.T. and D.S.; writing—review and editing, N.L., P.T., D.S., D.B. and M.P.; supervision, D.S., D.B. and M.P. All authors have read and agreed to the published version of the manuscript.

Funding: This research received no external funding.

Institutional Review Board Statement: Not applicable.

Informed Consent Statement: Not applicable.

Data Availability Statement: Data are available at https://www.arpalombardia.it/Pages/Aria/Richiesta-Dati.aspx.

Conflicts of Interest: The authors declare no conflict of interest.

References

1. Poli, P.; Boaga, J.; Molinari, I.; Cascone, V.; Boschi, L. The 2020 coronavirus lockdown and seismic monitoring of anthropic activities in Northern Italy. *Sci. Rep.* **2020**, *10*, 9404. [CrossRef]
2. Piccinini, D.; Giunchi, C.; Olivieri, M.; Frattini, F.; Di Giovanni, M.; Prodi, G.; Chiarabba, C. COVID-19 lockdown and its latency in Northern Italy: Seismic evidence and socio-economic interpretation. *Sci. Rep.* **2020**, *10*, 16487. [CrossRef]
3. IQAir and United Nations. 2020 World Air Quality Report. Available online: https://www.iqair.com/world-air-quality-report (accessed on 10 August 2020).
4. Malpede, M.; Percoco, M. Lockdown measures and air quality: Evidence from Italian provinces. *Lett. Spat. Resour. Sci.* **2021**, *14*, 101–110. [CrossRef]
5. Merico, E.; Grasso, F.M.; Cesari, D.; Decesari, S.; Belosi, F.; Manarini, F.; De Nuntiis, P.; Rinaldi, M.; Gambaro, A.; Morabito, E.; et al. Characterisation of atmospheric pollution near an industrial site with a biogas production and combustion plant in southern Italy. *Sci. Total Environ.* **2020**, *717*, 137220. [CrossRef]
6. Istituto superiore per la protezione ambientale. *La Qualità Dell'aria in Italia*; Edizione 2020; Istituto Superiore Per la Protezione Ambientale: Ispra, Italy, 2020.
7. Bassani, C.; Vichi, F.; Esposito, G.; Montagnoli, M.; Giusto, M., Ianniello, A. Nitrogen dioxide reductions from satellite and surface observations during COVID-19 mitigation in Rome (Italy). *Environ. Sci. Pollut. Res.* **2021**, *28*, 22981–23004. [CrossRef]
8. Scerri, M.M.; Kandler, K.; Weinbruch, S. Disentangling the contribution of Saharan dust and marine aerosol to PM10 levels in the Central Mediterranean. *Atmos. Environ.* **2016**, *147*, 395–408. [CrossRef]

9. European Environmental Agency. *Air Quality in Europe—2020 Report*; European Environmental Agency: Luxembourg, 2020. [CrossRef]
10. Rossi, R.; Ceccato, R.; Gastaldi, M. Effect of Road Traffic on Air Pollution. Experimental Evidence from COVID-19 Lockdown. *Sustainability* **2020**, *12*, 8984. [CrossRef]
11. Piccoli, A.; Agresti, V.; Balzarini, A.; Bedogni, M.; Bonanno, R.; Collino, E.; Colzi, F.; Lacavalla, M.; Lanzani, G.; Pirovano, G.; et al. Modeling the Effect of COVID-19 Lockdown on Mobility and NO_2 Concentration in the Lombardy Region. *Atmosphere* **2020**, *11*, 1319. [CrossRef]
12. Sofia, D.; Giuliano, A.; Gioiella, F. Air quality monitoring network for tracking pollutants: The case study of salerno city center. *Chem. Eng. Trans.* **2018**, *68*, 67–72. [CrossRef]
13. Sofia, D.; Lotrecchiano, N.; Giuliano, A.; Barletta, D.; Poletto, M. Optimization of number and location of sampling points of an air quality monitoring network in an urban contest. *Chem. Eng. Trans.* **2019**, *74*, 277–282. [CrossRef]
14. Rovetta, A. The Impact of COVID-19 Lockdowns on Particulate Matter Emissions in Lombardy and Italian Citizens' Consumption Habits. *Front. Sustain.* **2021**, *2*, 44. [CrossRef]
15. Lotrecchiano, N.; Gioiella, F.; Giuliano, A.; Sofia, D. Forecasting Model Validation of Particulate Air Pollution by Low Cost Sensors Data. *J. Model. Optim.* **2019**, *11*, 63–68. [CrossRef]
16. Sofia, D.; Lotrecchiano, N.; Cirillo, D.; Villetta, M.L.; Sofia, D. NO2 Dispersion model of emissions of a 20 kwe biomass gasifier. *Chem. Eng. Trans.* **2020**, *82*, 451–456. [CrossRef]
17. Lotrecchiano, N.; Sofia, D.; Giuliano, A.; Barletta, D.; Poletto, M. Pollution dispersion from a fire using a Gaussian plume model. *Int. J. Saf. Secur. Eng.* **2020**, *10*, 431–439. [CrossRef]
18. Sofia, D.; Lotrecchiano, N.; Trucillo, P.; Giuliano, A.; Terrone, L. Novel air pollution measurement system based on ethereum blockchain. *J. Sens. Actuator Netw.* **2020**, *9*, 49. [CrossRef]
19. Sofia, D.; Gioiella, F.; Lotrecchiano, N.; Giuliano, A. Mitigation strategies for reducing air pollution. *Environ. Sci. Pollut. Res.* **2020**, *27*, 19226–19235. [CrossRef] [PubMed]
20. Sofia, D.; Gioiella, F.; Lotrecchiano, N.; Giuliano, A. Cost-benefit analysis to support decarbonization scenario for 2030: A case study in Italy. *Energy Policy* **2020**, *137*, 111137. [CrossRef]
21. Squizzato, S.; Masiol, M.; Brunelli, A.; Pistollato, S.; Tarabotti, E.; Rampazzo, G.; Pavoni, B. Factors determining the formation of secondary inorganic aerosol: A case study in the Po Valley (Italy). *Atmos. Chem. Phys.* **2013**, *13*, 1927–1939. [CrossRef]
22. Pecorari, E.; Squizzato, S.; Masiol, M.; Radice, P.; Pavoni, B.; Rampazzo, G. Using a photochemical model to assess the horizontal, vertical and time distribution of PM2.5 in a complex area: Relationships between the regional and local sources and the meteorological conditions. *Sci. Total Environ.* **2012**, *443C*, 681–691. [CrossRef] [PubMed]
23. Lotrecchiano, N.; Sofia, D.; Giuliano, A.; Barletta, D.; Poletto, M. Real-time on-road monitoring network of air quality. *Chem. Eng. Trans.* **2019**, *74*, 241–246. [CrossRef]
24. Lotrecchiano, N.; Sofia, D.; Giuliano, A.; Barletta, D.; Poletto, M. Spatial Interpolation Techniques For innovative Air Quality Monitoring Systems. *Chem. Eng. Trans.* **2021**, *86*, 391–396. [CrossRef]
25. Arnold, M.; Seghaier, A.; Martin, D.; Buat-Ménard, P.; Chesselet, R. *Géochimie de L'aérosol Marin de la Méditerranée Occidentale*; CIESM: Cannes, France, 1982; pp. 2–4.
26. Bonasoni, P.; Cristofanelli, P.; Calzolari, F.; Bonafè, U.; Evangelisti, F.; Stohl, A.; Sajani, S.Z.; van Dingenen, R.; Colombo, T.; Balkanski, Y. Aerosol-ozone correlations during dust transport episodes. *Atmos. Chem. Phys. Discuss.* **2004**, *4*, 1201–1215. [CrossRef]
27. Lotrecchiano, N.; Capozzi, V.; Sofia, D. An Innovative Approach to Determining the Contribution of Saharan Dust to Pollution. *Int. J. Environ. Res. Public Health* **2021**, *18*, 6100. [CrossRef] [PubMed]

Article

Numerical Simulation of the Novel Coronavirus Spread in Commercial Aircraft Cabin

Mengya Zhang, Nu Yu *, Yao Zhang, Xin Zhang and Yu Cui

Department of Air Traffic, College of Civil Aviation, Nanjing 211106, China; zmy197319@nuaa.edu.cn (M.Z.); yaozhangnuaa@nuaa.edu.cn (Y.Z.); zx160705@nuaa.edu.cn (X.Z.); cuiyu1999@nuaa.edu.cn (Y.C.)
* Correspondence: nuyu@nuaa.edu.cn; Tel.: +86-18949642736

Abstract: Passengers carrying the severe acute respiratory syndrome coronavirus 2 (SARS-CoV-2) in a commercial aircraft cabin may infect other passengers and the cabin crew. In this study, a cabin model of the seven-row Airbus A320 aircraft is constructed and meshed for simulating the SARS-CoV-2 spread in the cabin with a virus carrier using the Computational Fluid Dynamics (CFD) modeling tool. The passengers' infection risk is also quantified with the susceptible exposure index (SEI) method. The results show that the virus spreads to the ceiling of the cabin within 50 s of the virus carrier's normal breathing. Coughing makes the virus spread to the front three rows with a higher mass fraction. While the high mass fraction areas always stay on the same side of the aisle as the virus carrier, the adjacent passengers and the passengers in the back two rows are affected more than the others when the virus carrier breathes normally. Spread patterns under the carrier's two breath conditions, normal breath and cough, were numerically simulated.

Keywords: CFD; SARS-CoV-2; aircraft cabin; cough; SEI

Citation: Zhang, M.; Yu, N.; Zhang, Y.; Zhang, X.; Cui, Y. Numerical Simulation of the Novel Coronavirus Spread in Commercial Aircraft Cabin. *Processes* **2021**, *9*, 1601. https://doi.org/10.3390/pr9091601

Academic Editors: Daniele Sofia and Paolo Trucillo

Received: 19 July 2021
Accepted: 2 September 2021
Published: 7 September 2021

Publisher's Note: MDPI stays neutral with regard to jurisdictional claims in published maps and institutional affiliations.

1. Introduction

In late December 2019, a novel coronavirus disease (COVID-19) broke out. The severe acute respiratory syndrome coronavirus 2 (SARS-CoV-2) is a highly infectious pathogen that can spread quickly. This pandemic not only impacts human health and community but also impacts global economics severely [1]. As of 13 July 2021, the total number of confirmed COVID-19 cases has exceeded 188 million, and the total number of deaths has exceeded 4 million, among which many cases have been defined as air travel infections.

There are more than two billion people traveling by air every year [2]. The microenvironment in the commercial aircraft cabin is enclosed, which provides a good place for the virus to spread among passengers [3,4]. Long-distance air travel makes the virus spread broadly and rapidly. When a virus carrier boards onto a commercial airliner, the surfaces inside the cabin might be contaminated by his/her breath, saliva, and mucus. The in-cabin air can also be contaminated by his/her exhalation. Other passengers may be infected if they touch the contaminated surfaces or breathe in the virus from the contaminated air. For example, the severe acute respiratory syndrome (SARS) pandemic caused by the SARS coronavirus (SARS-CoV) broke out in China in 2003. During that pandemic, 20 passengers on the Air China F112 flight traveling from Hong Kong to Beijing were infected by a virus carrier [5]. In most cases of virus spread, the fomite route plays the dominant role. The China National Health Commission (CNHC) has issued a policy in 2020, which defines the passengers sitting three rows in front of or behind a SARS-CoV-2 carrier as "close contacts". According to this policy, all "close contacts" shall be quarantined for at least 14 days after traveling [6].

Lei et al. indicate that passengers in aisle seats had a higher norovirus infection risk than others when traveling with a virus carrier in the same aircraft cabin, and the predicted infection risk from the fomite route for aisle seat passengers is 2.2 times higher than that for non-aisle seat passengers [7]. Viruses can be transmitted through either aerosols (≤ 5 μm) or droplets (>5–10 μm) exhaled by the carriers [8,9]. Larger respiratory virus droplets deposit to surfaces quicker due to gravity sedimentation, which may lead to contact transmission. Aerosols ($\leq$5 μm) evaporate faster than settling, so they are buoyant. These aerosols will be transported with the airflow in the aircraft cabin [9]. The particle size of a SARS-CoV-2 molecule is about 65–125 nm [10], which is in the ultrafine particle range (<100 nm). This is slightly smaller than SARS-CoV (80–220 nm). Studies have confirmed that SARS-CoV-2 can spread by aerosol [8,9,11,12]. The exposure risk of passengers traveling with the SARS-CoV-2 carrier inside an aircraft cabin increases with the cabin ventilation rate's increase [13].

There are two commonly used methods to simulate the aerosol systems with virus particles, or virus-carrying particles. One is to treat the system as one continuous substance and use Euler's method to solve a convective diffusion equation to obtain the virus concentration and distribution [14–16]. The other is to treat solid particles or droplets as a discrete phase material and use the Lagrangian's method to calculate the motion characteristics of each particle [14–16]. Some in-cabin studies use the discretization method to obtain the trajectory of virus particles [4,17,18]. In addition, Sandro et al. replaced respiratory pathogens with smoke particles and concluded that 60% of the particles sink to the surfaces of the cabin due to gravity during the propagation process [19]. Research studies have been conducted to calculate the virus concentrations inside commercial aircraft cabins with a virus carrier using the computational fluid dynamics (CFD) technique. Mazumdar et al. found that the seats and passengers tended to obstruct the lateral transport of the contaminants and confined their spread to the aisle of the cabin [20]. Gupta et al. found that the bulk airflow pattern in the cabin played the most important role in droplet transport [21]. Davis et al. used the CFD tool to calculate the propagation paths of particles released by the coughing passengers in different seats in a Boeing B737 aircraft [17].

To simplify the aerosol transmission process in the commercial aircraft cabin, N_2O is used as a tracer gas to establish a continuous system, and Euler's method is applied in the CFD tool to simulate the SARS-CoV-2 concentration and distribution in this study. The virus distribution changes in the cabin under the carrier's normal breathing and coughing are compared based on the simulation data. When a virus carrier presents in a commercial aircraft cabin, these results can provide guidance for risk-reducing policies, which is of great significance to the operation of commercial aircraft during a pandemic.

2. Methods

2.1. Cabin Physical Model and Meshing

As the CNHC requires, when a SARS-CoV-2 carrier is found in a commercial aircraft cabin, the passengers sitting three rows in front of and behind the virus carrier are defined as "close contacts" [6]. Based on this policy, this study builds a physical model with seven-row seats inside an Airbus A320 aircraft cabin to study the SARS-CoV-2 mass fraction around the virus carrier. The dimensions are from the A320 maintenance manual (Table 1) [22].

Table 1. Dimensions of the seven-row cabin model.

Dimension	Size
Length	6.3 m
Width	3.7 m
Height	2.26 m
Passenger breathing zone height	1.11 m
Inlet width	0.04 m
Outlet width	0.04 m
Passenger's mouth area	0.03 m^2
Row distance	0.85 m

The seats are numbered from 1A to 7F in this seven-row aircraft cabin model. We assume the virus carrier is sitting in the 4th row middle seat (Seat 4E; Figure 1a). To facilitate the analysis of the virus mass fraction distribution in the passengers' breathing zone, a plane at 1.11 m height (y = 1.11 m; Plane 1 on Figure 1b) from the cabin floor is set up. Since the virus carrier is sitting at a distance of 1.04 m from the midline of the cabin aisle, a vertical plane is set up parallel to the midline of the cabin (x = 1.04 m; Plane 2 on Figure 1b). The cabin physical model was meshed with the ANSYS Meshing (Pittsburgh, PA, USA) to achieve an accurate and efficient solution [23]. The maximum grid size is set to 0.05 m, in order to achieve the best resolution with the capacity of a regular personal computer. The total number of grids was finally calculated to be 2,822,402.

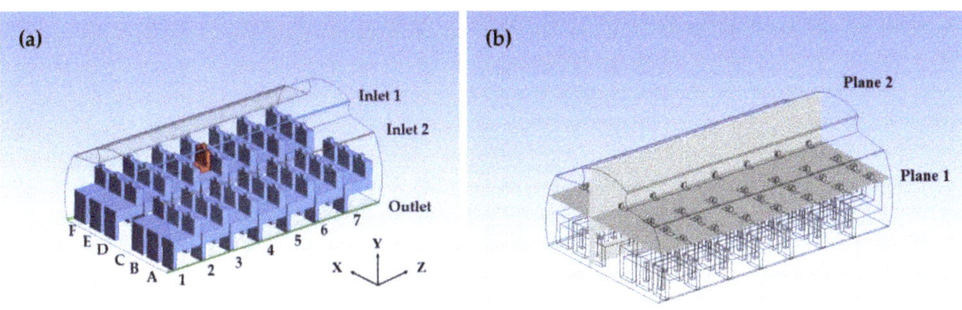

Figure 1. The physical model of the seven-row A320 cabin: (**a**) virus carrier seat in red; (**b**) plane 1 and plane 2.

2.2. Boundary Condition Setting

The cabin air supply and outlet are simplified to rectangles. The temperature (T) of the air inlet is set to 18 °C, the air velocity (v) near the ceiling and on both side walls are set to 1 m·s⁻¹, and the temperature of the wall and the cabin air is set to 22 °C to meet the requirements of ASHRAE 161-2018 [24].The airflow velocity generated by human coughing is 6–22 m·s⁻¹ [25]. The initial airflow speed from the virus carrier's mouth after coughing is set to 10 m·s⁻¹ and the initial air temperature is set to 37.5 °C. The initial airflow speed is set to 1 m·s⁻¹ for the virus carrier's normal breathing. The body temperature of other passengers is set to 36.5 °C. Table 2 summarizes the boundary conditions of the seven-row Airbus A320 Aircraft cabin model constructed in this study, including the boundary location, name, type and parameter.

Table 2. Boundary condition settings.

Location	Name	Parameter
Ceiling air inlet	Inlet 1	v = 1 m·s⁻¹, T = 18 °C
Side wall air inlet	Inlet 2	v = 1 m·s⁻¹, T = 18 °C
Near-ground air outlet	Outlet	P = 84,475.3 Pa
Cabin wall	Wall 1	T = 22 °C
Virus carrier	Virus	Normal breath: v = 1 m·s⁻¹, T = 37.5 °C; Cough: v = 10 m·s⁻¹, T = 37.5 °C
Non-ill passengers	Wall 2	T = 36.5 °C

2.3. SARS CoV-2 Mass Fraction Numerical Simulation

The commercial CFD software ANSYS Fluent 19.1 (Pittsburgh, PA, USA) is used for numerical simulation of the dynamic propagation process of viruses inside the aircraft cabin [18]. The initial scenarios are the time points when the SARS-CoV-2 is introduced to the cabin air by the virus carrier's normal breathing (initial airflow velocity = 1 m·s⁻¹) or coughing (initial airflow velocity = 10 m·s⁻¹).

Most of the droplets from coughing change to droplet nuclei (≤5 μm) in less than 1 s, and the diameter becomes about 1/3 of the original [26]. Because of the tiny size of the nuclei, tracer gases have been widely used to simulate the aerosol dynamics in recent studies. For example, Liu et al. simulated the aerosol transmission using nitrous oxide (N_2O) as a trace gas, and Tsoukias et al. established a human exhalation profile using N_2O to represent the tiny droplet nuclei [27,28]. The volume fraction of (N_2O) in the exhaled air is about 4 %, which is about the same volume fraction of CO_2 [27,29,30]. Therefore, in this study, N_2O is used as the tracer gas to simulate the virus's distribution after being exhaled by the virus carrier in the A320 aircraft cabin model.

The initial mass fraction of N_2O needs to be set in the Fluent software. First, suppose the volume of the mixed gas exhaled by the virus carrier is V_0, then the volume of N_2O is $V_{N_2O} = 4\%V_0$ and the mass of N_2O is:

$$m_{N_2O} = \rho_{N_2O} \cdot V_{N_2O} \tag{1}$$

The total mixed gas mass is:

$$m_{V_0} = \rho_{air} \cdot V_0 \tag{2}$$

The N_2O mass fraction is:

$$N_2O \; mass \; fraction = \frac{m_{N_2O}}{m_{V_0}} \tag{3}$$

where m_{N_2O}, ρ_{N_2O} and V_{N_2O} represent, respectively, the mass, density and volume of N_2O, where $\rho_{N_2O} = 1.23 \; g \cdot cm^{-3}$. ρ_{air} represents the density of aircraft cabin air, and $\rho_{air} = 1.29 \; g \cdot cm^{-3}$. m_{V_0} represents the mass of V_0. The initial mass fraction of N_2O calculated from Equations (1)–(3) is 0.038.

More particularly, the numerical solution of Reynolds Averaged Navier-Stokes equations with Re-Normalization Group (RNG) $k - \varepsilon$ turbulence closure is employed to simulate the continuum phase [31]. According to Zhang et al., the RNG $k - \varepsilon$ model fits the propagation and diffusion of pollutants in the cabin, so this study chooses the RNG $k - \varepsilon$ turbulence model [32–36].

2.4. Susceptible Exposure Index (SEI) Calculation

To quantify the exposure risks of other passengers in the aircraft cabin, this study utilizes the Susceptible Exposure Index (SEI) to quantify the passengers' exposure to the virus [27,37]. The SEI is defined with Formula (4) [27]:

$$\text{SEI} = \frac{C_i - C_S}{C_r - C_S} \tag{4}$$

where C_r, C_i, and C_S are the virus mass fraction at the air outlet, in the inhaled air of the receiving individual, and at air supply, respectively.

The virus mass fraction of the supply air is assumed to be zero in this study, then the SEI calculation formula can be simplified to $\text{SEI} = \frac{C_i}{C_r}$.

A high SEI means a high exposure of the receiving individual to the airborne substances exhaled by the virus carrier, and if the SEI is greater than 1, it means that the passenger is at risk of exposure [27].

3. Results and Discussion

3.1. Virus Mass Fraction Dynamics

Virus spread patterns on plane 1 and plane 2 are numerically simulated. Figures 2 and 3 show the virus mass fraction patterns on the two planes at 5 s, 50 s, 100 s, 150 s and 200 s after the virus is exhaled under normal breathing and coughing, respectively.

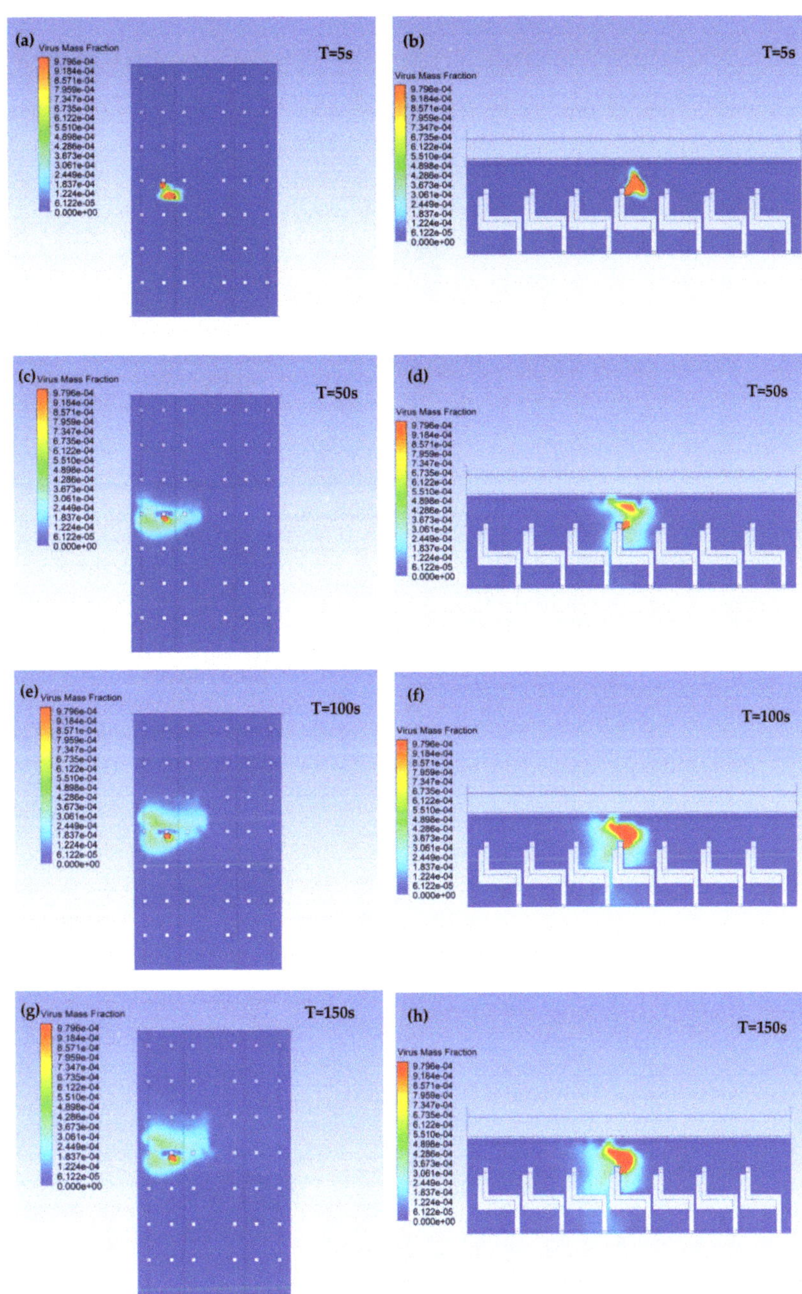

Figure 2. *Cont.*

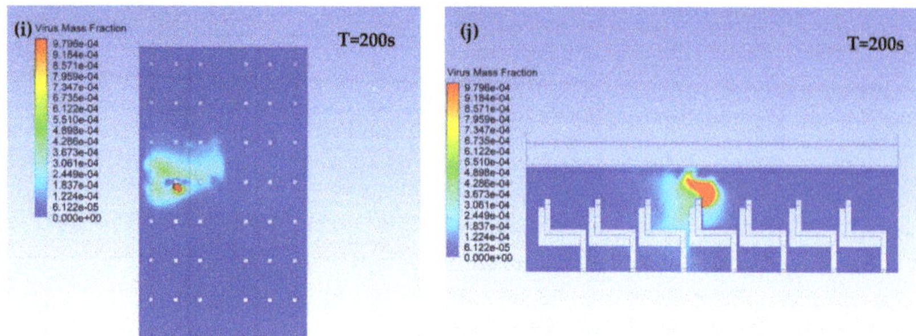

Figure 2. The dynamic propagation process of pathogens 5 s, 50 s, 100 s, 150 s 200 s after the virus carrier's single normal breath (left: (**a,c,e,g,i**)-Plane 1; right: (**b,d,f,h,j**)-Plane 2).

As shown in Figure 2a,b, when the virus carrier breathes normally, the high mass fraction ($>9 \times 10^{-4}$) virus area stays near the virus carrier at T = 5 s (Figure 2a,b). At T = 50 s, the boundary of contaminated air extends to the cabin ceiling, the adjacent seat (seats 4D and 4F) breathing zone, and the aisle area. However, the high mass fraction areas shrink on both planes (Figure 2c,d). At T = 100 s, the low mass fraction contaminant disperses to the cabin floor, the same side back row breathing zone (seats 5D, 5E and 5F), and front row breathing zone (seats 3E and 3F), except for the aisle seat (seat 3D). At T = 150 s and T = 200 s, the spread patterns are similar, and the very low mass fraction ($<1 \times 10^{-4}$) contaminant spreads to the other side of the aisle (seats 4A–4C and seats 5A–5C) (Figure 2e–j).

This result shows the vertical spread is much faster than the horizontal spread due to the fixed lateral rows of the passenger seats, which partially block the horizontal airflow. The aisle makes the virus travel farther, and the virus spread after normal breathing affects the back rows more than the front rows due to the ventilation of the airliner cabin. However, the higher initial airflow velocity after coughing overcomes the cabin airflow generated by the ventilation, which makes the virus spread farther to the front rows. This is consistent with the results of Mazumdar et al. [20].

As shown in Figure 3a,b, when the virus carrier coughs, the high mass fraction ($>9 \times 10^{-4}$) of the virus instantly spreads to the same side front row (seat 3D) breathing zone at T = 5 s, but the adjacent passengers' breathing zones in seats 4D and 4F stay unaffected at that moment (Figure 3a,b). The high virus mass fraction area continues to spread forward at T = 50 s, and reaches the cabin ceiling and the 2nd front row (seats 2D-2F) breathing zone on the same side. At the same time, the low mass fraction area ($<1 \times 10^{-4}$) extends to the cabin floor, the 3rd front row (seats 1D–1F) and the back row (seats 5D-5F) on the same side. The low mass fraction of the virus also spreads to the other side of the aisle in rows 4 and 5 (Figure 3c,d). At T = 100 s, the high mass fraction area covers rows 2–4 on the same side, while the low mass fraction area reaches rows 1–6 on both sides (Figure 3e,f). At T = 150 s and T = 200 s, the high mass fraction area stays at rows 2–4 on the same side, and the low mass fraction area stays in rows 1–6 on both sides (Figure 3e–j). Row 7 (seats 7A–7F) stays uncontaminated all the time (Figure 3).

Comparing Figures 2 and 3, we can find that when the virus carrier breathes normally, the virus spreads to the back seats more than the front seats. However, when the virus carrier coughs, the virus spreads to the front seats more than the back seats. This is due to the higher initial airflow velocity from coughing (Table 2). For the same reason, the spread of the virus is much faster after coughing than normal breathing. However, the passengers in the 7th row stay away from the virus contamination under both situations. This is not necessarily implying that the 7th row passengers, who are three rows back from the virus carrier, could be exempted from the "close contacts" group and the mandatory quarantine. This is because this study only simulates an ideal situation that the virus carrier only breathes or coughs once (single breath or cough), and all passengers and cabin crew members stay still in the cabin. According to other studies, when people are moving in the aisle, the wake can carry the virus to the far most in the cabin [20,38]. Although the 7th row in this model stays safe and clean under both breathing situations, in a realistic environment it cannot stay uncontaminated all the time.

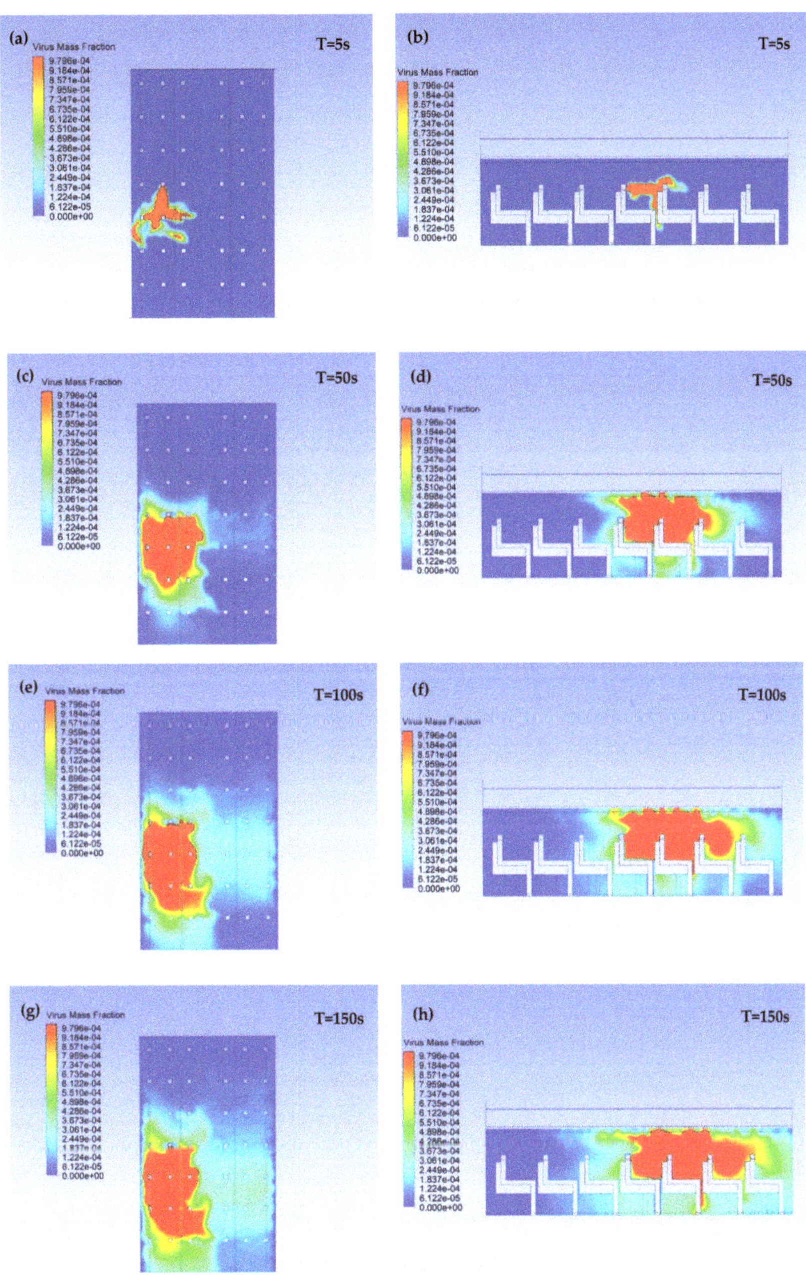

Figure 3. *Cont.*

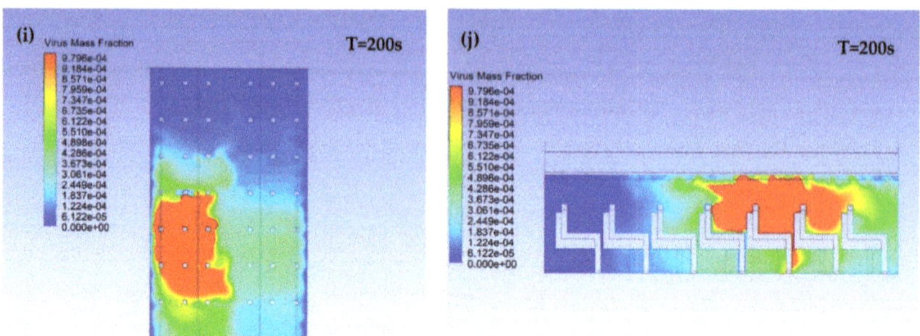

Figure 3. The dynamic 5 s, 50 s, 100 s, 150 s, 200 s after the virus carrier's single cough (left: (**a,c,e,g,i**)-Plane 1; right: (**b,d,f,h,j**)-Plane 2).

In general, no matter whether the virus carrier breathes or coughs in the aircraft cabin, the virus spreads to the front seats easier than the back seats. This implies that arranging the virus carriers in front of other passengers might be of help for reducing the infection risks.

The complete simulation results can be found in the Supplemental Information Section.

3.2. SEI

The virus mass fraction after a single breath or cough was calculated for each time point at the virus carrier breathing zone, and the results are shown in Figure 4.

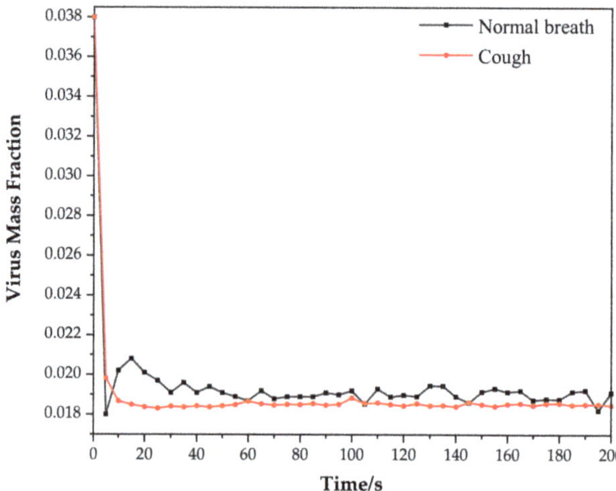

Figure 4. Virus mass fraction at the virus carrier's breathing zone after a single breath or cough.

As shown in Figure 4, the initial virus mass fraction at the virus carrier's mouth is 0.038 at T = 0 s. When the virus carrier breathes normally, the virus mass fraction decreases sharply to 0.018 within 5 s, and then goes up to 0.021 at 5–15 s, then drop again at 15–25 s. After T = 25 s, the virus mass fraction tends to get stabilized with little fluctuation. When the passenger coughs, the virus mass fraction decreases from the initial 0.038 at T = 0 s sharply to 0.019 within 10 s but stabilized much faster than normal breathing. After that, the fluctuations are also milder than normal breathing (Figure 4). The simulation data of a Boeing 737 cabin show that the virus concentration around the cougher experienced an exponential decay, and 95% of the particles were removed in 2.3 to 4.5 min [17]. The simulation results in this study show a similar decay; however, the decay is much faster in the A320 model, and the fraction curve becomes flat after 20 s.

During 5–15 s, the virus mass fraction at the virus carrier's mouth shows the opposite trends between normal breathing and coughing, because the curve under normal breathing bounces up, while the curve after coughing goes flat. The virus mass fraction stays higher under normal breathing during 10–190 s than under coughing. This is because the initial airflow velocity of coughing is higher ($10\ \mathrm{m\cdot s^{-1}}$) than the air velocity during normal breath ($1\ \mathrm{m\cdot s^{-1}}$), although the initial value is the same. This is because in the higher volume of air blown out from the virus carrier's mouth during coughing, while counting the total number of the virus molecules, more viruses are released from the virus carrier's mouth with the airflow.

Because of the high initial airflow velocity blowing away the virus aerosols and droplets, the virus carrier who coughs reduces the virus mass fraction at his/her mouth. At the same time, because the high volume of contaminated air is spread out into the cabin, this cough may increase the other passengers' chances of exposure.

To quantify the exposure risk of other passengers sitting around the virus carrier, the SEI is calculated for each passenger in this physical cabin model at T = 140 s, 160 s, 180 s and 200 s, and the results are shown in Figure 5.

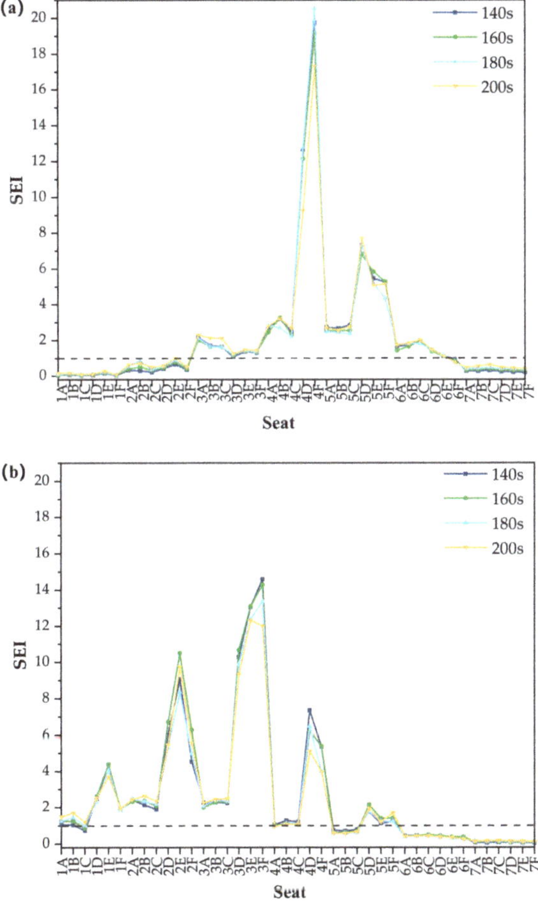

Figure 5. SEI results under when T = 140 s, 160 s, 180 s and 200 s: (**a**) normal breath; (**b**) cough.

Based on the SEI definition, we assume when the passenger calculated SEI is higher than 1, he/she has a risk for virus infection. It can be seen from Figure 5 that, when the virus carrier breathes normally, the SEI of the passengers in rows 3–6 are all greater than 1 except for seat 6F. This indicates that the passengers in rows 3–6 have exposure risks (Figure 5a). When the virus carrier coughs, the SEI of passengers in rows 1–4 and passengers 5D–5F is greater than 1 (Figure 5b). Through comparison, it can be found that the passengers in rows 3 and 4 (front row and same row of the virus carrier) and passengers in seats 5D and 5F in the rear seats always have exposure risks under the two conditions (Figure 5a,b).

When the virus carrier breathes normally, the tiny droplets exhaled from his/her mouth spread to the passengers in the back two rows with the airflow in the cabin. However, when the virus carrier coughs, the passengers sitting in the three rows in front of the virus carrier have higher exposure risks. This is due to the higher initial airflow velocity, as discussed. This is consistent with previous studies; for example, Liu et al. found that when a virus carrier coughs, the SEI remained greater than 1, even if the exposed person is 3 m away from the carrier, in an indoor environment [27]. In this study, the row spacing is set to 85 cm, and it is found that, when a virus carrier coughs, it can cause the SEI of the passengers in the three rows of seats in front to be greater than 1, the total virus travel distance is about 2.55 m inside this cabin model, which is a little shorter than the 3 m in a regular indoor environment. This is because the fixed lateral seat setup blocks the airflow to some extent.

Each passenger's SEI ratio after the virus carrier's coughing over normal breathing is calculated, and the results are shown in Figure 6 of each passenger in the cabin under the two breath conditions of the virus carrier to provide guidance for passengers to choose their seats.

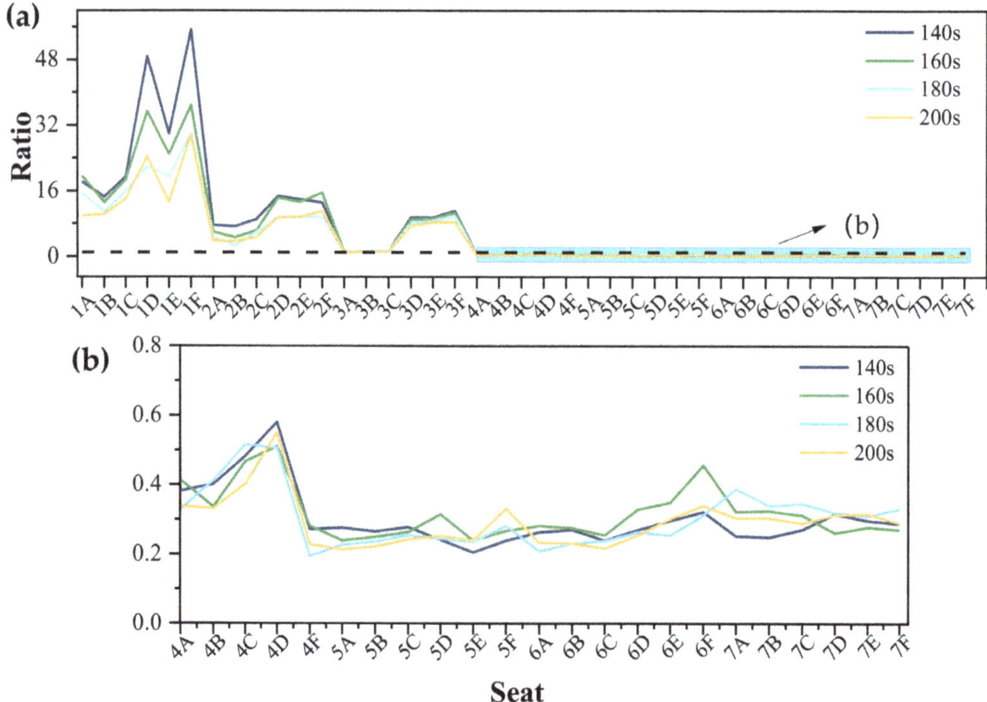

Figure 6. Passenger SEI ratio under virus carrier's cough over normal breath ((**a**)SEI ration; (**b**) the results for passengers in rows 4 to 7 are enlarged in (**b**)).

As shown in Figure 6, the SEI of passenger 1F can reach up to 55 at T = 140 s (Figure 6a). The calculated ratio is always higher than 1 for rows in front of the virus carrier, while the ratio for the same row and the back row passengers is always lower than 1. The ratio for the seats on the same side of the virus carrier is always higher than the ratio for the seats on the other side of the aisle in the same row (Figure 6). This shows that when the virus carrier coughs, the passengers behind him are relatively safer than the situation when he/she breathes normally, but the passengers in front are at a much greater risk of exposure compared with his/her normal breathing. These results indicate that, while a coughing passenger is observed in an airliner cabin, sitting on the opposite side of the aisle or in the rear of the passenger might be effective to lower the risk of infection.

As mentioned, due to the limitations and the restrictions of this research, we only studied one single breath or cough of one virus carrier in this seven-row A320 airliner cabin model. Realistic situations can be much more complicated, such as continuous breathing or coughing, and the wake introduced by moving people in the cabin is not considered. However, we build, for the first time, this seven-row airliner model with the CFD tool that complies with the policy issued by the CNHC, and numerically simulate the SARS-CoV-2 spread effectively with this model. Further studies can be projected for more realistic situations with this or a similar CFD method.

4. Conclusions

Based on our numerical simulation results calculated within the seven-row A320 airliner cabin model, the vertical spread of the virus is always faster than the horizontal spread due to the lateral seat organization in the cabin, and the high virus mass fraction area ($>9 \times 10^{-4}$) always stays in the same side of the aisle with the virus carrier. When the virus carrier breathes normally, the virus can spread to the seats in the front row, the same row and to the back two rows. When the virus carrier coughs, more viruses are carried into the air with higher initial airflow velocity, and the passengers sitting in the three rows in front, in the same row with the virus carrier, and in the two rows behind might have an SEI greater than 1, which indicates the risk for infection. In general, when a virus carrier coughs, more surrounding passengers are at risk than the situation when the virus carrier breathes normally. Avoiding sitting in front of the coughing virus carrier might lower the risk for infection.

Supplementary Materials: The following is available at https://doi.org/10.6084/m9.figshare.1657 0599.v6, Simulation video of the spread of the SARS-CoV-2 in an A320 cabin model.

Author Contributions: Conceptualization, M.Z. and N.Y.; methodology, N.Y.; software, M.Z.; validation, N.Y., M.Z., Y.Z., X.Z. and Y.C.; formal analysis, M.Z.; investigation, Y.Z., X.Z. and Y.C.; resources, M.Z.; data curation, M.Z. and X.Z; writing—original draft preparation, M.Z.; writing—review and editing, N.Y. and Y.Z.; visualization, M.Z. and N.Y.; supervision, N.Y.; project administration, N.Y.; funding acquisition, N.Y. and M.Z. All authors have read and agreed to the published version of the manuscript.

Funding: This research was funded by the Nanjing University of Aeronautics and Astronautics New Faculty Start-up fund, grant number 90YAH19018; This research was also funded by the Postgraduate Research & Practice Innovation Program of Jiangsu Province, grant number SJCX20_0067.

Institutional Review Board Statement: Not applicable.

Informed Consent Statement: Not applicable.

Data Availability Statement: Not applicable.

Acknowledgments: The authors sincerely thank Yao Zhang for helping with the field measurement and preliminary data processing. Special thanks should also be given to Nu Yu for providing the help of model processing and writing original paper.

Conflicts of Interest: The authors declare no conflict of interest.

References

1. World Health Organization. *Report of the WHO—China Joint Mission on Coronavirus Disease 2019 (COVID-19)*; WHO: Geneva, Switzerland, 2020.
2. Mboreha, C.A.; Abdallah, C.S.; Kumar, G. Risk and prevention of COVID-19 in a commercial aircraft cabin: An overview. *Int. J. Eng. Appl. Sci. Technol.* **2020**, *5*, 661–670. [CrossRef]
3. Mangili, A.; Gendreau, M.A.J.L. Transmission of infectious diseases during commercial air travel. *Lancet* **2005**, *365*, 989–996. [CrossRef]
4. Yan, Y.; Li, X.; Fang, X.; Yan, P.; Tu, J. Transmission of COVID-19 virus by cough-induced particles in an airliner cabin section. *Eng. Appl. Comput. Fluid Mech.* **2021**, *15*, 934–950. [CrossRef]
5. Olsen, S.J.; Chang, H.-L.; Cheung, T.Y.-Y.; Tang, A.F.-Y.; Fisk, T.L.; Ooi, S.P.-L.; Kuo, H.-W.; Jiang, D.D.-S.; Chen, K.-T.; Lando, J.; et al. Transmission of the severe acute respiratory syndrome on aircraft. *N. Engl. J. Med.* **2003**, *349*, 2416–2422. [CrossRef]
6. National Health Commission. *Novel Coronavirus Pneumonia Prevention and Control Plan*, 5th ed.; National Health Commission: Beijing, China, 2020. (In Chinese)
7. Lei, H.; Li, Y.; Xiao, S.; Yang, X.; Lin, C.H.; Norris, S.L.; Wei, D.; Hu, Z.; Ji, S.J.R. Logistic growth of a surface contamination network and its role in disease spread. *Sci. Rep.* **2017**, *7*, 14826. [CrossRef] [PubMed]
8. Ahlawat, A.; Wiedensohler, A.; Mishra, S.K. An Overview on the Role of Relative Humidity in Airborne Transmission of SARS-CoV-2 in Indoor Environments. *Aerosol Air Qual. Res.* **2020**, *20*, 1856–1861. [CrossRef]
9. Prather, K.A.; Wang, C.C.; Schooley, R.T.J.S. Reducing transmission of SARS-CoV-2. *Science* **2020**, *368*, eabc6197. [CrossRef] [PubMed]
10. Astuti, I.; Diabetes, Y.J. Severe Acute Respiratory Syndrome Coronavirus 2 (SARS-CoV-2): An overview of viral structure and host response. *Diabetes Metab. Syndr. Clin. Res. Rev.* **2020**, *14*, 407–412. [CrossRef]
11. Morawska, L.; Milton, D.K. It Is Time to Address Airborne Transmission of Coronavirus Disease 2019 (COVID-19). *Clin. Infect. Dis.* **2020**, *71*, 2311–2313. [CrossRef] [PubMed]
12. Asadi, S.; Bouvier, N.; Wexler, A.S.; Ristenpart, W.D. The coronavirus pandemic and aerosols: Does COVID-19 transmit via expiratory particles? *Aerosol Sci. Technol.* **2020**, *54*, 635–638. [CrossRef]
13. To, G.S.; Wan, M.P.; Chao, C.; Fang, L.; Melikov, A. Experimental Study of Dispersion and Deposition of Expiratory Aerosols in Aircraft Cabins and Impact on Infectious Disease Transmission. *Aerosol Sci. Technol.* **2009**, *43*, 466–485. [CrossRef]
14. Wang, G.Q.; Ji, S.C. Summary of the analysis of the spread of germs in the cabin. *Acta Aeronautica et Astronautica Sinica.* **2015**, *36*, 2577–2587. (In Chinese) [CrossRef]
15. Mazumdar, S.; Chen, Q.J. Influence of cabin conditions on placement and response of contaminant detection sensors in a commercial aircraft. *J. Environ. Monit.* **2008**, *10*, 71–81. [CrossRef]
16. Zhang, Z.; Chen, Q. Comparison of the Eulerian and Lagrangian methods for predicting particle transport in enclosed spaces. *Atmos. Environ.* **2007**, *41*, 5236–5248. [CrossRef]
17. Davis, A.C.; Zee, M.; Clark, A.D.; Wu, T.; Olson, N.A. Computational Fluid Dynamics Modeling of Cough Transport in an Aircraft Cabin. *bioRxiv* **2021**, *2*, 15. [CrossRef]
18. Talaat, K.; Abuhegazy, M.; Mahfoze, O.A.; Anderoglu, O.; Poroseva, S.V. Simulation of aerosol transmission on a Boeing 737 airplane with intervention measures for COVID-19 mitigation. *Phys. Fluids* **2021**, *33*, 033312. [CrossRef] [PubMed]
19. Conceição, S.T.; Pereira, M.L.; Tribess, A. A Review of Methods Applied to Study Airborne Biocontaminants inside Aircraft Cabins. *Int. J. Aerosp. Eng.* **2011**, *2011*, 824591. [CrossRef]
20. Mazumdar, S.; Poussou, S.B.; Lin, C.-H.; Isukapalli, S.S.; Plesniak, M.W.; Chen, Q. Impact of scaling and body movement on contaminant transport in airliner cabins. *Atmos. Environ.* **2011**, *45*, 6019–6028. [CrossRef]
21. Gupta, J.K.; Lin, C.H.; Chen, Q. Transport of expiratory droplets in an aircraft cabin. *Indoor Air* **2011**, *21*, 3–11. [CrossRef]
22. AIRBUS. *Aircraft Characteristics Airport and Maintenance Plannin*; AIRBUS: Blagnac, France, 2005.
23. Available online: https://www.ansys.com/products/meshing (accessed on 2 July 2021).
24. ASHRAE. *Standard 161-2018. Air Quality within Commercial Aircraft (ANSI Approved)*; ASHRAE: Atlanta, GA, USA, 2021.
25. Gupta, J.K.; Lin, C.H.; Chen, Q. Flow dynamics and characterization of a cough. *Indoor Air* **2010**, *19*, 517–525. [CrossRef]
26. Nicas, M.; Nazaroff, W.W.; Hubbard, A. Toward understanding the risk of secondary airborne infection: Emission of respirable pathogens. *J. Occup. Environ. Hyg.* **2005**, *2*, 143–154. [CrossRef] [PubMed]
27. Liu, L.; Li, Y.; Nielsen, P.V.; Wei, J.; Jensen, R.L. Short-range airborne transmission of expiratory droplets between two people. *Indoor Air* **2016**, *27*, 452–462. [CrossRef]
28. Tsoukias, N.M.; Tannous, Z.; Wilson, A.F.; George, S.C. Single-exhalation profiles of NO and CO_2 in humans: Effect of dynamically changing flow rate. *Jpn. J. Physiol.* **1998**, *85*, 642–652. [CrossRef]
29. Chang, T.B.; Sheu, J.J.; Huang, J.W.; Lin, Y.S.; Chang, C.C. Development of a CFD model for simulating vehicle cabin indoor air quality. *Transp. Res. Part D Transp. Environ.* **2018**, *62*, 433–440. [CrossRef]
30. Scott, J.L.; Kraemer, D.G.; Keller, R.J. Occupational hazards of carbon dioxide exposure. *ACS Chem. Health Saf.* **2009**, *16*, 18–22. [CrossRef]
31. Yakhot, V.; Orszag, S.A.; Thangam, S.; Gatski, T.B.; Speziale, C.G. Development of turbulence models for shear flows by a double expansion technique. *Phys. Fluids* **1992**, *4*, 1510–1520. [CrossRef]

32. Acikgoz, M.B.; Akay, B.; Miguel, A.F.; Aydin, M. Airborne Pathogens Transport in an Aircraft Cabin. *Defect Diffus. Forum* **2011**, *312–315*, 865–870. [CrossRef]
33. Zhang, Z. Modeling of Airflow and Contaminant Transport in Enclosed Environments. Ph.D. Thesis, Purdue University, West Lafayette, OH, USA, 2007.
34. Shengcheng, J.; Jinglei, X.; Shanggui, C.; Zhongmin, H. Numerical Study of Virus Droplet Transport in Civil Aircraft Cabin. *Int. J. Astronaut. Aeronaut. Eng.* **2018**, *3*, 1–13. [CrossRef]
35. Chen, Q. Comparison of different k-ε models for indoor air flow computations. *Numer. Heat Transf. Part B Fundam.* **1995**, *28*, 353–369. [CrossRef]
36. Liu, W.; Wen, J.; Lin, C.H.; Liu, J.; Long, Z.; Chen, Q.J.B. Evaluation of various categories of turbulence models for predicting air distribution in an airliner cabin. *Build. Environ.* **2013**, *65*, 118–131. [CrossRef]
37. Qian, H.; Li, Y. Removal of exhaled particles by ventilation and deposition in a multibed airborne infection isolation room. *Indoor Air* **2010**, *20*, 284–297. [CrossRef] [PubMed]
38. Han, Z.Y.; To, G.N.S.; Fu, S.C.; Chao, C.Y.H.; Weng, W.G.; Huang, Q.Y. Effect of human movement on airborne disease transmission in an airplane cabin: Study using numerical modeling and quantitative risk analysis. *BMC Infect. Dis.* **2014**, *14*, 434. [CrossRef] [PubMed]

Article

Particulate Matter Exposures under Five Different Transportation Modes during Spring Festival Travel Rush in China

Yao Zhang [1,†], **Nu Yu** [1,*,†], **Mengya Zhang** [1] and **Quan Ye** [2]

1   Department of Air Traffic, College of Civil Aviation, Nanjing 211106, China; yaozhangnuaa@nuaa.edu.cn (Y.Z.); zmy197319@nuaa.edu.cn (M.Z.)
2   Department of Computer Science, School of Electronics and Information Engineering, Lanzhou Jiaotong University, Lanzhou 730070, China; 11200758@stu.lzjtu.edu.cn
*   Correspondence: nuyu@nuaa.edu.cn; Tel.: +86-189-4964-2736
†   These two authors contributed equally to this work.

Abstract: Serious traffic-related pollution and high population density during the spring festival (Chinese new year) travel rush (SFTR) increases the travelers' exposure risk to pollutants and biohazards. This study investigates personal exposure to particulate matter (PM) mass concentration when commuting in five transportation modes during and after the 2020 SFTR: China railway high-speed train (CRH train), subway, bus, car, and walking. The routes are selected between Nanjing and Xuzhou, two major transportation hubs in the Yangtze Delta. The results indicate that personal exposure levels to PM on the CRH train are the lowest and relatively stable, and so it is recommended to take the CRH train back home during the SFTR to reduce the personal PM exposure. The exposure level to $PM_{2.5}$ during SFTR is twice as high as the average level of Asia, and it is higher than the WHO air quality guideline (AQG).

Keywords: Chinese spring festival travel rush; COVID-19; exposure; particulate matter

Citation: Zhang, Y.; Yu, N.; Zhang, M.; Ye, Q. Particulate Matter Exposures under Five Different Transportation Modes during Spring Festival Travel Rush in China. *Processes* **2021**, *9*, 1133. https://doi.org/10.3390/pr9071133

Academic Editor: Daniele Sofia

Received: 31 May 2021
Accepted: 27 June 2021
Published: 29 June 2021

Publisher's Note: MDPI stays neutral with regard to jurisdictional claims in published maps and institutional affiliations.

1. Introduction

In-vehicle air quality can affect the exposure level of commuters. Those pollutants related to traffic emissions may include ultrafine particles (particles with aerodynamic diameter $\leq$ 100 nm, UFPs), fine particles (particles with aerodynamic diameter $\leq$ 2.5 μm, $PM_{2.5}$), black carbon (BC), carbon monoxide (CO), nitrogen dioxide (NO_2), and volatile organic compounds (VOCs) [1]. Previous studies indicate that people spend about 5.5% and 7.6% of their day in vehicles and outdoors, respectively [2]. About 26.7% of commuters have the preference of choosing public transportations in Nanjing in recent years [3]. There is a population of 1.4 billion in China. During the Chinese spring festival travel rush (SFTR), workers and college students buy tickets to return home, which is the largest human migration on earth happening annually. The investigated SFTR started on 10 January 2020, with an estimate of 3 billion passengers over 40 days. Public transportation was fully loaded during the SFTR. In late December 2019, a novel coronavirus disease (COVID-19) broke out in Wuhan, Hubei province, China. Before the suddenly announced strict quarantine starting from 23 January 2020, many unknowns remain. COVID 19 is both deadly and highly transmissible. Recently, the airborne mode of COVID-19 transmission occurring primarily in indoor places has been recognized by many countries and research communities, suggesting proper mask use in transportation facilities such as cars and trains [4–8]. As of June 2021, the number of global confirmed cases is over 190 million, and the death toll is 3.9 million. Air pollution during transportation can have other adverse health effects, including respiratory and cardiovascular diseases and death [9]. A study by Goel et al. shows changes in traffic density for multiple weeks may not induce immediate $PM_{2.5}$ or BC exposure changes, but some toxic elements such as P, S, As, Cu, and Pb may change significantly [10]. Despite the exposure time in a vehicle being relatively limited, the chemical species and their concentrations can represent a significant health risk to commuters [11].

Large proportions of aerosol exposure are experienced during daily commuting trips due to heavy traffic, which contributes to a serious problem for public health [12,13]. High levels of $PM_{2.5}$ in traffic exposure are associated with systemic inflammatory response and respiratory injury [14,15]. The study of Xu et al. indicates that the carbonyl compounds emitted by private cars in Nanjing are harmful to passengers, and most people prefer to take public transports for convenience [16]. Commuting using light rail and the subway have potential health benefits when compared to driving on roadways for the general population [17]. Commuters' $PM_{2.5}$ exposure level in public transport modes is always impacted by wind speed and the number of passengers [18]. In addition, commuting via walking and bus has more exposure to a larger particle concentration and $PM_{2.5}$ mass concentrations for commuters in Beijing [19].

Traffic-related air pollution is a serious health risk, especially for susceptible people [20]. Heavy traffic and various public transport modes also increase the risk of cross-infection during the SFTR. However, in recent years, there have been no articles about the exposure risk of passengers on public transport during the SFTR. Public transportation is always fully or overloaded during the SFTR, as well as the roads and highways being congested. This study was conducted during the worst transportation scenario in Yangtze Delta, which is the most populated area in China. In this study, (1) the personal exposure to PM mass concentrations of five transportation modes was investigated. (2) PM mass concentrations were analyzed during and after the Chinese SFTR with the impact of the COVID-19 outbreak.

2. Methods

2.1. Sampling Routes

Nanjing Railway Station, servicing approximately 210,000 passengers daily during the SFTR, is the center of transportation and tourism in the Yangtze Delta [21]. Passengers are mostly made up of workers and students, and public transport is their first choice. Figure 1 shows the four routes of commuter monitoring data, namely NO.1, 2, 3, 4. The travel routes are connecting Nanjing Railway Station, Nanjing South Railway Station, Xuzhou Railway Station, Xuzhou East Railway Station, and Guanyin International Airport. Xuzhou Railway Station, as the main railway station, services approximately 33,200 passengers daily during the SFTR. Meanwhile, the Xuzhou Subway Line 1 is also selected as the typical route to approach Xuzhou East Railway Station for convenience, and it services approximately 85,700 passengers every day. A total of 1.06 million people is the highest single-day passenger volume of Nanjing Subway Line 3.

Table 1 shows the specific segments of the four routes during and after the SFTR. Table 2 summarizes the basic information of four selected travel routes, such as ventilation type, duration, and distance. Five transport modes are included, which are China railway high-speed train (CRH train), subway, bus, passenger car, and walking. Air conditioning (AC) is used in CRH trains, subways, and busses. The passenger car did not use AC (Non-AC), and the windows were opened (WO) when NO.1–3 routes are selected, but the car AC mode was on when NO.4 route is selected. The carriages of the CRH train and bus are fully loaded. In addition, walking is also considered open-air. There is a distance between Xuzhou Railway Station and Xuzhou Subway Station, and passengers walk along the street for about 24 min for the transfer. The NO.4 route was operated after the strictest quarantine was lifted, and students and workers were back in their original places. The other routes were operated during the SFTR with the beginning of the COVID-19 outbreak.

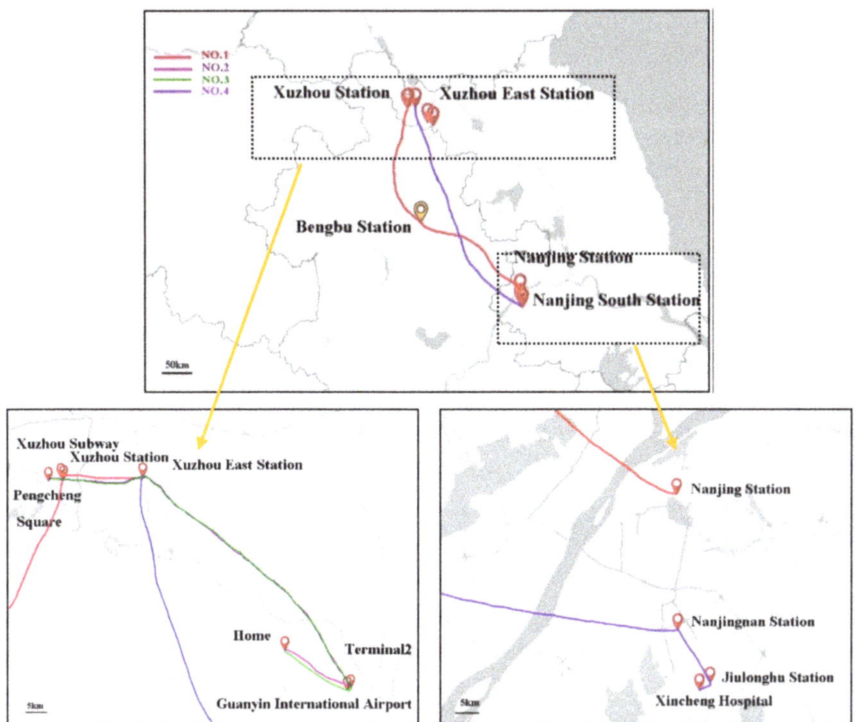

Figure 1. The selected route for different transport modes.

Table 1. The segments of travel routes during and after the SFTR.

Transportation Modes	NO.1 (19 January 2020)	NO.2 (21 January 2020)	NO.3 (21 January 2020)	NO.4 (6 May 2020)
CRH train 1	Nanjing-Xuzhou	-	-	Xuzhou East-Nanjing South
Walk	Xuzhou-Xuzhou Subway	-	-	Jiulonghu-Xincheng Hospital
Subway	Xuzhou Subway-Xuzhou East	Xuzhou East-Pengcheng Square	Pengcheng Square-Xuzhou East	Nanjing South-Jiulonghu
CRH train 2	Xuzhou East-Guanyin Airport	Guanyin Airport-Xuzhou East	Xuzhou East-Guanyin Airport	Guanyin Airport-Xuzhou East
Bus	Guanyin Airport-Terminal 2	-	Guanyin Airport-Terminal 2	-
Car	Terminal 2-Home	Home-Guanyin Airport	Guanyin Airport-Home	Home-Guanyin Airport

2.2. Measurements

In this study, personal exposure monitoring began when the researcher left school for winter vacation. The first experiment was performed on 19 January 2020 from 11:03 to 17:25 (NO.1). The second experiment was performed on 21 January 2020, from 10:06 to 18:23 (NO.2, 3). The third experiment was performed on 6 May 2020, from 09:58 to 13:39 (NO.4). The PM mass concentrations were measured using a portable air quality monitor (BoHu model BH1-B3, China), and the instrument's detailed information is shown in Table 3. This instrument uses the light scattering method to measure the PM from 0 to 1999 $\mu g \cdot m^{3}$ (PM_1, $PM_{2.5}$, PM_{10}). This instrument has been calibrated against a TSI DustTrak 8532 for $PM_{2.5}$ indoor and outdoor use before the experiment (R2 = 0.89). Weather parameters such as temperature (T) and relative humidity (RH) were also measured by the same instrument. All collected data are stored in the memory card inside the instrument with a 1 min sampling interval.

Table 2. The information summary about travel routes.

Transportation Modes	Ventilation Type	Routes	Duration (min)	Distance (km)	Note
CRH train 1	AC	NO.1	150	283.5	-
		NO.4	222	283.5	-
Walk	Open air	NO.1	24	0.4	8 lanes in both directions
		NO.4	9	0.2	Single direction
Subway	AC	NO.1	27	9.0	-
		NO.2	30	9.0	-
		NO.3	33	9.0	-
		NO.4	20	8.9	-
CRH train 2	AC	NO.1	16	35.3	-
		NO.2	15	35.3	-
		NO.3	16	35.3	-
		NO.4	17	35.3	-
Bus	AC	NO.1	9	2.0	2 lanes in both directions
		NO.3	9	2.0	
Car	Non-AC+WO	NO.1	15	9.2	
	Non-AC+WO	NO.2	20	9.2	6 lanes in both directions
	Non-AC+WO	NO.3	23	9.2	
	AC	NO.4	34	9.2	

Table 3. Instrument information.

Instrument	Parameters Collected	Interval	Range	Accuracy
BoHu model BH1-B3	$PM_1/PM_{2.5}/PM_{10}$	1 min	$0\text{--}1999\ \mu g\cdot m^{-3}$	$\pm 15\%$
	RH		0–100%	$\pm 5\%$
	T		$-20\text{--}99\ ^{\circ}C$	$\pm 2\ ^{\circ}C$

The ambient PM mass concentration is usually affected by meteorological conditions. However, the in-cabin microenvironment in trains, passenger cars, subway cabins, and buses was mainly affected by the air conditioning systems when the doors and windows were tightly closed. The ambient relative humidity (RH) and temperature (T) mostly affected the walking exposure in our study. The air quality index (AQI) has been set up based on the ambient environment. However, in terms of personal exposure, the ambient and in-cabin exposure levels were both important because nowadays, people spend more time in microenvironments, such as buildings and cabins. The average temperature (T) and relative humidity (RH) of four transportation routes are summarized in Table 4. The outdoor T and RH during the walking mode were similar to the in-cabin conditions in this study. The cabin microenvironment in different transportation modes was affected by the air conditioning/ventilation systems, mostly when the cabin doors and windows were tightly closed.

Table 4. Average temperature (T) and relative humidity (RH).

Route	NO.1		NO.2		NO.3		NO.4	
	T/°C	RH/%	T/°C	RH/%	T/°C	RH/%	T/°C	RH/%
CRH train 1	23.0	41.0	-	-	-	-	25.7	47.9
Walk	20.9	31.5	-	-	-	-	26.2	48.7
Subway	20.1	38.6	13.5	52.1	15.3	53.7	25.9	50.5
CRH train 2	22.1	36.3	13.6	64.4	19.5	44.1	25.1	50.3
Bus	21.2	35.8	-	-	17.3	54.1	-	-
Car	19.0	43.7	12.9	52.0	18.0	49.2	25.3	48.6

The maximum speed of the CRH train is about 250 km/h. The Xuzhou Subway Line 1 started operation on 28 September 2019, and Nanjing Subway Line 3 has operated from 1 January 2015. The bus is free for the airport to shuttle passengers who get off from the Guanyin international airport. The private passenger car is a Mazda-6 four-door sedan with a 2.0 L engine (gasoline) and high-performance cabin air filter VF2018, which was used during SFTR. After the strictest quarantine was lifted, a new (less than 1 year) Lynkco-03 four-door SUV with a 1.5 L engine (gasoline) and a high-efficiency particulate air filter (HEPA) was driven from home to the Guanyin international airport. During exposure measurement, the instrument was positioned outside a laptop backpack and fixed with tape, which has no threat to the security check during the SFTR.

3. Results and Discussion

3.1. During the SFTR

The World Health Organization (WHO) air quality guideline (AQG) for 24-h mean $PM_{2.5}$ and PM_{10} are 25 $\mu g \cdot m^{-3}$ and 50 $\mu g \cdot m^{-3}$, respectively. The AQG is the lowest level at which total cardiopulmonary and lung cancer mortality have been shown to increase with more than 95% confidence in response to $PM_{2.5}$ in the American Cancer Society study [22]. There is no related guideline of PM for the health of commuters in public transportations in China. Figure 2 shows the box plot of PM mass concentrations during the SFTR. Average PM_{10} mass concentrations of five different transport modes range from 23.4 to 202.3 $\mu g \cdot m^{-3}$.

When passengers commute using the subway and car, the exposure levels of PM_{10} mass concentrations are all higher and about twice as high as the results of Qiu et al. (subway: 58.2 $\mu g \cdot m^{-3}$; car: 95.9 $\mu g \cdot m^{-3}$; walking: 110.0 $\mu g \cdot m^{-3}$) [9]. However, when commuting on the bus, the exposure level to PM_{10} mass concentration in this study is similar to the results of Qiu et al. (123.6 $\mu g \cdot m^{-3}$) and Chan et al.(184.0 $\mu g \cdot m^{-3}$) [9,23]. The exposure level of passengers in a car and by walking is higher than other transport modes, and the mean PM_{10} mass concentrations of car and walking are 202.3 $\mu g \cdot m^{-3}$ and 164.5 $\mu g \cdot m^{-3}$, respectively. However, the average PM_{10} mass concentration of commuter exposure in CRH train1 is the lowest, which is one ninth of the commuting in car level. Average $PM_{2.5}$ mass concentrations for six segments are 20.2 $\mu g \cdot m^{-3}$, 137.5 $\mu g \cdot m^{-3}$, 97.7 $\mu g \cdot m^{-3}$, 43.8 $\mu g \cdot m^{-3}$, 118.9 $\mu g \cdot m^{-3}$, and 166.7 $\mu g \cdot m^{-3}$, when commuters travel by CRH train1, walk, subway, CRH train2, bus, and car, respectively. During the SFTR, the exposure levels of $PM_{2.5}$ mass concentrations in five tested transport modes are 2 to 4 times the results of other researchers, as shown in Figure 3 [9,19,24–28].

As Figure 3a shows the exposure level to particles when passengers take the CRH train, due to less research on pollution in CRH train cabins, this study makes a comparison with other electronic trains and diesel trains. According to Figure 3a, the exposure level in diesel trains are two times that in electronic trains [29]. Overall, the average exposure levels of electronic trains in Europe, China, and America are close, and they are 33.7 $\mu g \cdot m^{-3}$, 38.3 $\mu g \cdot m^{-3}$, and 22.5 $\mu g \cdot m^{-3}$, respectively [30–38]. In this study, the PM mass concentrations in the long-haul CRH train (CRH train1) are half of the short-haul CRH train (CRH train2). The fast speed (no more than 250 km/h) and frequent air exchange of the air conditioner might decrease the PM mass concentration.

Figure 3b indicates that the $PM_{2.5}$ mass concentration is close to 2.4 and 2 times higher in this study (141.1 $\mu g \cdot m^{-3}$) when commuting via walking during the SFTR compared to the usual level in China (59.4 $\mu g \cdot m^{-3}$) and Asia (71.7 $\mu g \cdot m^{-3}$), respectively [9,13,19,24,25,39,40]. According to the study of Ozgen et al., the mean exposure level of SFTR in China is seven times that of Italy [40]. In India, PM pollution is more serious (234 $\mu g \cdot m^{-3}$), which is 1.7 times that of China during the SFTR [41]. In addition, the pollution level of China's surrounding environment is getting worse compared with this study when commuting via walking, and 5.3 times that of the study of Yan et al. in Beijing (26.7 $\mu g \cdot m^{-3}$) [19].

(a)

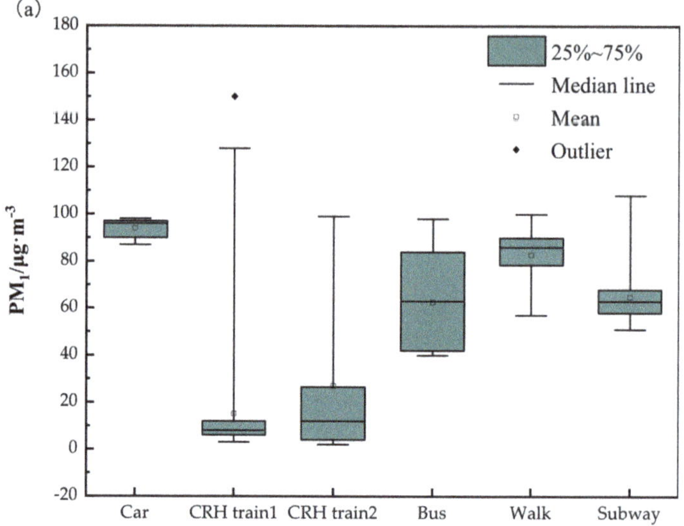

(b)

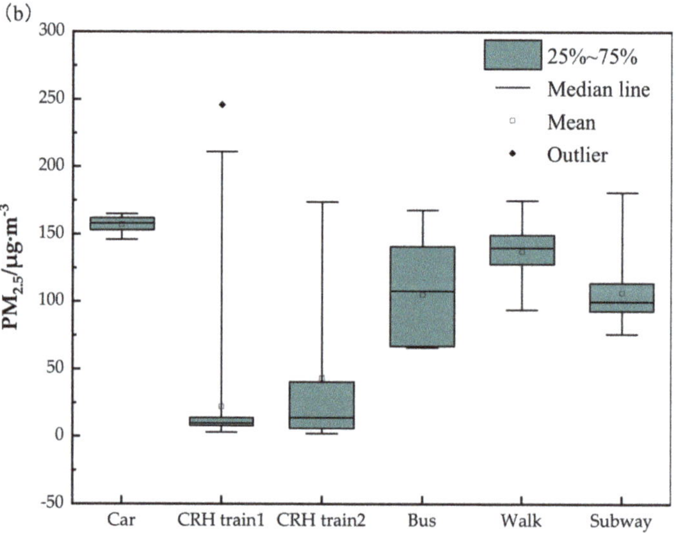

Figure 2. *Cont.*

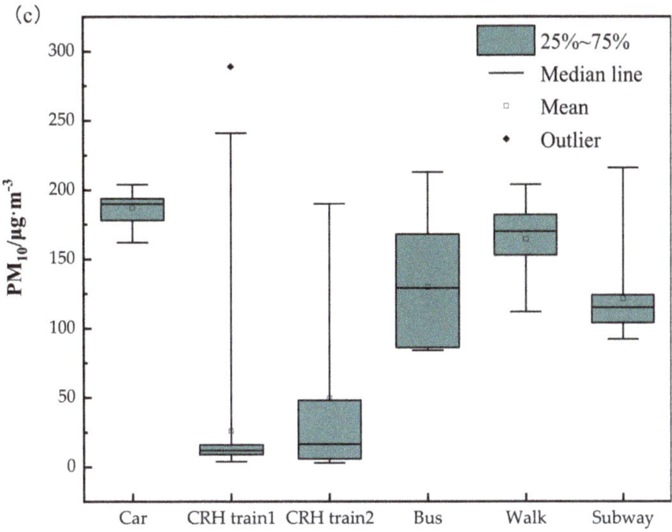

Figure 2. The box plot of PM mass concentrations during the SFTR: (**a**) PM$_1$; (**b**) PM$_{2.5}$; (**c**) PM$_{10}$.

Figure 3c shows that the mean PM$_{2.5}$ mass concentration is 97.7 µg·m^{-3} when commuting on the subway during the SFTR, which is about 2.2, 2, and 1.6 times higher than the usual level in China (43.7 µg·m^{-3}), Asia (47.95 µg·m^{-3}, and Chile (62.4 µg·m^{-3}), respectively [9,13,19,24,25,28,38–40,42]. When commuting on the subway, the exposure level to PM$_{2.5}$ mass concentration in this study is similar to the results of Ozgen et al. in Italy (91.1 µg·m^{-3}), and the PM pollution on the subway is the most serious compared to other public travel modes in the study by Ozgen et al. in Italy [40]. In particular, in Milan, a well-known European hot-spot for PM pollution, which may contribute to the high and similar PM exposure level with China during the SFTR. As shown in Figure 3c, the PM pollution in China is more and more serious, and in particular during the SFTR. In addition, the subway is the primary choice for commuters' transit plans, which may do harm to the health of commuters and increase the exposure risk to COVID-19.

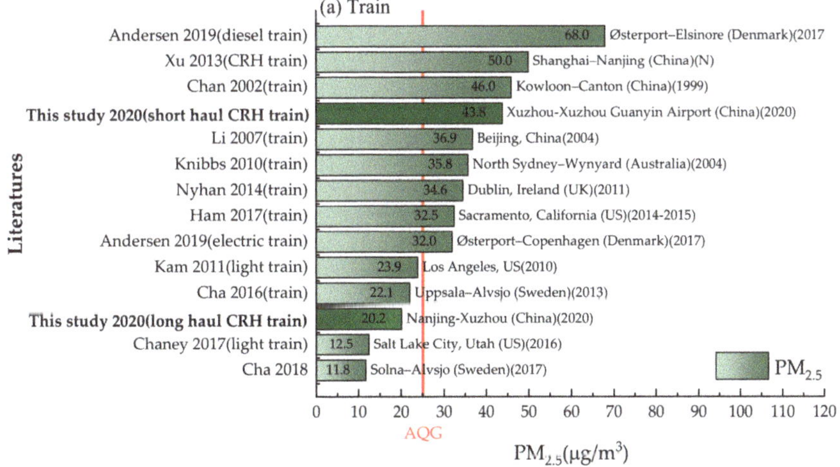

Figure 3. *Cont.*

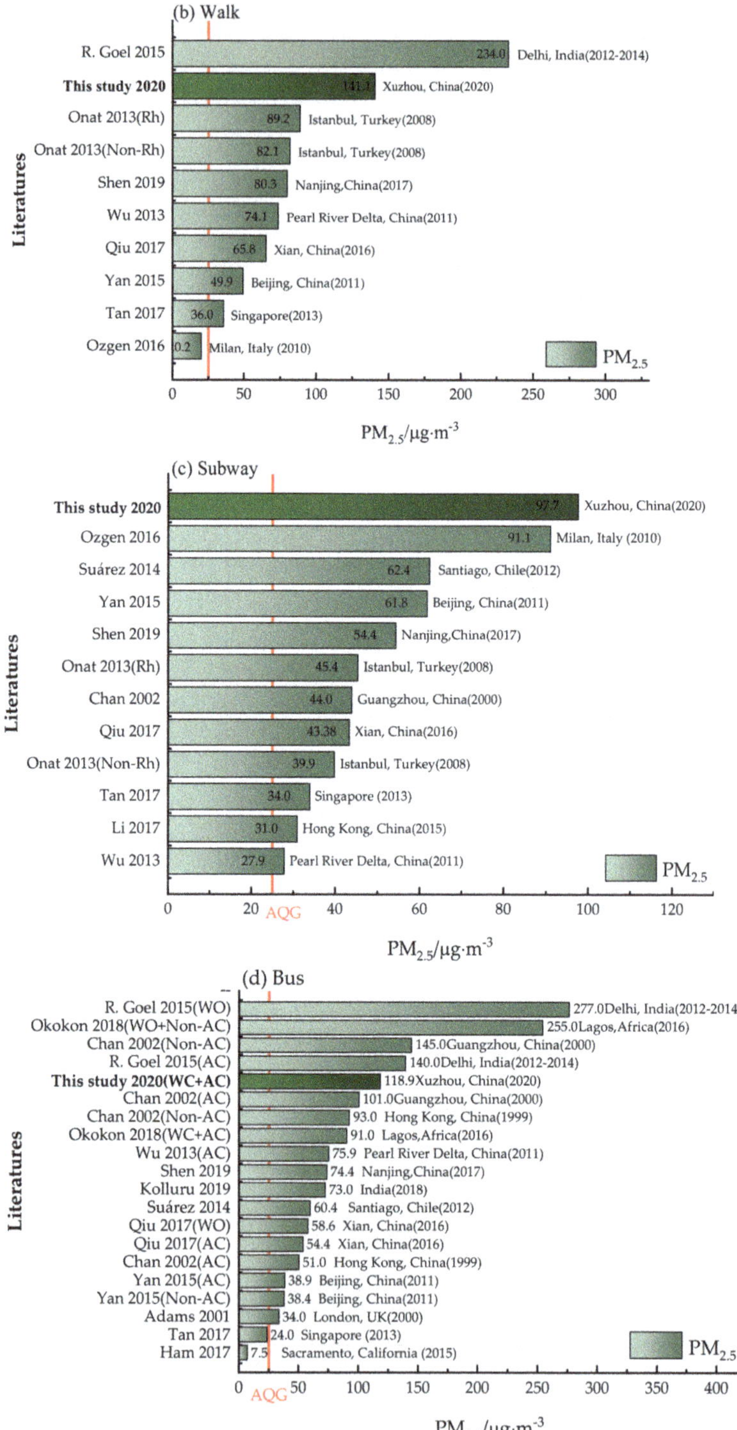

Figure 3. *Cont.*

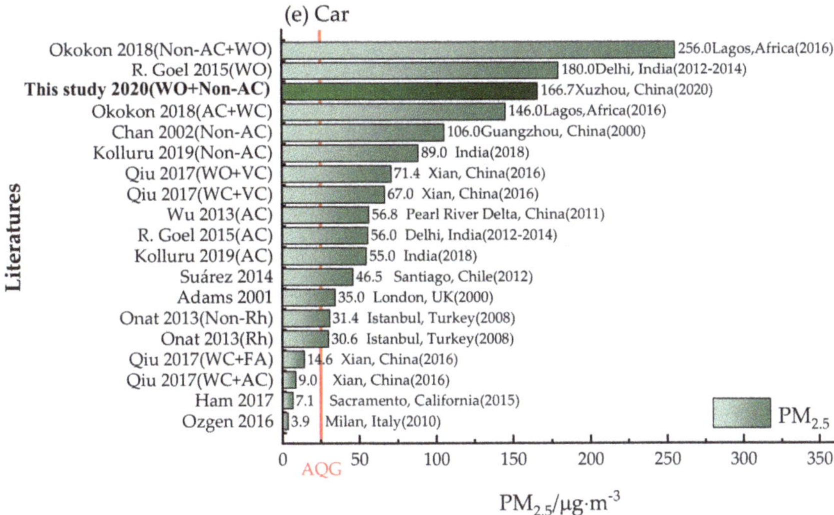

Figure 3. The $PM_{2.5}$ mass concentration of different transportations in the literature: (**a**) Train; (**b**) Walk; (**c**) Subway; (**d**) Bus; (**e**) Car. Rh stands for rush hours. Non-Rh stands for non-rush hours. WC + VC stands for all windows and vents being closed. AC + WC stands for all windows closed, air condition on recirculation. WC + FA stands for all windows closed, AC system on fresh air mode.

When commuting on the bus, the windows are all closed with the air conditioner working. The mean level of $PM_{2.5}$ mass concentration is 118.9 $\mu g \cdot m^{-3}$ during the SFTR in this study. As shown in Figure 3d, the measured mean $PM_{2.5}$ mass concentration during the SFTR is 1.6 times that of the regular time in India and the non-SFTR in China, 1.5 times that of the regular time in Europe, and 3.5 times that of the United States [9,19,23–25,28,34,38,39,43,44]. According to Yan et al., whether working or not, the air conditioner is not a critical factor influencing the exposure level of passengers on the bus (AC: 38.9 $\mu g \cdot m^{-3}$; Non-AC:38.4 $\mu g \cdot m^{-3}$) [19]. However, the study of Chan et al. issued in 2002 indicates that the exposure level of a Non-AC bus (145.0 $\mu g \cdot m^{-3}$) is 1.4 times that of an AC bus (101.0 $\mu g \cdot m^{-3}$) in Guangzhou [23]. In addition, the exposure level to $PM_{2.5}$ mass concentration on the bus is similar to the results of Okokon et al.'s study, which was researched in Africa when commuting on the bus with the AC on and the windows closed (91.0 $\mu g \cdot m^{-3}$) [45]. When the bus windows were opened in Non-AC mode, higher exposure levels to $PM_{2.5}$ mass concentration were found in Okokon et al.'s study (255.0 $\mu g \cdot m^{-3}$), which are 2.1 and 2.8 times that of this study and Okokon et al.'s study in AC mode [45], respectively. However, similar exposure levels were obtained in the WO mode bus (58.6 $\mu g \cdot m^{-3}$) and AC mode bus (54.4 $\mu g \cdot m^{-3}$) in the study of Qiu et al. [9]. Due to more serious ambient pollution in Africa compared to China, which contributes to more difference between the inside and outside of the bus.

In this study, when commuting in a car with WO and VC mode during the SFTR in 2020, the mean exposure level to $PM_{2.5}$ mass concentration is 166.7 $\mu g \cdot m^{-3}$, which is 2.3 times that of Qiu et al.'s study [9]. However, when the car windows are all closed with FA, VC, or AC modes, a lower exposure level was found in Qiu et al.'s study, as shown in Figure 3e [9]. Furthermore, the mean exposure level of commuting in a car is similar to commuting via walking, which is 3, 4.3, 24, and 2.3 times that of Non-SFTR, UK, US, and India, respectively [9,23–25,28,34,41,43–45].

In summary, the exposure levels to $PM_{2.5}$ on public transport are apparently higher than AQG. Short-term exposure to $PM_{2.5}$ can trigger cardiovascular disease-related mortality and nonfatal events, which also can affect healthy human lymphocyte subsets [46,47]. Therefore, it is critical for commuters, for instance, to wear a facial mask to decrease the risk of pollutants exposure and avoid inhalation of PM into the lungs. In this study, workers and students were found to have higher exposure levels to PM mass concentrations when commuting via subway, bus, car, and walking during the SFTR. Although the average PM mass concentrations of commuting by car are the highest, the standard deviations (SDs) of them are the lowest. Car windows are opened during the non-AC mode, which can provide a relatively stable in-cabin microenvironment. The mean exposure level to particles is lowest in the CRH train, and following is the subway, bus, walking, and car. While the SDs of PM mass concentration are higher than the average, which is contributed to by the speed of the CRH train suddenly slowing down when approaching the destination, the pollutants in the ambient air will enter the cabin via AC system or by air pressure. Therefore, it is recommended that the passengers choose the better combination of CRH train and subway during the SFTR.

The time series of the CRH train1, CRH train2, subway, and the car started from the closing of the cabin door to reopening. When walking, the researchers are exposed to the open environment. According to Figure 4a,d time series PM mass concentrations in CRH trains are similar. When CRH trains approached destinations, about 100 times PM mass concentrations were monitored in the cabin microenvironment. The study of Adams et al. indicates that wind speed negatively affects the PM mass concentration [43].

As shown in Figure 4, the gaps of PM_1, $PM_{2.5}$, PM_{10} in CRH train1 and CRH train2 are the smallest when the cabin door is closed compared to other transportation modes. As shown in Figure 4b, the mean PM mass concentrations decrease when walking into the subway and preparing to check out under the ground from 13:52 to 13:54. After the safety check, the mean PM mass concentrations increased from 13:54 to 13:58, which might be affected by passengers' activities. Figure 4b,c,e indicate that the exposure levels to particles via subway and bus are similar to commuting via walking during the SFTR. In addition, there are more fluctuations in the PM mass concentrations when commuting via subway, which may be contributed to by frequent stopping (9 stops) during operation. Figure 4e shows a little higher PM mass concentration from 17:01 to 17:04, which might be affected by the fresh air mode of the air conditioner and the number of passengers, which can introduce more particles into the bus. When the bus door opened, passengers aboard the bus experience a personal exposure level to PM mass concentrations similar to commuting via walking (Figure 4b) and car (Figure 4f).

The WHO recommends the use of $PM_{2.5}$ as an indicator and a $PM_{2.5}/PM_{10}$ ratio of 0.5 (50%) is used to derive an appropriate PM_{10} guideline value [22]. This ratio of 0.5 is close to that typically observed in urban areas in developing countries and at the bottom of the range (0.5–0.8) found in urban areas in developed countries [22]. For the five transport modes, PM ratios consist of PM_1/PM_{10} and $PM_{2.5}/PM_{10}$, as shown in Figure 5. Figure 5 indicates $PM_{2.5}/PM_{10}$ are all higher than the typical guideline value of the WHO, which shows more $PM_{2.5}$ on public transportation in China. In this study, the PM_1/PM_{10} and $PM_{2.5}/PM_{10}$ in CRH train1 are the highest, and they are 68.2% and 87.4%, respectively. These ratios on CRH trains and the subway are significantly higher than that of the other three transport modes, which indicates that there are more fine particles on CRH trains and the subway. When the cabin door is closed, the air conditioner exchanges cabin air for passengers. According to the study on traffic environmental pollution exposure of Yu Nu et al., the exposure levels of road traffic pollution sources $PM_{2.5}$, and ultrafine particles (UFPs) were significantly correlated with the increase of the body's oxidation marker malondialdehyde among taxi driver groups in Los Angeles [48]. The PM_1/PM_{10} and $PM_{2.5}/PM_{10}$ on the bus are the lowest, and they are 48.2% and 80.8%, respectively. Although the ventilation on the bus is supported by an air conditioner, the air filtration performance is not good. Considering the health of passengers, the bus driver should change the air conditioning filter regularly. Overall, the $PM_{2.5}/PM_{10}$ in the microenvironment of the subway, bus, car, and walking are all higher than PM_1/PM_{10}, and the exposure levels to fine particles are most serious on CRH trains and the subway, which is similar to the study of Qiu et al. [9].

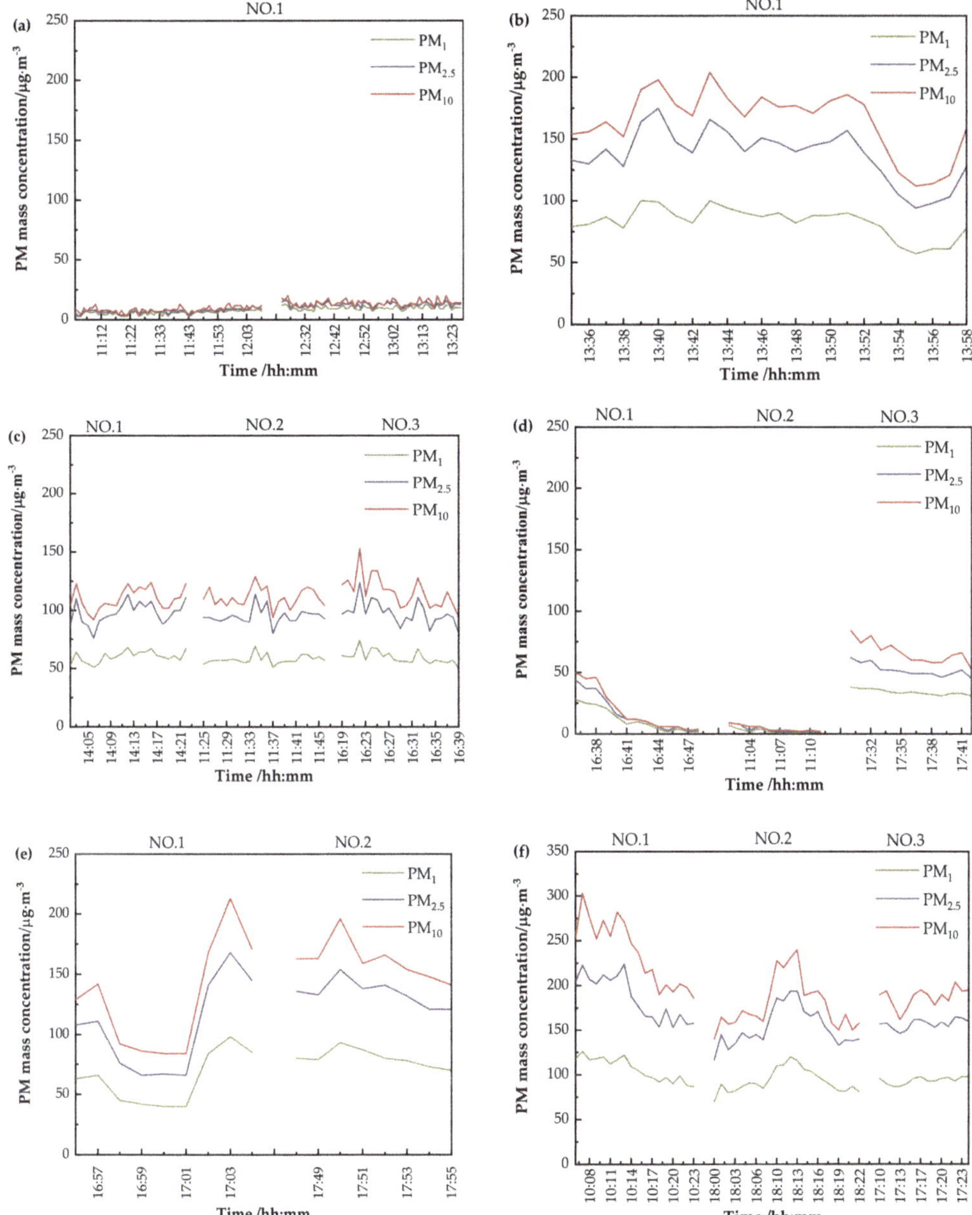

Figure 4. The PM mass concentrations in time series for different transport modes during the SFTR: (**a**) CRH train1; (**b**) Walk; (**c**) Subway; (**d**) CRH train2; (**e**) Bus; (**f**) Car.

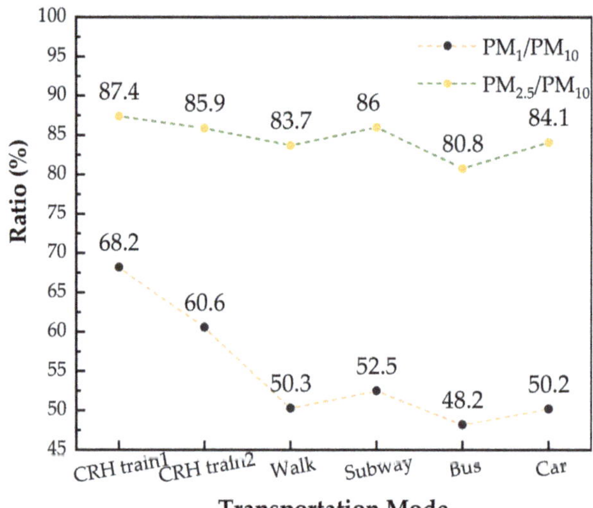

Figure 5. The PM ratios with PM_{10}.

3.2. After the SFTR

Figure 6a is the radar chart of the mass concentration of PM_1, $PM_{2.5}$, and PM_{10} on different transportation from Nanjing to Xuzhou during the SFTR. Figure 6b is the return route from Xuzhou to Nanjing after the SFTR. The results show that the mass concentrations of PM_1, $PM_{2.5}$, and PM_{10} are reduced by varying degrees in different modes except for the CRH train1 after the SFTR. For instance, the $PM_{2.5}$ mass concentration is reduced by 2.6, 2.2, and 5.9 times for walking, subway, and car, respectively. However, the PM_1, $PM_{2.5}$, and PM_{10} mass concentrations did not show much variation for the CRH train1 and CRH train2 during and after the SFTR. Although the strictest quarantine was removed and passengers returned, the factories in China have not yet been fully restored. Therefore, the exposure level is apparently lower than during the SFTR, and NASA stated: the industrial shutdown caused a significant improvement in global air quality affected by COVID-19.

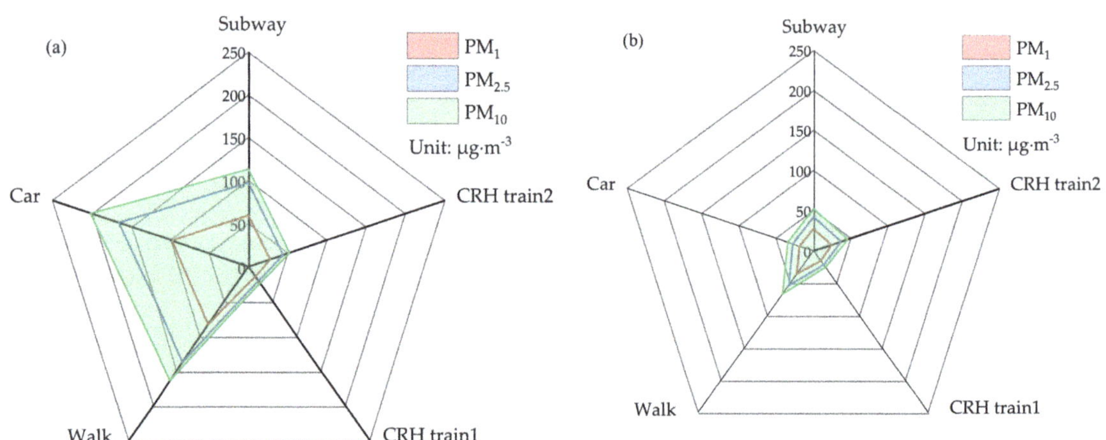

Figure 6. PM mass concentration in different transport modes: (**a**) during the SFTR; (**b**) after the SFTR.

The results of *t*-tests among different modes during and after the SFTR are shown in Table 5. There are no significant differences among the PM_1, $PM_{2.5}$, and PM_{10} mass concentrations on the CRH train1 and the CRH train2 during and after the SFTR (Sig > 0.05). However, there are significant differences among the PM_1, $PM_{2.5}$, and PM_{10} mass concentrations for the subway, car, and walking, during and after the SFTR (Sig < 0.05). For the long-haul CRH train (> 1 h), the exposure level to $PM_{2.5}$ on the CRH train1 (21.77 µg m^{-3}) is lower than the AQG. For the short-haul CRH train (< 20 min), the exposure level to $PM_{2.5}$ in CRH train2 (36.41 µg·m^{-3}) is similar to the AQG. As mentioned above, the CRH trains are always fully loaded during and after the SFTR, and it maintains high-speed operation ($\leq$250 km/h) for the long term. The microenvironment of the CRH train is more stable than subway, car, and walking. Therefore, it is recommended for passengers to take the CRH train during the SFTR to lower the PM exposure level.

Table 5. The results of *t*-tests among different modes during and after the SFTR.

	PM_1			$PM_{2.5}$			PM_{10}		
	t	df	Sig	*t*	df	Sig	*t*	df	Sig
CRH train1	−0.19	194.25	0.85	0.15	194.98	0.88	0.12	204.45	0.91
CRH train2	0.36	17.46	0.72	0.45	17.33	0.66	0.22	18.21	0.83
Subway	13.46	45	0.00	12.81	45	0.00	10.78	45	0.00
Car	21.46	40.24	0.00	42.68	45.47	0.00	34.13	45.94	0.00
Walking	8.40	42	0.00	17.87	29.39	0.00	16.61	30.64	0.00

4. Conclusions

This study monitors personal exposure using different transport modes during and after the SFTR. Commuters have the lowest personal exposure to PM on the CRH trains. The $PM_{2.5}$ mass concentration is reduced by 2.6, 2.2, and 5.9 times for walking, subway, and car, respectively, while there is no change on CRH train1 and CRH train2 during and after the SFTR. The results of *t*-tests indicate that personal exposure level is relatively stable in the CRH trains, and it is no more than the WHO guideline (AQG). Personal exposure to particles is similar to commuting by walking if the air conditioning filter is not replaced in time. Therefore, to reduce personal exposure for passengers, the CRH train is the best choice during the SFTR.

Author Contributions: Conceptualization, N.Y. and Y.Z.; methodology, N.Y.; software, Y.Z.; validation, N.Y., Y.Z. and M.Z.; formal analysis, Q.Y.; investigation, Y.Z.; resources, Y.Z.; data curation, Y.Z.; writing—original draft preparation, Y.Z.; writing—review and editing, N.Y.; visualization, Y.Z. and Q.Y.; supervision, N.Y.; project administration, N.Y.; funding acquisition, N.Y. and M.Z. All authors have read and agreed to the published version of the manuscript.

Funding: This research was funded by the Nanjing University of Aeronautics and Astronautics New Faculty Start-up fund, grant number 90YAH19018; this research was also funded by the Postgraduate Research and Practice Innovation Program of Jiangsu Province, grant number SJCX20_0067.

Data Availability Statement: Not applicable.

Acknowledgments: The authors sincerely thank Mengya Zhang and Quan Ye for helping with the field measurement and preliminary data processing. Special thanks should also be given to Nu Yu for providing the needed instruments during experiments.

Conflicts of Interest: The authors declare no conflict of interest.

References

1. Weichenthal, S.; Van Ryswyk, K.; Kulka, R.; Sun, L.; Wallace, L.; Joseph, L. In-Vehicle Exposures to Particulate Air Pollution in Canadian Metropolitan Areas: The Urban Transportation Exposure Study. *Environ. Sci. Technol.* **2015**, *49*, 597–605. [CrossRef]
2. Klepeis, N.E.; Nelson, W.C.; Ott, W.R.; Robinson, J.P.; Tsang, A.M.; Switzer, P.; Behar, J.V.; Hern, S.C.; Engelmann, W.H. The National Human Activity Pattern Survey (NHAPS): A resource for assessing exposure to environmental pollutants. *J. Expo. Sci. Environ. Epidemiol.* **2001**, *11*, 231–252. [CrossRef] [PubMed]
3. Bi, X.; Luo, W. Analysis of travel characteristics and transportation improvement strategies of small and medium-sized cities. *Technol. Econ. Area Commun.* **2018**, *20*, 28–31. [CrossRef]

4. Asadi, S.; Bouvier, N.; Wexler, A.S.; Ristenpart, W.D. The coronavirus pandemic and aerosols: Does COVID-19 transmit via expiratory particles? *Aerosol Sci. Technol.* **2020**, *54*, 635–638. [CrossRef] [PubMed]

5. Prather, K.A.; Wang, C.C.; Schooley, R.T. Reducing transmission of SARS-CoV-2. *Science* **2020**, *368*, 1422–1424. [CrossRef] [PubMed]

6. Morawska, L.; Milton, D.K. It Is Time to Address Airborne Transmission of Coronavirus Disease 2019 (COVID-19). *Clin. Infect. Dis.* **2020**, *71*, 2311–2313. [CrossRef]

7. Morawska, L.; Tang, J.W.; Bahnfleth, W.; Bluyssen, P.M.; Boerstra, A.; Buonanno, G.; Cao, J.; Dancer, S.; Floto, A.; Franchimon, F.; et al. How can airborne transmission of COVID-19 indoors be minimised? *Environ. Int.* **2020**, *142*, 105832. [CrossRef]

8. Ahlawat, A.; Wiedensohler, A.; Mishra, S.K. An Overview on the Role of Relative Humidity in Airborne Transmission of SARS-CoV-2 in Indoor Environments. *Aerosol Air Qual. Res.* **2020**, *20*, 1856–1861. [CrossRef]

9. Qiu, Z.; Song, J.; Xu, X.; Luo, Y.; Zhao, R.; Zhou, W.; Xiang, B.; Hao, Y. Commuter exposure to particulate matter for different transportation modes in Xi'an, China. *Atmos. Pollut. Res.* **2017**, *8*, 940–948. [CrossRef]

10. Goel, V.; Mishra, S.K.; Ahlawat, A.; Sharma, C.; Vijayan, N.; Radhakrishnan, S.R.; Dimri, A.P.; Kotnala, R.K. Effect of Reduced Traffic Density on Characteristics of Particulate Matter Over Delhi. *Curr. Sci.* **2018**, *115*, 315–319. [CrossRef]

11. Yang, T.; Zhang, P.; Xu, B.; Xiong, J. Predicting VOC emissions from materials in vehicle cabins: Determination of the key parameters and the influence of environmental factors. *Int. J. Heat Mass Transf.* **2017**, *110*, 671–679. [CrossRef]

12. Maji, K.J.; Arora, M.; Dikshit, A.K. Burden of disease attributed to ambient $PM_{2.5}$ and PM_{10} exposure in 190 cities in China. *Environ. Sci. Pollut. Res.* **2017**, *24*, 11559–11572. [CrossRef] [PubMed]

13. Tan, S.H.; Roth, M.; Velasco, E. Particle exposure and inhaled dose during commuting in Singapore. *Atmos. Environ.* **2017**, *170*, 245–258. [CrossRef]

14. Zhao, J.; Bo, L.; Gong, C.; Cheng, P.; Kan, H.; Xie, Y.; Song, W. Preliminary study to explore gene-$PM_{2.5}$ interactive effects on respiratory system in traffic policemen. *Int. J. Occup. Med. Environ. Health* **2015**, *28*, 971–983. [CrossRef] [PubMed]

15. Zhao, J.; Gao, Z.; Tian, Z.; Xie, Y.; Xin, F.; Jiang, R.; Kan, H.; Song, W. The biological effects of individual-level $PM_{2.5}$ exposure on systemic immunity and inflammatory response in traffic policemen. *Occup. Environ. Med.* **2013**, *70*, 426–431. [CrossRef]

16. Xu, H.; Song, N.; Ji, G.; Shi, L.; Zhang, Q.; Guo, M. Personal exposure and health risk assessment of carbonyls in family cars and public transports—a comparative study in Nanjing, China. *Environ. Sci. Pollut. Res.* **2017**, *24*, 26111–26119. [CrossRef]

17. Kam, W.; Delfino, R.J.; Schauer, J.J.; Sioutas, C. A comparative assessment of $PM_{2.5}$ exposures in light-rail, subway, freeway, and surface street environments in Los Angeles and estimated lung cancer risk. *Environ. Sci. Process. Impacts* **2012**, *15*, 234–243. [CrossRef] [PubMed]

18. Gómez-Perales, J.; Colvile, R.; Nieuwenhuijsen, M.; Fernández-Bremauntz, A.; Gutiérrez-Avedoy, V.; Páramo-Figueroa, V.; Blanco-Jiménez, S.; Bueno-López, E.; Mandujano, F.; Bernabé-Cabanillas, R.; et al. Commuters' exposure to $PM_{2.5}$, CO, and benzene in public transport in the metropolitan area of Mexico City. *Atmos. Environ.* **2004**, *38*, 1219–1229. [CrossRef]

19. Yan, C.; Zheng, M.; Yang, Q.; Zhang, Q.; Qiu, X.; Zhang, Y.; Fu, H.; Li, X.; Zhu, T.; Zhu, Y. Commuter exposure to particulate matter and particle-bound PAHs in three transportation modes in Beijing, China. *Environ. Pollut.* **2015**, *204*, 199–206. [CrossRef]

20. Ruttens, D.; Verleden, S.E.; Bijnens, E.M.; Winckelmans, E.; Gottlieb, J.; Warnecke, G.; Meloni, F.; Morosini, M.; Van Der Bij, W.; Verschuuren, E.A.; et al. An association of particulate air pollution and traffic exposure with mortality after lung transplantation in Europe. *Eur. Respir. J.* **2016**, *49*, 1600484. [CrossRef]

21. Television, C.C. Nanjing Railway Station Sent 210,000 Passengers on the First Day of Spring Festival Transport. Available online: https://haokan.baidu.com/v?pd=wisenatural&vid=13281730742068025235 (accessed on 11 January 2021).

22. World Health Organization. *Air Quality Guidelines: Global Update 2005: Particulate Matter, Ozone, Nitrogen Dioxide, and Sulfur Dioxide*; World Health Organization: Geneva, Switzerland, 2006; pp. 277–279.

23. Chan, L.; Lau, W.; Zou, S.; Cao, Z.; Lai, S. Exposure level of carbon monoxide and respirable suspended particulate in public transportation modes while commuting in urban area of Guangzhou, China. *Atmos. Environ.* **2002**, *36*, 5831–5840. [CrossRef]

24. Onat, B.; Stakeeva, B. Personal exposure of commuters in public transport to $PM_{2.5}$ and fine particle counts. *Atmos. Pollut. Res.* **2013**, *4*, 329–335. [CrossRef]

25. Wu, D.-L.; Lin, M.; Chan, C.-Y.; Li, W.-Z.; Tao, J.; Li, Y.-P.; Sang, X.-F.; Bu, C.-W. Influences of Commuting Mode, Air Conditioning Mode and Meteorological Parameters on Fine Particle ($PM_{2.5}$) Exposure Levels in Traffic Microenvironments. *Aerosol Air Qual. Res.* **2013**, *13*, 709–720. [CrossRef]

26. Kumar, P.; Patton, A.; Durant, J.L.; Frey, H.C. A review of factors impacting exposure to $PM_{2.5}$, ultrafine particles and black carbon in Asian transport microenvironments. *Atmos. Environ.* **2018**, *187*, 301–316. [CrossRef]

27. Kumar, P.; Rivas, I.; Singh, A.P.; Ganesh, V.J.; Ananya, M.; Frey, H.C. Dynamics of coarse and fine particle exposure in transport microenvironments. *npj Clim. Atmos. Sci.* **2018**, *1*, 11. [CrossRef]

28. Suárez, L.; Mesías, S.; Iglesias, V.; Silva, C.; Cáceres, D.D.; Ruiz-Rudolph, P. Personal exposure to particulate matter in commuters using different transport modes (bus, bicycle, car and subway) in an assigned route in downtown Santiago, Chile. *Environ. Sci. Process. Impacts* **2014**, *16*, 1309–1317. [CrossRef] [PubMed]

29. Andersen, M.H.G.; Johannesson, S.; Fonseca, A.S.; Clausen, P.A.; Saber, A.T.; Roursgaard, M.; Loeschner, K.; Koponen, I.K.; Loft, S.; Vogel, U.; et al. Exposure to Air Pollution inside Electric and Diesel-Powered Passenger Trains. *Environ. Sci. Technol.* **2019**, *53*, 4579–4587. [CrossRef] [PubMed]

30. Cha, Y.; Tu, M.; Elmgren, M.; Silvergren, S.; Olofsson, U. Factors affecting the exposure of passengers, service staff and train drivers inside trains to airborne particles. *Environ. Res.* **2018**, *166*, 16–24. [CrossRef]
31. Cha, Y.; Abbasi, S.; Olofsson, U. Indoor and outdoor measurement of airborne particulates on a commuter train running partly in tunnels. *Proc. Inst. Mech. Eng. Part F J. Rail Rapid Transit* **2016**, *232*, 3–13. [CrossRef]
32. Chaney, R.A.; Sloan, C.D.; Cooper, V.C.; Robinson, D.R.; Hendrickson, N.R.; Mccord, T.A.; Johnston, J.D. Personal exposure to fine particulate air pollution while commuting: An examination of six transport modes on an urban arterial roadway. *PLoS ONE* **2017**, *12*, e0188053. [CrossRef]
33. Kam, W.; Cheung, K.; Daher, N.; Sioutas, C. Particulate matter (PM) concentrations in underground and ground-level rail systems of the Los Angeles Metro. *Atmos. Environ.* **2011**, *45*, 1506–1516. [CrossRef]
34. Ham, W.; Vijayan, A.; Schulte, N.; Herner, J.D. Commuter exposure to $PM_{2.5}$, BC, and UFP in six common transport microenvironments in Sacramento, California. *Atmos. Environ.* **2017**, *167*, 335–345. [CrossRef]
35. Nyhan, M.; McNabola, A.; Misstear, B. Comparison of particulate matter dose and acute heart rate variability response in cyclists, pedestrians, bus and train passengers. *Sci. Total. Environ.* **2014**, *468-469*, 821–831. [CrossRef]
36. Knibbs, L.; De Dear, R.J. Exposure to ultrafine particles and $PM_{2.5}$ in four Sydney transport modes. *Atmos. Environ.* **2010**, *44*, 3224–3227. [CrossRef]
37. Li, T.-T.; Bai, Y.-H.; Liu, Z.-R.; Li, J.-L. In-train air quality assessment of the railway transit system in Beijing: A note. *Transp. Res. Part D Transp. Environ.* **2007**, *12*, 64–67. [CrossRef]
38. Chan, L.; Lau, W.; Lee, S.-C.; Chan, C. Commuter exposure to particulate matter in public transportation modes in Hong Kong. *Atmos. Environ.* **2002**, *36*, 3363–3373. [CrossRef]
39. Shen, J.; Gao, Z. Commuter exposure to particulate matters in four common transportation modes in Nanjing. *Build. Environ.* **2019**, *156*, 156–170. [CrossRef]
40. Ozgen, M.S.; Ripamonti, G.; Malandrini, A.; Ragettli, M.S.; Lonati, G. Particle number and mass exposure concentrations by commuter transport modes in Milan, Italy. *AIMS Environ. Sci.* **2016**, *3*, 168–184. [CrossRef]
41. Goel, R.; Gani, S.; Guttikunda, S.; Wilson, D.; Tiwari, G. On-road $PM_{2.5}$ pollution exposure in multiple transport microenvironments in Delhi. *Atmos. Environ.* **2015**, *123*, 129–138. [CrossRef]
42. Li, Z.; Che, W.; Frey, H.C.; Lau, A.K.; Lin, C. Characterization of $PM_{2.5}$ exposure concentration in transport microenvironments using portable monitors. *Environ. Pollut.* **2017**, *228*, 433–442. [CrossRef]
43. Adams, H.; Nieuwenhuijsen, M.; Colvile, R. Determinants of fine particle ($PM_{2.5}$) personal exposure levels in transport microenvironments, London, UK. *Atmos. Environ.* **2001**, *35*, 4557–4566. [CrossRef]
44. Kolluru, S.S.R.; Patra, A.K.; Kumar, P. Determinants of commuter exposure to $PM_{2.5}$ and CO during long-haul journeys on national highways in India. *Atmos. Pollut. Res.* **2019**, *10*, 1031–1041. [CrossRef]
45. Okokon, E.O.; Taimisto, P.; Turunen, A.W.; Amoda, O.A.; Fasasi, A.E.; Adeyemi, L.G.; Juutilainen, J.; Lanki, T. Particulate air pollution and noise: Assessing commuter exposure in Africa's most populous city. *J. Transp. Health* **2018**, *9*, 150–160. [CrossRef]
46. Brook, R.D.; Rajagopalan, S.; Pope, I.C.A.; Brook, J.R.; Bhatnagar, A.; Diez-Roux, A.V.; Holguin, F.; Hong, Y.; Luepker, R.V.; Mittleman, M.; et al. Particulate Matter Air Pollution and Cardiovascular Disease. *Circulation* **2010**, *121*, 2331–2378. [CrossRef] [PubMed]
47. Yuan, Z.H.; Hou, Y.J.; Zhao, C.; Li, Y. Dynamicly Observing Influence of Ambient $PM_{2.5}$ on Human Lymphocyte Subsets. *Appl. Mech. Mater.* **2014**, *508*, 286–289. [CrossRef]
48. Yu, N.; Shu, S.; Lin, Y.; She, J.; Ip, H.S.S.; Qiu, X.; Zhu, Y. High efficiency cabin air filter in vehicles reduces drivers' roadway particulate matter exposures and associated lipid peroxidation. *PLoS ONE* **2017**, *12*, e0188498. [CrossRef] [PubMed]

MDPI

St. Alban-Anlage 66

4052 Basel

Switzerland

www.mdpi.com

Processes Editorial Office

E-mail: processes@mdpi.com

www.mdpi.com/journal/processes

www.ingramcontent.com/pod-product-compliance
Lightning Source LLC
LaVergne TN
LVHW071010170726
843515LV00006B/1128